Classes of Modules

MONOGRAPHS AND TEXTBOOKS IN PURE AND APPLIED MATHEMATICS

Recent Titles

G. S. Ladde and M. Sambandham, Stochastic versus Deterministic Systems of Differential Equations (2004)

B. J. Gardner and R. Wiegandt, Radical Theory of Rings (2004)

J. Haluska, The Mathematical Theory of Tone Systems (2004)

C. Menini and F. Van Oystaeyen, Abstract Algebra: A Comprehensive Treatment (2004)

E. Hansen and G. W. Walster, Global Optimization Using Interval Analysis, Second Edition, Revised and Expanded (2004)

M. M. Rao, Measure Theory and Integration, Second Edition, Revised and Expanded (2004)

W. J. Wickless, A First Graduate Course in Abstract Algebra (2004)

R. P. Agarwal, M. Bohner, and W-T Li, Nonoscillation and Oscillation Theory for Functional Differential Equations (2004)

J. Galambos and I. Simonelli, Products of Random Variables: Applications to Problems of Physics and to Arithmetical Functions (2004)

Walter Ferrer and Alvaro Rittatore, Actions and Invariants of Algebraic Groups (2005)

Christof Eck, Jiri Jarusek, and Miroslav Krbec, Unilateral Contact Problems: Variational Methods and Existence Theorems (2005)

M. M. Rao, Conditional Measures and Applications, Second Edition (2005)

A. B. Kharazishvili, Strange Functions in Real Analysis, Second Edition (2006)

Vincenzo Ancona and Bernard Gaveau, Differential Forms on Singular Varieties: De Rham and Hodge Theory Simplified (2005)

Santiago Alves Tavares, Generation of Multivariate Hermite Interpolating Polynomials (2005)

Sergio Macías, Topics on Continua (2005)

Mircea Sofonea, Weimin Han, and Meir Shillor, Analysis and Approximation of Contact Problems with Adhesion or Damage (2006)

Marwan Moubachir and Jean-Paul Zolésio, Moving Shape Analysis and Control: Applications to Fluid Structure Interactions (2006)

Alfred Geroldinger and Franz Halter-Koch, Non-Unique Factorizations: Algebraic, Combinatorial and Analytic Theory (2006)

Kevin J. Hastings, Introduction to the Mathematics of Operations Research with *Mathematica*®, Second Edition (2006)

Robert Carlson, A Concrete Introduction to Real Analysis (2006)

John Dauns and Yiqiang Zhou, Classes of Modules (2006)

Classes of Modules

John Dauns

Tulane University
New Orleans, Louisiana, U.S.A.

Yiqiang Zhou

Memorial University of Newfoundland
St. John's, Canada

Chapman & Hall/CRC
Taylor & Francis Group
Boca Raton London New York

Chapman & Hall/CRC is an imprint of the
Taylor & Francis Group, an informa business

Published in 2006 by
Chapman & Hall/CRC
Taylor & Francis Group
6000 Broken Sound Parkway NW, Suite 300
Boca Raton, FL 33487-2742

Printed in the United States of America on acid-free paper
10 9 8 7 6 5 4 3 2 1

International Standard Book Number-10: 1-5848-8660-9 (Hardcover)
International Standard Book Number-13: 978-1-5848-8660-0 (Hardcover)

Library of Congress Cataloging-in-Publication Data

Dauns, John.
Classes of modules / John Dauns, Yiqiang Zhou.
p. cm.
Includes bibliographical references and index.
ISBN-13: 978-1-5848-8660-0 (acid-free paper)
ISBN-10: 1-5848-8660-9 (acid-free paper)
1. Set theory. 2. Modules (Algebra) 3. Rings (Algebra) I. Zhou, Yiqiang. II. Title.
QA248.D275 2006
512'.42--dc22 2006045438

informa
Taylor & Francis Group
is the Academic Division of Informa plc.

Visit the Taylor & Francis Web site at
http://www.taylorandfrancis.com

and the CRC Press Web site at
http://www.crcpress.com

Contents

Preface v

Note to the Reader vii

List of Symbols ix

Chapter 1. Preliminary Background **1**

1.1. Notation and Terminology 1
1.2. Lattices 5

Chapter 2. Important Module Classes and Constructions **7**

2.1. Torsion Theory 7
2.2. Module Class $\sigma[M]$ 12
2.3. Natural Classes 19
2.4. M-Natural Classes 24
2.5. Pre-Natural Classes 29

Chapter 3. Finiteness Conditions **33**

3.1. Ascending Chain Conditions 33
3.2. Descending Chain Conditions 50
3.3. Covers and Ascending Chain Conditions 65

Chapter 4. Type Theory of Modules: Dimension **71**

4.1. Type Submodules and Type Dimensions 71
4.2. Several Type Dimension Formulas 80
4.3. Some Non-Classical Finiteness Conditions 87

Chapter 5. Type Theory of Modules: Decompositions **107**

5.1. Type Direct Sum Decompositions 108
5.2. Decomposability of Modules 117
5.3. Unique Type Closure Modules 132
5.4. TS-Modules 142

Chapter 6. Lattices of Module Classes **149**

6.1. Lattice of Pre-Natural Classes 149
6.2. More Sublattice Structures 153
6.3. Lattice Properties of $\mathcal{N}_r^p(R)$ 163
6.4. More Lattice Properties of $\mathcal{N}_r^p(R)$ 177
6.5. Lattice $\mathcal{N}_r(R)$ and Its Applications 190
6.6. Boolean Ideal Lattice 200

References **205**

Index **215**

Preface

The main theme of the book is in two concepts and how they pervade and structure much of ring and module theory. They are a natural class, and a type submodule. A natural class $\mathcal{K}$ of right modules over an arbitrary associative ring R with identity is one that is closed under isomorphic copies, submodules, arbitrary direct sums, and injective hulls. A submodule N of a right R-module M is a type submodule if there exists some natural class $\mathcal{K}$ such that $N \subseteq M$ is a submodule maximal with respect to the property that $N \in \mathcal{K}$. There are also equivalent but somewhat technical ways of defining this notion totally internally in terms of the given module M without reference to any outside classes $\mathcal{K}$. Equivalently a submodule $N \leq M$ is a type submodule if and only if N is a complement submodule of M such that, for some submodule $C \leq M$, N and C do not have nonzero isomorphic submodules and $N + C = N \oplus C$ is essential in M.

An attempt is made to make this book self contained and accessible to someone who either has some knowledge of basic ring theory, such as beginning graduate and advanced undergraduate students, or to someone who is willing to acquire the basic definitions along the way.

A brief description of the contents of the book is given next. For some readers such a description will only be useful after they have started reading the book. The beginning Chapter 1 defines the more or less standard notation used, and lists a few useful facts mostly without proof. Chapter 2 presents all the module classes that will be used, among which are torsion, torsion free classes, $\sigma[M]$, natural classes, and pre-natural classes. Chapter 3 utilizes chain conditions relative to some module class $\mathcal{K}$ of right R-modules. These chain conditions guarantee that direct sums of injective modules in $\mathcal{K}$ are injective, or that nil subrings are nilpotent in certain rings. Chapter 4 develops the basic theory of type submodules, and the type dimension of a module, which is analogous to the finite uniform dimension. Here new chain conditions, the type ascending and descending chain conditions, are explained and used to obtain structure theorems for modules and rings.

Chapter 5 shows that the collection $\mathcal{N}(R)$ of all natural classes of right R-modules ordered by class inclusion is a Boolean lattice. By use of this lattice, new natural classes are defined and used to give module decompositions of a module M as a direct sum of submodules belonging to certain natural classes. A module M is a TS-module if every type submodule of M is a direct summand of M. Extending or CS-modules are a very special case of a TS-module. A decomposition theory for TS-modules is developed which parallels the much studied theory of CS-modules. Chapter 6 studies the collection $\mathcal{N}^p(R)$ of all pre-natural classes of right R-modules. This complete lattice is significant in several ways. First of all, it contains almost all the well and lesser known lattices of module classes as sublattices. Many sublattices of $\mathcal{N}^p(R)$ are identified. Several connections between ring theoretic properties of the ring R, and purely lattice theoretic properties of $\mathcal{N}^p(R)$, or some of its many sublattices are proved. Thus, $\mathcal{N}(R)$ is a sublattice of $\mathcal{N}^p(R)$, but in general not a complete sublattice. The interaction between a ring R and the lattices associated to R is explored.

If ring theory is to progress, it seems that some new kind of finiteness or chain conditions will be required, such as the type ascending and descending chain conditions. Also, placing restrictive hypotheses on all submodules of a module, or even only on all complement submodules, is too restrictive. However, putting restrictions on the type submodules only is more reasonable.

At the time of this writing, this is the only book on the present subject. Previously it was very inaccessibly scattered throughout the literature. Moreover, some results in the literature have been either improved or extended, or proofs have been simplified and made more elegant.

We view this book not merely as a presentation of a certain theory, but believe that it gives more. It gives tools or new methods and concepts to do ring and module theory. So we regard the book as a program, a path, a direction or road, on which so far we have only started to travel with still a long way ahead.

The second author expresses his gratitude to his wife Hongwa and their son David for their support and patience during this project, and he gratefully acknowledges the support by the Natural Sciences and Engineering Research Council of Canada.

John Dauns
Yiqiang Zhou

Note to the Reader

Lemmas, corollaries, theorems, etc. are labeled by three integers separated by two periods. Thus (4.2.10) stands for Chapter 4, section 2, and item 10.

Chapter 1 and Chapter 2, section 1 (section 2.1) list many facts without proof. In order to read the book, knowledge of the proofs of these facts is not needed. Section 2.1 serves to give an overview of several previously used module classes or categories so that the reader can see where the present material fits into the broader picture.

One can read only certain parts of the book without going through everything. However, Chapter 2, section 3 on natural classes and Chapter 4, section 1 on type submodules is basic for everything and should not be omitted. After reading Chapters 2 and 4, the reader could go anywhere else in the book.

List of Symbols

ACC	ascending chain condition
DCC	descending chain condition
t-ACC	type ascending chain condition
t-DCC	type descending chain condition
$E(N)$	injective hull of a module N
$E_M(N)$	M-injective hull of a module N
$\mathrm{Hom}_R(M, N)$	group of R-homomorphisms from M to N
$\mathrm{End}(N_R)$	ring of R-endomorphisms of a module N
$Gdim(M)$	Gabriel dimension of a module M
$u.dim(M)$	uniform dimension of a module M
$t.dim(M)$	type dimension of a module M
Mod-R	category of all right R-modules
R-Mod	category of all left R-modules
$\sigma[M]$	category of M-subgenerated modules
$\mathbb{N}$	set of positive integers
$\mathbb{Z}$	ring of integers
$\mathbb{Q}$	field of rationals
$\mathbb{R}$	field of real numbers
$\mathbb{Z}_n$	ring of integers modulo n
$\mathbb{Z}_{p^\infty}$	Prüfer group
$M^{(I)}$	direct sum of I copies of M
$M^{(n)}$	direct sum of n copies of M
$M_n(R)$	n by n matrix ring over a ring R

$\mathcal{N}(R), \mathcal{N}_r(R)$	lattice of natural classes of right R-modules
$\mathcal{N}^p(R), \mathcal{N}_r^p(R)$	lattice of pre-natural classes of right R-modules
$\mathcal{N}(R, M)$	lattice of M-natural classes of right R-modules
$N \leq_e M$	N is an essential submodule of a module M
$N \subseteq^{\oplus} M$	N is a direct summand of a module M
$N \leq_t M$	N is a type submodule of a module M
$J(R)$	Jacobson radical of a ring R
$Soc(N)$	socle of a module N
$Z(N)$	singular submodule of a module N
$Z_2(N)$	Goldie torsion (or second singular) submodule of a module N
$Z_M(N)$	sum of M-singular submodules of a module N
$\tau(N)$	torsion submodule of a module N
$\text{tr}(\mathcal{K}, N)$	trace of a module class $\mathcal{K}$ in a module N
$\mathcal{F}_r(R)$	set of hereditary torsion free classes of right R-modules
$\mathcal{T}_r(R)$	set of hereditary torsion classes of right R-modules
$\mathcal{T}_r^p(R)$	set of hereditary pretorsion classes of right R-modules
fil-R	lattice of right linear topologies of R
Tor-R	lattice of hereditary torsion theories of R
$\subseteq, \subset$	inclusion, proper inclusion
$X \backslash Y$	set-theoretic difference
$2^X, \mathcal{P}(X)$	power set of a set X
$\aleph_0, \aleph^+$	first infinite cardinal, successor of the infinite cardinal $\aleph$
$\varinjlim M_a$	direct limit
$i_M(\mathcal{F})$	M-injective modules in $\mathcal{F}$

Chapter 1

Preliminary Background

Section 1.1 gives notation and terminology about rings and modules that are used throughout. Then section 1.2 briefly outlines some material about lattices.

1.1 Notation and Terminology

Some notation is introduced that will be used throughout.

1.1.1. Unless said otherwise modules M are right unital over a ring R with identity. Submodules N are denoted by $<, \leq, \subset, \subseteq$ as $N \leq M, N \subset M$, etc. If it is understood that we have a submodule and we wish to emphasize containment, $\subset$ and $\subseteq$ are used. **Essential** or **large** submodules are denoted by "$\leq_e$" or "$<_e$." A nonzero module M is **uniform** if $N \leq_e M$ for every nonzero submodule N of M.

For a module A and index set I, as usual, $A^{(I)} = \bigoplus_I A$ denotes the direct sum $\bigoplus\{A_i : i \in I\}$ with $A_i = A$ for all $i \in I$. Similarly, $A^I = \Pi_I A$.

The **injective hull** of M is written as $E(M) = EM$. The latter is particularly useful if M is given by a complex formula. Note that $M \leq_e E(M)$.

For sets X and Y, the complement of Y in X is $X \setminus Y = \{z \in X : z \notin Y\}$. Note that $Y \subseteq X$ is not assumed. Sets are denoted as $\{x : x \text{ satisfies } P\}$. The cardinality of X is denoted by $|X|$, and $\mathcal{P}(X) = \{A : A \subseteq X\} = 2^X$ denotes the set of all subsets of X. Note that $|\mathcal{P}(X)| = |2^X| = 2^{|X|}$.

For a function $f : A \longrightarrow B$ of sets and for $X \subseteq A$, $f|_X : X \longrightarrow B$ denotes the restriction of f to X. For $Y \subseteq B$, $f^{-1}(Y) = f^{-1}Y = \{a \in A : f(a) \in Y\}$; it is not required that $Y \subseteq f(A)$. For a singleton $\{y\}$, write $f^{-1}(y) = f^{-1}(\{y\})$. If $f : A \longrightarrow B$ is an R-module homomorphism or more shortly an R-map, then $Ker(f) = f^{-1}(0) \subseteq A$, $Im(f) = f(A) \subseteq B$ are submodules. If clear from context, we just say "f is a map" in place of "f is an R-homomorphism of right R-modules". For modules A and B, we write $A \twoheadrightarrow B$ to mean that B is a homomorphic image of A.

Throughout, Mod-R and R-Mod denote the categories of right and left R-modules. The notation "M_R" and "${}_RM$" mean that M is a right and a left R-module, respectively.

1.1.2. For a given module M, a module V is M-**generated** if $V \cong M^{(I)}/N$ for some index set I and submodule $N \leq M^{(I)}$. A module W is M-**cogenerated** or M cogenerates W if W is isomorphic to a submodule of M^I for some I (notation $W \hookrightarrow M^I$). A class $\mathcal{F}$ of modules is cogenerated by a module M if every W in $\mathcal{F}$ is cogenerated by M.

A module N is M-**injective** if for any submodule $X \leq M$, every R-homomorphism $f : X \longrightarrow N$ can be extended to a homomorphism $\hat{f} : M \longrightarrow N$. A module M is **quasi-injective** if M is M-injective. It is known that M is quasi-injective if and only if $\mathrm{Hom}_R(EM, EM)M \subseteq M$ where M is viewed as a submodule $M \leq E(M)$.

For a right R-module M_R and a subset $X \subseteq M$, define the annihilator of X in R by $X^\perp = \{r \in R : xr = 0, \forall x \in X\}$ and write $x^\perp$ for $\{x\}^\perp$ for all $x \in M$. For a left R-module ${}_RM$, if $X \subseteq M$ and $x \in M$, then ${}^\perp X$ and ${}^\perp x$ are defined similarly. So if I is a subset of the ring R, then $I^\perp = \{r \in R : ar = 0, \forall a \in I\}$ and ${}^\perp I = \{r \in R : ra = 0, \forall a \in I\}$. For $N \leq M$ and $x \in M$, $(x + N)^\perp = x^{-1}N = \{r \in R : xr \in N\}$.

1.1.3. For a module M_R, its **singular submodule** is $Z(M) = ZM = \{x \in M : x^\perp \leq_e R_R\}$, and its **Goldie torsion (or second singular) submodule** $Z_2(M) = Z_2M$ is defined by $Z[M/Z(M)] = Z_2(M)/Z(M)$. Then $Z(M) \leq_e Z_2(M)$. A module M is **singular** if $M = Z(M)$, and **nonsingular** if $Z(M) = 0$; it is not singular if $Z(M) \neq M$. Thus nonsingular $\neq$ not singular. The ring R is said to be nonsingular if R_R is nonsingular.

1.1.4. A submodule $X \leq M$ is said to be a **complement submodule** of M if it satisfies one of the following two equivalent conditions: (i) There exists a submodule $Y \leq M$ such that X is maximal among the submodules of M such that $X \cap Y = 0$ (such an X is also called a **complement** of Y in M); (ii) For any $X \leq_e N \leq M$, $N = X$. In words, X does not have a proper essential extension inside M. Because of the latter, complement submodules are also called **closed submodules**.

For a submodule $X \leq M$, any maximal essential extension C in M of X is called a **complement closure** of X. One way to obtain it is to start with any $Y \leq M$ such that $X \oplus Y \leq_e M$, and then take $C \leq M$ to be maximal such that $X \subseteq C$ but $C \cap Y = 0$. In general, complement closures are not unique.

Some useful properties of complement submodules are listed.

1.1.5. Let $X \leq N \leq M$ be modules.

1. **TRANSITIVITY.** If X is a complement submodule of N and N is a complement submodule of M then X is a complement submodule of M (see [66, Prop.1.5, p.15]).

2. N is a complement submodule of M if and only if for any $Y \leq_e M$, $(N+Y)/N \leq_e M/N$ (see [30, Prop.3-3.5, p.41]).

3. For $x \in M$, $N \leq_e N + xR \Longrightarrow x^{-1}N \leq_e R$. The converse holds when $ZM \subseteq N$. □

1.1.6. Let M be nonsingular. Then $N \leq M$ is a complement submodule if and only if $Z(M/N) = 0$. To see this, take any complement closure C of N and any $X \leq M$ with $X \oplus C \leq_e M$. "$\Longrightarrow$". If $N = C$, then $Z[(X \oplus N)/N] = (ZX \oplus N)/N = 0$, and M/N contains an essential nonsingular submodule $(X \oplus N)/N$ by (1.1.5)(2). "$\Longleftarrow$". If $N \neq C$, then $0 \neq C/N = Z(C/N) \leq M/N$, a contradiction. □

Since the following useful fact is often used and stated but seldom proved, we give a particularly short and simple proof.

1.1.7. LEMMA. Let M be nonsingular and $N \leq M$. Then N has a unique complement closure in M.

PROOF. Let C_1, C_2 be two complement closures of N in M. Then $N \subseteq C_1 \cap C_2$. Consequently $M/(C_1 \cap C_2) \hookrightarrow M/C_1 \oplus M/C_2$. Since $M/C_1 \oplus M/C_2$ is nonsingular by (1.1.6), also $M/(C_1 \cap C_2)$ is nonsingular. Hence again by (1.1.6), $C_1 \cap C_2$ is a complement submodule of M. So $C_1 = C_1 \cap C_2 = C_2$. □

In view of (1.1.5)(3), for any modules $ZM \subseteq N \leq M$, the unique complement closure of N in M is the module $\{x \in M : x^{-1}N \leq_e R\}$.

1.1.8. An extension of modules $N \leq M$ is **rational** if it satisfies any one of the following equivalent properties:

1. For any $N \subset A \leq M$, and any homomorphism $f : A \longrightarrow M$ such that the restriction $f|_N = 0$, necessarily $f = 0$.

2. $\text{Hom}_R(M/N, EM) = 0$.

3. For any $m_1, 0 \neq m_2 \in M$, $m_2(m_1^{-1}N) \neq 0$.

If $N < M$ is rational, upon taking $m = m_1 = m_2 \in M \backslash N$, $0 \neq m(m^{-1}N) \leq N$ shows that $N \leq_e M$. If $ZM = 0$ and $N \leq_e M$, then in (3), $m_1^{-1}N \leq_e R$ for all $m_1 \in M$, and since $0 \neq m_2 \notin ZM = 0$, $m_2(m^{-1}N) \neq 0$. So for a nonsingular module M, the rational and essential submodules of M coincide. A right ideal I of R is called **dense** if I_R is rational in R_R.

1.1.9. A **right quotient ring** of a given ring R is a ring S that contains R as a subring so that the identity element of R is also the identity element of S

and such that $R_R \leq S_R$ is a rational extension. A **maximal right quotient ring** of R is a right quotient ring Q of R such that for any other right quotient ring S of R, the inclusion map $R \longrightarrow Q$ lifts to a monic ring homomorphism $S \longrightarrow Q$. It is known that every ring R has a maximal right quotient ring ([66, Thm.2.29, p.58]).

As usual, we write $\mathbb{Z}, \mathbb{Q}, \mathbb{R}$ for the integers, the rationals, and the real numbers. For a positive integer n, $\mathbb{Z}_n = \mathbb{Z}/n\mathbb{Z}$ denotes the ring of integers modulo n or the Abelian group of order n.

1.2 Lattices

Standard lattice definitions and concepts are used.

1.2.1. DEFINITION. A **lattice** L is a partially ordered set in which any two elements $x, y \in L$ have a least upper bound $x \vee y \in L$ and a greatest lower bound $x \wedge y \in L$. Similarly for subsets $\{x_\gamma : \gamma \in \Gamma\}$, $\bigvee\{x_\gamma : \gamma \in \Gamma\} = \bigvee_{\gamma\in\Gamma} x_\gamma = \bigvee x_\gamma$ denotes the least upper bound provided it exists at all, and similarly for the greatest lower bound $\bigwedge_{\gamma\in\Gamma} x_\gamma$.

1.2.2. DEFINITION. A lattice L with largest element $1 \in L$ and least element $0 \in L$ is **complemented** if for every $x \in L$ there exists $x^c \in L$ with $x \vee x^c = 1$ and $x \wedge x^c = 0$.

Terms such as the following are used: **complete** lattice, **modular** lattice, **distributive** lattice, lattice homomorphism.

1.2.3. DEFINITION. A **Boolean** lattice is a lattice L with least element $0 \in L$ and largest element $1 \in L$, which is distributive and complemented. "Boolean lattice" and "Boolean algebra" are synonyms.

1.2.4. DEFINITION. A ring (with or without identity) is a **Boolean** ring if all of its elements x satisfy $x^2 = x$. A Boolean lattice and a Boolean ring are known to be equivalent concepts [71]. Therefore, for example, if L_1 and L_2 are Boolean lattices and $f : L_1 \longrightarrow L_2$ is a lattice homomorphism which preserves zeros, but not necessarily ones, then f is also, equivalently, a homomorphism of the Boolean rings L_1 and L_2, or of L_1 and $f(L_1)$. Note that if $f(1) \neq 1 \in L_2$, then in general f does not preserve complements $x^c \in L_1$; in fact, always $f(x^c) \leq (f(x))^c$, where $f(x^c) \neq (f(x))^c$ holds for $x = 1$.

1.2.5. DEFINITION. A Boolean lattice L can be specified by listing some or most of its ingredients $L = \langle L, \vee, \wedge, c, 0, 1\rangle$. For any set X, $\mathcal{P}(X) = \langle \mathcal{P}(X), \cup, \cap, \setminus, 0 = \emptyset, 1 = X\rangle$.

The following is not used except a couple of times at the end of Chapter 6.

1.2.6. COMPLETION. Let R be a Boolean ring with $1 \in R$. Then R is nonsingular. For $x, y \in R$, $x \leq y$ means that $x = yx$; $x \wedge y = xy$, $x \vee y = x + y$, and $1 - x$ is the complement of x. The Boolean ring R is complete if every subset of R has a least upper bound in this order (equivalently, a greatest lower bound). Every Boolean ring R has a minimal completion Λ, where Λ is a complete Boolean ring containing R densely ($\forall\, \lambda_1 < \lambda_2 \in R$, $[\lambda_1, \lambda_2] \cap R \neq \emptyset$) with Λ being unique over R ([71, pp.92-93]).

The completion Λ of R can be equivalently characterized in three ways ([13, Thm.5, Cor.4, pp.26-27] and [71, Thm.11, p.93]):

1. $\Lambda = ER$ is the injective hull of R_R.

2. Λ is the so-called Dedekind–MacNeille completion of the partially ordered set R.

3. Λ is the maximal ring of quotients of R.

Let X be a Boolean space, i.e., a compact Hausdorff space where the closed and open (clopen) sets form a base for the topology. A **field of sets** on X is a sublattice of the lattice $\mathcal{P}(X)$ in (1.2.5). By the Stone Representation Theorem ([71, Thm.6, p.78]), every Boolean ring R is (isomorphic to) the field of all clopen sets in some Boolean space X. An open set $P \subseteq X$ is called **regular** if P is equal to the interior of its closure $\overline{P}$. For any set P, its interior is $X\backslash\overline{(X\backslash P)}$ where the upper bar denotes closure. For example, for any $x_0 \in X$, $X\backslash\{x_0\}$ is open but not regular. The completion Λ of the above R can be viewed as the field of all regular open sets in X by [71, Thm.11, p.93].

Chapter 2

Important Module Classes and Constructions

Section 2.1 outlines the main facts about torsion theories. Proofs are not given in section 2.1, because it is intended merely for giving definitions. Section 2.2 develops the basic properties of the category $\sigma[M]$ subgenerated by an arbitrary module M. It includes the Goldie torsion class in $\sigma[M]$. The next section 2.3 covers natural classes, which is a central topic here. Then section 2.4 is about M-natural classes, which can be regarded as natural classes relativized to the subcategory $\sigma[M]$ of Mod-R. The last section 2.5 on pre-natural classes gives a generalization of natural classes. A natural class, or an M-natural class, is a special case of a pre-natural class. Their importance and usefulness will not become fully apparent until Chapter 6, where it will be shown that almost all lattices of module classes associated to a ring are sublattices of the lattice of pre-natural classes.

2.1 Torsion Theory

A brief outline of pretorsion, hereditary pretorsion, torsion, and hereditary torsion classes is given, and how each of these corresponds to preradicals with additional properties. Here in this section, unless said otherwise, all classes of modules are in Mod-R and are closed under isomorphic copies.

Although all definitions here are in the category Mod-R, they can be extended and used in any full subcategory of Mod-R which has direct products and sums, injective hulls, and is closed under extensions. A class of modules will be said to be closed under $\mathrm{Ext}^1_R(\cdot,\cdot)$, or closed under extensions, if whenever A and C belong to the class, and $0 \longrightarrow A \longrightarrow B \longrightarrow C \longrightarrow 0$ is a short exact sequence, then also B belongs to the class in question.

2.1.1. DEFINITION. The fact that a class of modules is closed under quotient modules, arbitrary direct sums, submodules, extensions, arbitrary products, and injective hulls is indicated respectively by: $/, \bigoplus, \leq, \mathrm{Ext}^1_R(\cdot,\cdot), \Pi$, and $E(\cdot)$. The following table defines seven classes of modules by defining what they are closed under:

Class	Closed Under
Pretorsion	$/, \bigoplus$
Hereditary pretorsion	$/, \bigoplus, \leq$
Torsion	$/, \bigoplus, \mathrm{Ext}_R^1(\cdot,\cdot)$
Hereditary torsion	$/, \bigoplus, \leq, \mathrm{Ext}_R^1(\cdot,\cdot)$
Pretorsion free	$\leq, \Pi$
Torsion free	$\leq, \Pi, \mathrm{Ext}_R^1(\cdot,\cdot)$
Hereditary torsion free	$\leq, \Pi, \mathrm{Ext}_R^1(\cdot,\cdot), E(\cdot)$

A module class is called **stable** if it is closed under injective hulls. Some classes of modules are uniquely defined by certain functors, and also by sets of right ideals called filters. Both are defined next.

2.1.2. DEFINITION. For functors ρ, σ : Mod-$R \longrightarrow$ Mod-R, ρ is a **subfunctor** of σ if for any morphism $f : M \longrightarrow N$, first $\rho M = \rho(M) \subseteq \sigma(M)$ and secondly $\rho(f) = \sigma(f)|_{\rho(M)}$ is the restriction of $\sigma(f)$ to $\rho(M)$. A **preradical** ρ is a subfunctor of the identity functor id : Mod-$R \longrightarrow$ Mod-R.

For a preradical ρ and a module M, M is ρ-**torsion** if $\rho(M) = M$, and ρ-**torsion free** if $\rho(M) = 0$. A submodule N of M is called ρ-**dense** in M if M/N is ρ-torsion, and N is called ρ-**closed** in M if M/N is ρ-torsion free. A preradical ρ is **idempotent** if $\rho^2 = \rho$, i.e., $\rho(\rho M) = \rho(M)$ for all modules M. A preradical ρ is called a **radical** if $\rho(M/\rho(M)) = 0$ for all modules M, and is said to be **left exact** if the sequence $0 \longrightarrow \rho(A) \longrightarrow \rho(B) \longrightarrow \rho(C)$ is exact for every short exact sequence $0 \longrightarrow A \longrightarrow B \longrightarrow C \longrightarrow 0$ in Mod-R.

Associated to a preradical ρ, there are two classes of modules in Mod-R, namely $\mathcal{T}_\rho = \{N \in \text{Mod-}R : \rho(N) = N\}$ and $\mathcal{F}_\rho = \{N \in \text{Mod-}R : \rho(N) = 0\}$. Note that $\mathcal{T}_\rho$ is a pretorsion class and $\mathcal{F}_\rho$ is a pretorsion free class. Let $\mathcal{T}$ be a pretorsion class. For any $N \in$ Mod-R, let $\tau(N) = \Sigma\{X \leq N : X \in \mathcal{T}\}$, and then $\tau(N)$ is the unique largest submodule that belongs to $\mathcal{T}$. Thus, $\mathcal{T}$ gives rise to a preradical τ which is clearly idempotent. The next proposition can be found in Stenström [112].

2.1.3. PROPOSITION. The following hold in the category Mod-R:

1. The correspondence $\rho \longmapsto \mathcal{T}_\rho$ between idempotent preradicals and pretorsion classes is bijective.

2. The correspondence $\rho \longmapsto \mathcal{T}_\rho$ between left exact preradicals and hereditary pretorsion classes is bijective.

3. The correspondence $\rho \longmapsto \mathcal{T}_\rho$ between idempotent radicals and torsion classes is bijective.

4. The correspondence $\rho \longmapsto \mathcal{T}_\rho$ between left exact radicals and hereditary torsion classes is bijective.

2.1.4. DEFINITION. A **torsion theory** for Mod-R consists of a pair $(\mathcal{T}, \mathcal{F})$ of classes of R-modules satisfying the following conditions:

1. $T \in \mathcal{T} \iff \forall F \in \mathcal{F}, \mathrm{Hom}_R(T, F) = 0$, and
2. $F \in \mathcal{F} \iff \forall T \in \mathcal{T}, \mathrm{Hom}_R(T, F) = 0$.

2.1.5. DEFINITION. For a class $\mathcal{K}$ of modules, set

$$\mathcal{F} = \{F \in \text{Mod-}R : \mathrm{Hom}_R(K, F) = 0, \forall K \in \mathcal{K}\},$$
$$\mathcal{T} = \{T \in \text{Mod-}R : \mathrm{Hom}_R(T, F) = 0, \forall F \in \mathcal{F}\};$$

and dually set

$$\mathcal{G} = \{G \in \text{Mod-}R : \mathrm{Hom}_R(G, K) = 0, \forall K \in \mathcal{K}\},$$
$$\mathcal{H} = \{H \in \text{Mod-}R : \mathrm{Hom}_R(G, H) = 0, \forall G \in \mathcal{G}\}.$$

Then $(\mathcal{T}, \mathcal{F})$ is a torsion theory, called the **torsion theory generated** by $\mathcal{K}$, and $\mathcal{T}$ is the smallest torsion class containing $\mathcal{K}$. Dually, $(\mathcal{G}, \mathcal{H})$ is also a torsion theory, called the **torsion theory cogenerated** by $\mathcal{K}$, and $\mathcal{H}$ is the smallest torsion free class containing $\mathcal{K}$.

2.1.6. Let $\mathcal{T}$ and $\mathcal{F}$ be classes of modules. Then the following hold:

1. There exists a class $\mathcal{D}$ such that $(\mathcal{T}, \mathcal{D})$ is a torsion theory if and only if $\mathcal{T}$ is a torsion class.
2. There exists a class $\mathcal{C}$ such that $(\mathcal{C}, \mathcal{F})$ is a torsion theory if and only if $\mathcal{F}$ is a torsion free class.
3. If $(\mathcal{T}, \mathcal{F})$ is a torsion theory, then $\mathcal{T}$ is a hereditary torsion class iff $\mathcal{F}$ is a hereditary torsion free class. In this case, we call $(\mathcal{T}, \mathcal{F})$ a **hereditary torsion theory**. It is known ([112, Prop.3.2, p.141]) that a torsion theory $(\mathcal{T}, \mathcal{F})$ is hereditary if and only if $\mathcal{F}$ is closed under injective hulls.
4. For a preradical ρ and the associated classes of modules $\mathcal{T}_\rho$ and $\mathcal{F}_\rho$ as above, ρ is a left exact radical if and only if $\mathcal{T}_\rho$ is a hereditary torsion class if and only if $\mathcal{F}_\rho$ is a hereditary torsion free class if and only if $(\mathcal{T}_\rho, \mathcal{F}_\rho)$ is a hereditary torsion theory. In this case, we simply write that $\rho = (\mathcal{T}_\rho, \mathcal{F}_\rho)$ is a hereditary torsion theory.

A torsion theory also defines a torsion submodule. For a discussion and clear proof of the following, see Bland [11, pp.11-14].

2.1.7. TORSION SUBMODULE. Let $(\mathcal{T}, \mathcal{F})$ be any torsion theory such that τ is the left exact radical associated with $\mathcal{T}$ and M any module. Then $\tau(M) = \Sigma\{V : V \leq M, V \in \mathcal{T}\}$ is the unique largest submodule of M in $\mathcal{T}$, is fully invariant in M, and $M/\tau(M) \in \mathcal{F}$.

There are different equivalent ways of describing hereditary pretorsion classes and torsion theories. We state the next proposition mainly for general orientation purposes. It can be found in [112, Cor.1.8, p.138 and Prop.3.1, p.141]. Below, we do not use the associated torsion free classes.

2.1.8. DEFINITION. For a ring R, a **prefilter** $\mathfrak{A}$ is a nonempty set of right ideals satisfying the following:

1. $\forall A \subseteq B \leq R,\ A \in \mathfrak{A} \Longrightarrow B \in \mathfrak{A}$;
2. $\forall A \in \mathfrak{A}, B \in \mathfrak{A} \Longrightarrow A \cap B \in \mathfrak{A}$;
3. $\forall x \in R, A \in \mathfrak{A} \Longrightarrow x^{-1}A \in \mathfrak{A}$.

   A prefilter $\mathfrak{A}$ will be called a **hereditary filter** if also (4) below holds:

4. $\forall B \leq R, \forall A \in \mathfrak{A}$, if $\{a^{-1}B : a \in A\} \subseteq \mathfrak{A}$, then also $B \in \mathfrak{A}$.

Prefilters and hereditary filters are sometimes called **linear topologies** and **Gabriel topologies**, respectively.

For a prefilter $\mathfrak{A}$, set $\mathcal{T}_{\mathfrak{A}} = \{M \in \text{Mod-}R : x^{\perp} \in \mathfrak{A}, \forall x \in M\}$.

2.1.9. PROPOSITION. The following hold in the category Mod-R:

1. The correspondence $\mathfrak{A} \longmapsto \mathcal{T}_{\mathfrak{A}}$ between prefilters and hereditary pretorsion classes is bijective.
2. The correspondence $\mathfrak{A} \longmapsto \mathcal{T}_{\mathfrak{A}}$ between hereditary filters and hereditary torsion classes is bijective.

2.1.10. REMARK. For a module M, any prefilter $\mathfrak{A}$ defines a topology on M whose open subsets around the point $0 \in M$ are

$$\{V \leq M : m^{-1}V \in \mathfrak{A}, \forall m \in M\}.$$

In other words, $V \leq M$ is open iff $M/V \in \mathcal{T}_{\mathfrak{A}}$. For an arbitrary $v \in M$, the open neighborhood system of v is obtained from that of "0" by translation as $\{v + V : V \leq M, V \text{ is open}\}$. We make no further use of this remark. The **completion** of the topological space M is a module that so far has not been extensively studied.

2.1.11. EXAMPLE. (1) The semisimple modules form a hereditary pretorsion class.

(2) The singular modules form a hereditary pretorsion class whose associated left exact preradical is the functor Z, where $Z(M)$ is the singular submodule of M.

(3) The class of all **Goldie torsion** modules is, by definition, the smallest hereditary torsion class containing the singular modules. Its associated left exact radical is the functor Z_2, where $Z_2(M)$ is the Goldie torsion submodule of M. Note that both $Z(M) \leq Z_2(M)$ are fully invariant in M, while

$M/Z_2(M)$ is nonsingular. Let $\mathfrak{G}$ be the class of the Goldie torsion modules. Then $\mathfrak{G} = \{N \in \text{Mod-}R : Z(N) \leq_e N\}$ is indeed a stable hereditary torsion class; $\mathfrak{F} = \{N \in \text{Mod-}R : Z(N) = 0\}$ is a hereditary torsion free class; furthermore, $(\mathfrak{G}, \mathfrak{F})$ is a hereditary torsion theory, called the **Goldie torsion theory**.

2.1.12. EXAMPLE. The hereditary torsion theory $(\mathcal{T}, \mathcal{F})$ cogenerated by $E(R)$ is called the **Lambek torsion theory**. In this case, a module M is in $\mathcal{T}$ if and only if $\text{Hom}_R(M, E(R)) = 0$; and clearly $\text{Hom}_R(M, E(R)) = 0$ if and only if $\text{Hom}_R(N, R) = 0\ \forall N \leq M$. A right ideal I of R is dense if and only if $R/I \in \mathcal{T}$; and this is the case if and only if $\text{Hom}(J/I, R) = 0\ \forall I \leq J \leq R$ if and only if ${}^{\perp}(a^{-1}I) = 0\ \forall a \in R$ (see [112, Prop.6.4, p.149]). Every dense right ideal of R is essential in R; a ring R is right nonsingular if and only if every essential right ideal of R is dense. Thus, for a right nonsingular ring, the Goldie torsion theory and the Lambek torsion theory coincide.

2.2 Module Class $\sigma[M]$

For a ring R and an arbitrary R-module M, $\sigma[M]$ is the full subcategory of Mod-R subgenerated by M. As shown in Wisbauer [131], most of the known theorems in ring and module theory carry over to their analogues in $\sigma[M]$. Later in sections 2.4 and 2.5 and other places, facts about $\sigma[M]$ will be used for various purposes, in particular, to study the so called M-natural classes, and pre-natural classes.

2.2.1. DEFINITION. For any ring R and any fixed right R-module M, the category $\sigma[M]$ is the smallest full subcategory of Mod-R consisting of submodules of M-generated modules. That is,

$$\sigma[M] = \{N \in \text{Mod-}R : \exists I, \exists V \leq M^{(I)} \,\text{such that}\, N \hookrightarrow M^{(I)}/V\}.$$

Its morphisms are all possible module morphisms. Let $N \oplus (W/V) \leq_e M^{(I)}/V$ for $N \in \sigma[M]$ as above where W/V is a complement of N in $M^{(I)}/V$. Then $N \hookrightarrow M^{(I)}/W$ as an essential submodule. Consequently, $N \in \sigma[M]$ if and only if $N \leq_e M^{(I)}/W$ for some I and some $W \leq M^{(I)}$. Note that $\sigma[R_R] =$ Mod-R.

It will be useful later that if the sum of modules make sense in $\sigma[M]$, then that sum also belongs to $\sigma[M]$. That is, $\sigma[M]$ is closed under arbitrary sums.

2.2.2. LEMMA. If Y is an R-module and $X_i, i \in I$, are submodules of Y with all $X_i \in \sigma[M]$ $(i \in I)$, then $\Sigma_{i\in I} X_i \in \sigma[M]$.

PROOF. If g is the sum map $g : \oplus_{i\in I} X_i \longrightarrow \Sigma_{i\in I} X_i \leq Y$, and $P = Ker(g)$, then $\Sigma_{i\in I} X_i \cong (\oplus_{i\in I} X_i)/P \in \sigma[M]$. □

The next lemma shows how categories of the form $\sigma[M]$ arise naturally.

2.2.3. DEFINITION AND LEMMA. For any class or set $\mathcal{F}$ of right R-modules, select any set $\{X_i : i \in I\}$ of representatives of isomorphy classes of cyclic submodules of modules in $\mathcal{F}$. That is, for any $A \in \mathcal{F}$ and $a \in A$ (where aR need not be in $\mathcal{F}$), there exists $i \in I$ for which $aR \cong X_i$. Then define "$M_{\mathcal{F}}$" by $M_{\mathcal{F}} = \oplus\{X_i : i \in I\}$. Note that the cardinal $|I|$ is unique, and $M_{\mathcal{F}}$ is unique up to an isomorphism of the direct summands X_i, after a permutation of the index set. Then the following hold for a class $\mathcal{F}$ of modules:

1. $\sigma[M_{\mathcal{F}}]$ is the unique smallest hereditary pretorsion class containing $\mathcal{F}$.

2. $\mathcal{F}$ is a hereditary pretorsion class iff $\mathcal{F} = \sigma[M_{\mathcal{F}}]$.

PROOF. (1). It is straightforward to verify that $\sigma[M_{\mathcal{F}}]$ is a hereditary pretorsion class. The intersection of all hereditary pretorsion classes containing

$\mathcal{F}$ is also a hereditary pretorsion class, which, moreover, contains $\sigma[M_{\mathcal{F}}]$. It only remains to show that $\mathcal{F} \subseteq \sigma[M_{\mathcal{F}}]$. For $N \in \mathcal{F}$ and $x \in N$, $xR \hookrightarrow M_{\mathcal{F}}$ and so $xR \in \sigma[M_{\mathcal{F}}]$. Thus $N = \Sigma\{xR : x \in N\} \in \sigma[M_{\mathcal{F}}]$ by (2.2.2).

(2) follows from (1). □

2.2.4. COROLLARY. A class $\mathcal{F}$ of modules is a hereditary pretorsion class iff $\mathcal{F} = \sigma[M]$ for some module M. □

The next lemma will be used several times later.

2.2.5. Let M be an R-module and xR be a cyclic R-module. Then $xR \in \sigma[M]$ if and only if there exist $x_1, \cdots, x_n \in M$ such that $x_1^{\perp} \cap \cdots \cap x_n^{\perp} \subseteq x^{\perp}$.

PROOF. The implication "$\Longleftarrow$" follows because $\sigma[M]$ is a hereditary pretorsion class.

"$\Longrightarrow$". If $xR \in \sigma[M]$, then $xR \subseteq M^{(I)}/V$ where I is an index set and $V \subseteq M^{(I)}$. Write $x = (x_1, \cdots, x_n) + V$ where $(x_1, \cdots, x_n) \in M^n$. Then $x_1^{\perp} \cap \cdots \cap x_n^{\perp} \subseteq x^{\perp}$. □

2.2.6. DEFINITION. For R-modules N and M, the **trace** of M in N is defined to be

$$\operatorname{tr}(M, N) = \Sigma\{f(M) : f \in \operatorname{Hom}_R(M, N)\} \leq N.$$

Recall that a module N is M-**injective** if for any $X \leq M$ and any map $f : X \longrightarrow N$, f extends to $\overline{f} : M \longrightarrow N$. It is a consequence of these definitions that $\operatorname{tr}(M, EN)$ is an M-injective module. It is called the M-**injective hull** of N, and denoted by $E_M(N) = \operatorname{tr}(M, EN) \leq E(N)$. For any set or class $\mathcal{K}$ of modules, the trace of $\mathcal{K}$ in N is $\operatorname{tr}(\mathcal{K}, N) = \Sigma\{\operatorname{tr}(M, N) : M \in \mathcal{K}\}$. We say that a module N is $\mathcal{K}$-**injective** if N is M-injective for every $M \in \mathcal{K}$.

2.2.7. LEMMA. Let N and M be modules and I any index set. The following hold:

1. $\operatorname{tr}(M, N) \in \sigma[M]$.
2. $E_M(N) \in \sigma[M]$.
3. $\operatorname{tr}(M, EN) = \operatorname{tr}(M^{(I)}, EN) = \operatorname{tr}(\sigma[M], EN)$.

PROOF. Since $\sigma[M]$ is a hereditary pretorsion class and $\operatorname{tr}(M, N)$ is a quotient of a direct sum of copies of M, (1) follows. Similarly, (2) follows.

(3) For any module N, $\operatorname{tr}(M, EN) = \operatorname{tr}(M^{(I)}, EN) \subseteq \operatorname{tr}(\sigma[M], EN)$. For any $A \in \sigma[M]$ and any $f : A \longrightarrow EN$, we may assume $A \leq M^{(J)}/V$ for some J and V. Then f extends to $\overline{f} : M^{(J)}/V \longrightarrow EN$. Let $\pi : M^{(J)} \longrightarrow M^{(J)}/V$ be the canonical homomorphism. Then

$$f(A) \subseteq \overline{f}(M^{(J)}/V) = (\overline{f}\pi)(M^{(J)}) \leq \operatorname{tr}(M, EN).$$

□

2.2.8. LEMMA. Let N and M be modules. If N is M-injective then N is $\sigma[M]$-injective.

PROOF. For any $A \in \sigma[M]$, $A \leq M^{(I)}/V$ for some I and V. First, by Mohamed and Müller [87, Prop.1.5], $E_M(N)$ is $M^{(I)}$-injective. Then apply [87, Prop.1.3] two times to conclude that first $E_M(N)$ is $M^{(I)}/V$-injective, and secondly, it is A-injective. $\square$

2.2.9. LEMMA. For any module $N \in \sigma[M]$, $N \subseteq E_M(N)$.

PROOF. Let $N \leq M^{(I)}/V$ as in (2.2.1) for some I and some $V \leq M^{(I)}$. Then $N = L/V$ for some $L \leq M^{(I)}$. Let $\pi : L \longrightarrow L/V$ be the canonical homomorphism, and $\overline{\pi} : M^{(I)} \longrightarrow EN$ be the extension of π to $M^{(I)}$. But then $N = \pi(L) \subseteq \overline{\pi}(M^{(I)}) \subseteq \operatorname{tr}(\sigma[M], EN) = E_M(N)$ by (2.2.7)(3). $\square$

The next proposition is used frequently. It says that $E_M(\cdot)$ behaves the same as $E(\cdot)$.

2.2.10. PROPOSITION. For any module M, let $Y \leq X \in \sigma[M]$ and $N \in \sigma[M]$. Then the following hold:

1. If Y is M-injective, then Y is a summand of X.
2. $E_M(X) = E_M(Y) \oplus E_M(Z)$ for some $Z \leq X$.
3. If $N = N_1 \oplus \cdots \oplus N_m$, then $E_M(N) = E_M(N_1) \oplus \cdots \oplus E_M(N_m)$.
4. N is M-injective iff $N = E_M(N)$.
5. If N is M-injective, then N is quasi-injective.

PROOF. (1) If Y is M-injective, then Y is $\sigma[M]$-injective, and hence is X-injective. So Y is a summand of X.

(2) It follows from $Y \leq X \in \sigma[M]$ that $E_M(Y) \leq E_M(X) \in \sigma[M]$. Since $E_M(Y)$ is M-injective, $E_M(X) = E_M(Y) \oplus V$ for some V by (1). Let $Z = X \cap V$. Then $Z \leq V$ and so $E_M(Z) = E_M(V)$ (since $EZ = EV$). It follows that

$$\begin{aligned} E_M(X) =& E_M(Y) \oplus V \\ \leq& E_M(Y) \oplus E_M(V) \\ =& E_M(Y) \oplus E_M(Z) \leq E_M(X). \end{aligned}$$

(3) It suffices to show this for $m = 2$. Let $p_i : E(N_1 \oplus N_2) \longrightarrow E(N_i)$ $(i = 1, 2)$ be the projection. We have

$$E_M(N_1) \oplus E_M(N_2) \subseteq E_M(N_1 \oplus N_2) = E_M(N),$$

since $p_j(E_M(N_i)) = 0$ if $i \neq j$. Since $E_M(N_1) \oplus E_M(N_2)$ is M-injective and is essential in $E_M(N)$, it follows from (1) that $E_M(N) = E_M(N_1) \oplus E_M(N_2)$.

(4) $E_M(N)$ is clearly M-injective by its definition. Assume that N is M-injective. Since $N \subseteq E_M(N)$ by (2.2.9) and $N \leq_e E_M(N)$, it follows from (1) that N is a summand of $E_M(N)$, and hence $N = E_M(N)$.

(5) If N is M-injective, then N is $\sigma[M]$-injective, and hence is N-injective. □

It is essential that in the previous proposition the modules X and N are in $\sigma[M]$. The next proposition, which appeared in [101], explains the case where N may not be in $\sigma[M]$.

2.2.11. PROPOSITION. The following are equivalent for R-modules M and N:

1. N is M-injective.

2. $E_M(N) \cap N$ is M-injective.

3. $E_M(N) \subseteq N$.

PROOF. (1) $\Longrightarrow$ (2). Let $f : K \longrightarrow E_M(N) \cap N$ be an R-homomorphism, where $K \leq M$. Because N is M-injective, f extends to an R-homomorphism $g : M \longrightarrow N$. Note that $g(M) \subseteq E_M(N)$ and so $g(M) \subseteq E_M(N) \cap N$. Thus, $E_M(N) \cap N$ is M-injective.

(2) $\Longrightarrow$ (3). We may assume $E_M(N) \neq 0$. Let $T = E_M(N) \cap N$. Then $T \leq_e E_M(N) \subseteq E(N)$ and hence $E_M(N) \subseteq E(T) \subseteq E(N)$. It follows that $E_M(T) \subseteq E_M(N)$. But if $f \in \mathrm{Hom}(M, EN)$, then $f(M) \subseteq E_M(N) \subseteq E(T)$, implying $E_M(N) \subseteq E_M(T)$. So, $E_M(N) = E_M(T)$. By (2), T is M-injective. Note that $T \in \sigma[M]$. Then $T = E_M(T)$ by (2.2.10)(4). Therefore, $E_M(N) \subseteq N$.

(3) $\Longrightarrow$ (1). Let $f : K \longrightarrow N$ be an R-homomorphism, where $K \leq M$. Now f can be extended to $g : M \longrightarrow E(N)$. Since $g(M) \subseteq E_M(N) \subseteq N$, N is M-injective. □

2.2.12. DEFINITION. For $M, N \in$ Mod-R, N is defined to be M-**singular** if $N \cong L/K$ for some $L \in \sigma[M]$ and $K \leq_e L$; consequently, $N \in \sigma[M]$. The class of all M-singular modules is closed under submodules, quotient modules, and direct sums. It then follows from this that for any $N \in$ Mod-R, $\Sigma\{V : V \leq N, V \text{ is } M\text{-singular}\}$ is M-singular. This unique largest M-singular submodule of N is denoted by $Z_M(N) = Z_M N$. Note that $Z_M(N) \subseteq E_M(N) \in \sigma[M]$ even if N is not in $\sigma[M]$, and that $Z_M(N) \subseteq Z(N)$. N is called **non** M-**singular** if $Z_M(N) = 0$. For any $A, B \in$ Mod-R and any R-map $f : A \longrightarrow B$, since $f(Z_M A) \subseteq Z_M B$, Z_M is a subfunctor of the identity functor of Mod-R.

Recall that a module M is Goldie torsion if and only if $M = Z_2(M)$. Consequently, the class of Goldie torsion modules is precisely the class $\mathfrak{G} = \{M \in \text{Mod-}R : ZM \leq_e M\}$. The latter is clearly closed under extensions and injective hulls, and it is easy to see that the class of Goldie torsion modules is the unique smallest hereditary torsion class containing all the singular modules.

2.2.13. DEFINITION. For a given module M, the Goldie torsion class $\mathfrak{G}_M$ in $\sigma[M]$ is $\mathfrak{G}_M = \{N \in \sigma[M] : Z_M(N) \leq_e N\}$.

2.2.14. LEMMA. The above class $\mathfrak{G}_M$ is the smallest hereditary torsion class in the category $\sigma[M]$ containing all M-singular modules.

PROOF. Let $\mathcal{C}$ be a subclass of $\sigma[M]$ such that $\mathcal{C}$ contains all M-singular modules and is closed under extensions. Then for any $N \in \mathfrak{G}_M$, since $Z_M(N) \leq_e N \in \sigma[M]$, $N/Z_M(N)$ is M-singular. From the short exact sequence $0 \longrightarrow Z_M(N) \longrightarrow N \longrightarrow N/Z_2(N) \longrightarrow 0$ in $\sigma[M]$, it follows that $N \in \mathcal{C}$. Thus $\mathfrak{G}_M \subseteq \mathcal{C}$.

In order to show that $\mathfrak{G}_M$ actually is a hereditary torsion class in $\sigma[M]$, it has to be shown that it is closed under the following operations: $\leq, /, \oplus$, and $\mathrm{Ext}^1_R(\cdot,\cdot)$. For $C \leq A \in \mathfrak{G}_M$, we have $Z_M(A) \leq_e A$ and hence $Z_M(C) = Z_M(A) \cap C \leq_e A \cap C = C$. So $C \in \mathfrak{G}_M$. There exists a $B/C \leq A/C$ such that $Z_M(A/C) \oplus B/C \leq_e A/C$. Since $[(Z_M(A) \cap B) + C]/C \subseteq Z_M(B/C) = 0$, $Z_M(A) \cap B \subseteq C$. Since $Z_M(A) \leq_e A$, it follows that $Z_M(A) \cap B \leq_e B$ and hence $C \leq_e B$. So $B/C = Z_M(B/C) = 0$. Thus $Z_M(A/C) \leq_e A/C$, so $A/C \in \mathfrak{G}_M$.

The closure of $\mathfrak{G}_M$ under arbitrary direct sums is clear.

For a short exact sequence $0 \longrightarrow A \longrightarrow B \longrightarrow B/A \longrightarrow 0$ in Mod-R with $A, B/A \in \mathfrak{G}_M$, let $D \leq B$ such that $Z_M(B) \oplus D \leq_e B$. Thus $Z_M(D) = 0$. Since $Z_M(A) \subseteq Z_M(B)$ and $Z_M(A) \leq_e A$, $A \cap D = 0$. Thus $D \cong (D+A)/A \leq B/A$ and so $Z_M((D+A)/A) = 0$. Hence

$$Z_M(B/A) \cap (D+A)/A \subseteq Z_M((D+A)/A) = 0.$$

Since $Z_M(B/A) \leq_e B/A$, it follows that $(D+A)/A = 0$ and so $D = 0$. Thus $Z_M(B) \leq_e B$. So if also $B \in \sigma[M]$, then $B \in \mathfrak{G}_M$, and $\mathfrak{G}_M$ is closed under extensions. □

Some properties of the Goldie torsion theory in $\sigma[M]$ are described as follows. For any $N \in \sigma[M]$, by (2.2.14) N has a unique largest submodule in $\mathfrak{G}_M$, which is denoted by $\mathfrak{G}_M(N)$.

2.2.15. PROPOSITION. For any $N \in \mathfrak{G}_M$ and $A \in \sigma[M]$, the following hold:

1. $E_M(N) \in \mathfrak{G}_M$.

2. $\mathfrak{G}_M(A) = A \cap E_M(Z_M A) \in \mathfrak{G}_M$.

PROOF. (1) Since $N \in \sigma[M]$ and $N \leq_e E_M(N) \in \sigma[M]$, $E_M(N)/N \in \mathfrak{G}_M$. From the short exact sequence $0 \longrightarrow N \longrightarrow E_M(N) \longrightarrow E_M(N)/N \longrightarrow 0$, we conclude that $E_M(N) \in \mathfrak{G}_M$.

(2) First, $Z_M(A) \leq_e A \cap E_M(Z_M A) \in \mathfrak{G}_M$. Suppose that $Z_M(A) \leq_e L \leq A$. At this point the hypothesis that $A \in \sigma[M]$ is needed to conclude that $L \subseteq E_M(L) \cap A$. But since $Z_M(A) \leq_e L$, $E_M(L) = E_M(Z_M A)$ and hence

$L \subseteq A \cap E_M(Z_M A)$. Therefore, $A \cap E_M(Z_M A)$ is the unique largest submodule of A that belongs to $\mathfrak{G}_M$. So $\mathfrak{G}_M(A) = A \cap E_M(Z_M A)$. □

Whenever M is a generator of Mod-R, such as $M = R$, we have $\sigma[M] =$ Mod-R, and concepts in $\sigma[M]$ become the usual ones in Mod-R. In particular, $E_M(N) = E(N), Z_M(N) = Z(N), \mathfrak{G}_M = \mathfrak{G}$, and $\mathfrak{G}_M(N) = Z_2(N)$.

2.2.16. Recall that a nonzero submodule L of a module N is a **rational** submodule if for every submodule A of N with $L \subset A$ and every homomorphism $f : A \longrightarrow N$, $f(L) = 0$ implies that $f = 0$. The module N is **strongly uniform** if every nonzero submodule of N is a rational submodule (see Storrer [113, pp.621-622]). A synonym for strongly uniform is **monoform** (see Gordon and Robson [69, p.11]). Thus $0 \neq L \leq N$ is a rational submodule iff for any $m_1, m_2 \in M$ with $m_2 \neq 0$, $m_2(m_1^{-1}L) \neq 0$.

If M is strongly uniform, then M is uniform. For if $L \oplus mR \leq M$ with $L \neq 0$ and $m \neq 0$, take $m_1 = m_2 = m$. Then $m^{-1}L = m^{\perp}$. Hence $m(m^{-1}L) = mm^{\perp} = 0$, a contradiction.

It is known that if U is uniform and nonsingular, then U is strongly uniform. However, in the category $\sigma[M]$, U may be non M-singular but with $U = Z(U)$ singular.

2.2.17. LEMMA. Let $U \in \sigma[M]$ be uniform and non M-singular. Then U is strongly uniform.

PROOF. If not, there exists a nonzero essential submodule L of U such that L is not rational. Hence there exist $m_1 \in U, 0 \neq m_2 \in U$ with $m_2(m_1^{-1}L) = 0$. Then there is an epimorphism

$$(m_1 R + L)/L \longrightarrow m_2 R, \ m_1 r + L \longmapsto m_2 r, \ r \in R.$$

But $L \leq_e m_1 R + L \in \sigma[M]$ shows that $m_2 R$ is M-singular, a contradiction. □

2.2.18. For any module $N \in$ Mod-R, let $\overline{N} = (\text{End}_R EN)N \leq EN$ be its quasi-injective hull. Then

1. $E_M(N)$ is quasi-injective.
2. If $N \in \sigma[M]$, then $N \subseteq \overline{N} \subseteq E_M(N)$.

PROOF. (1) This follows, since $E_M(N)$ is $\sigma[M]$-injective, or see (2.2.10)(5).

(2) For $f \in \text{End}_R(EN)$, we may assume that $N \subseteq E_M(N) \leq M^{(I)}/V$ for some I and V, and $f|_N : N \longrightarrow EN$ extends to $g : M^{(I)}/V \longrightarrow EN$. Hence $f(N) \subseteq \text{tr}(\sigma[M], EN) = E_M(N)$. Thus $\overline{N} \subseteq E_M(N)$. □

Since strongly uniform modules occasionally arise in the study of natural and M-natural classes, we list a few of their known properties below.

2.2.19. Let U be any module and $\overline{U} = (\text{End}_R EU)U$ its **quasi-injective hull**. Then the following hold:

1. U is strongly uniform iff $\mathrm{End}_R\overline{U}$ is a division ring ([113, Prop.7.5, p.656] or [69, p.103]).

2. U is strongly uniform iff $\overline{U}$ is ([113, Prop.7.5, p.656 and pp.622-624]).

3. If U is strongly uniform, there is a natural embedding $\mathrm{End}_R U \subseteq \mathrm{End}_R\overline{U}$.

2.3 Natural Classes

A key feature here is the concept of a natural class. Historically, many results were first proven for the set $\mathcal{N}(R)$ of natural classes and then later extended to the set $\mathcal{N}^p(R)$ of pre-natural classes, since $\mathcal{N}(R) \subseteq \mathcal{N}^p(R)$.

When possible, we give the proofs for $\mathcal{N}^p(R)$, but in such a way that the relevant facts and conclusions for $\mathcal{N}(R)$ are transparent. For this reason, in this section, we will state two theorems, but postpone the proofs to a later section on pre-natural classes.

The operators "d" and "c", which can be applied to any nonempty set or class of modules, will be used frequently in this section as well as later on.

2.3.1. DEFINITION. For any nonempty class $\mathcal{F}$ of right R-modules, define $d(\mathcal{F})$ and a complementary class $c(\mathcal{F})$ of $d(\mathcal{F})$ as

$$d(\mathcal{F}) = \{N \in \text{Mod-}R : \forall 0 \neq W \leq N, \exists 0 \neq V \leq W, V \hookrightarrow A \,\text{for some}\, A \in \mathcal{F}\},$$
$$c(\mathcal{F}) = \{W \in \text{Mod-}R : \forall 0 \neq V \leq W, V \not\hookrightarrow A \,\text{for any}\, A \in \mathcal{F}\}.$$

In words, $d(\mathcal{F})$ consists of those modules N whose every nonzero submodule W contains a nonzero submodule V which embeds into some module A of $\mathcal{F}$, while $c(\mathcal{F})$ consists of all modules W which do not contain any nonzero submodules V embeddable in some module A of $\mathcal{F}$. Thus

$$\begin{aligned} c(c(\mathcal{F})) &= \{N \in \text{Mod-}R : \forall 0 \neq W \leq N, W \not\hookrightarrow A \,\text{for any}\, A \in c(\mathcal{F})\} \\ &= \{N \in \text{Mod-}R : \forall 0 \neq W \leq N, \exists 0 \neq V \leq W, V \hookrightarrow A \,\text{for some}\, A \in \mathcal{F}\} \\ &= d(\mathcal{F}). \end{aligned}$$

Let $\overline{\mathcal{F}} = \{N \in \text{Mod-}R : N \hookrightarrow A \text{ for some } A \in \mathcal{F}\}$. Most of the time we will think in terms of $\overline{\mathcal{F}}$, for which the above definitions simplify as

$$d(\mathcal{F}) = d(\overline{\mathcal{F}}) = \{N \in \text{Mod-}R : \forall 0 \neq W \leq N, \exists 0 \neq V \leq W, V \in \overline{\mathcal{F}}\},$$
$$c(\mathcal{F}) = c(\overline{\mathcal{F}}) = \{W \in \text{Mod-}R : \forall 0 \neq V \leq W, V \notin \overline{\mathcal{F}}\}.$$

An application of the very last formula to $\overline{\mathcal{F}}$ in place of $\mathcal{F}$ shows that $c(c(\mathcal{F})) = c(c(\overline{\mathcal{F}})) = d(\overline{\mathcal{F}})$. We already saw that $d(\mathcal{F}) = c(c(\mathcal{F}))$. For this reason, "$d$" and "$c$" are thought of as complementary operations. For example, if $\mathcal{F}$ is the class of all singular modules, then $c(\mathcal{F})$ is the class of all nonsingular modules and $d(\mathcal{F})$ is the class of all Goldie torsion modules.

Modules $N \in d(\mathcal{F})$ can be regarded as being "$\mathcal{F}$-dense", because every nonzero submodule of N contains a possibly smaller, but still nonzero submodule of $\overline{\mathcal{F}}$. Such modules N may be thought of as being locally in $\mathcal{F}$. Modules $W \in c(\mathcal{F})$ can be regarded as being "$\mathcal{F}$-free" (or free of $\mathcal{F}$), because they do not contain any nonzero submodules of $\overline{\mathcal{F}}$. Note that $0 \in d(\mathcal{F})$ and $0 \in c(\mathcal{F})$, and that $c(\mathcal{F}) \cap d(\mathcal{F}) = \{0\}$.

For a singleton set $\mathcal{F} = \{P\}$ for $P \in$ Mod-R, we sometimes abbreviate $c(\{P\}) = c(P)$ and $d(\{P\}) = c(P)$. If $\mathcal{F} = \emptyset$, then the above definitions require that $c(\emptyset) =$ Mod-R and $d(\emptyset) = \{0\}$.

The concept of a natural class will be a unifying thread running throughout.

2.3.2. DEFINITION. A nonempty class $\mathcal{K}$ of modules is a **natural class** (also called **saturated class** or **type**) if it is closed under submodules, arbitrary direct sums, and essential extensions (equivalently injective hulls). The collection of all natural classes of right R-modules will be denoted by $\mathcal{N}(R)$. When we will also be considering left modules, we will distinguish the natural classes of left R-modules from the right ones by $\mathcal{N}_l(R)$ and $\mathcal{N}_r(R)$.

In order to work with natural classes, the following argument will be used repeatedly later on, sometimes without mentioning that it is used.

2.3.3. PROJECTION ARGUMENT. Let W_α, $\alpha \in \Lambda$, be any indexed family of modules, and $0 \neq x \in E(\oplus_{\alpha\in\Lambda} W_\alpha)$. Then there exists $\alpha \in \Lambda, 0 \neq r \in R$, and $0 \neq w \in W_\alpha$ such that $0 \neq xrR \cong wR \leq W_\alpha$ with $(xr)^\perp = w^\perp$. In other words, every nonzero submodule of $E(\oplus_{\alpha\in\Lambda} W_\alpha)$ contains a nonzero submodule isomorphic to a submodule of some W_α.

2.3.4. The following hold for a nonempty class $\mathcal{F}$ of modules:

1. $d(\mathcal{F}), c(\mathcal{F})$ are natural classes, and $d(\mathcal{F}) \cap c(\mathcal{F}) = 0$.

2. $d(\mathcal{F}) = c(c(\mathcal{F}))$ is the unique smallest natural class containing $\mathcal{F}$.

3. $\forall \mathcal{K} \in \mathcal{N}(R)$, $\mathcal{K} = c(c(\mathcal{K}))$.

PROOF. (1) By (2.3.2), $d(\mathcal{F}) \in \mathcal{N}(R)$. The projection argument (2.3.3) shows that $c(\mathcal{F})$ is closed under injective hulls of arbitrary direct sums. Hence $c(\mathcal{F}) \in \mathcal{N}(R)$. It follows from (2.3.1) that $d(\mathcal{F}) \cap c(\mathcal{F}) = 0$.

(3) For any $0 \neq M \in c(c(\mathcal{K}))$, by (2.3.1) and Zorn's Lemma, there exists $\oplus_{\alpha\in\Lambda} V_\alpha \leq_e M$ with $0 \neq V_\alpha \in \mathcal{K}$. Hence by (2.3.2), $M \in \mathcal{K}$. Thus $\mathcal{K} = c(c(\mathcal{K}))$.

(2) By (2.3.1), $\mathcal{F} \subseteq c(c(\mathcal{F}))$. If $\mathcal{F} \subseteq \mathcal{K} \in \mathcal{N}(R)$, then $c(c(\mathcal{F})) \subseteq c(c(\mathcal{K})) = \mathcal{K}$. Moreover, $d(\mathcal{F}) = c(c(\mathcal{F}))$ by (2.3.1). □

2.3.5. COROLLARY. Every natural class $\mathcal{K}$ is closed under extensions.

PROOF. Suppose that $0 \longrightarrow A \longrightarrow B \longrightarrow C \longrightarrow 0$ is exact with $A, C \in \mathcal{K}$ but $B \notin \mathcal{K}$. So $B \notin c(c(\mathcal{K}))$ by (2.3.4)(3). Thus there exists a $0 \neq D \leq B$ with $D \in c(\mathcal{K})$. Then $A \cap D = 0$, and $0 \neq D \cong (D+A)/A \hookrightarrow C$, contradicting that $C \in \mathcal{K}$. □

2.3.6. REPRESENTATIVE OF A NATURAL CLASS.

1. For any module A, $d(A) = \{B \in \text{Mod-}R : \exists I, B \hookrightarrow E(A^{(I)})\}$.

2. For any class $\mathcal{F}$ of modules, $d(\mathcal{F}) = d(M_{\mathcal{F}})$ (see 2.2.3).

3. $\forall \mathcal{K} \in \mathcal{N}(R)$, $\mathcal{K} = d(M_{\mathcal{K}}) = c(M_{c(\mathcal{K})})$.

For $\mathcal{K}, \mathcal{L} \in \mathcal{N}(R)$, define $\mathcal{K} \leq \mathcal{L}$ if $\mathcal{K} \subseteq \mathcal{L}$. This defines a partial order on $\mathcal{N}(R)$ with smallest and largest elements $\mathbf{0} = \{0\}$ and $\mathbf{1} = \text{Mod-}R$. For $\mathcal{K}, \mathcal{L} \in \mathcal{N}(R)$, whenever their greatest lower and least upper bounds exist in the poset $\mathcal{N}(R)$, they are denoted as $\mathcal{K} \wedge \mathcal{L}$ and $\mathcal{K} \vee \mathcal{L}$, and similarly for infinite sups and infs of subsets of $\mathcal{N}(R)$. Later it will be shown that the collection $\mathcal{N}(R)$ actually is a set. We call $\mathcal{K} \in \mathcal{N}(R)$ an **atom** if $\mathcal{K}$ is an atom with respect to the partial order in $\mathcal{N}(R)$, i.e., $\mathcal{K} \neq \mathbf{0}$ and, whenever $\mathbf{0} \neq \mathcal{L} \leq \mathcal{K}$ with $\mathcal{L} \in \mathcal{N}(R)$, we have that $\mathcal{L} = \mathcal{K}$.

2.3.7. DEFINITION. A nonzero module A is **atomic** if $d(A)$ is an atom of $\mathcal{N}(R)$. It can easily be verified that a module A is atomic iff $A \neq 0$ and, for any $x, y \in A \backslash \{0\}$, xR and yR have isomorphic nonzero submodules.

2.3.8. DEFINITION. Let $\mathcal{F}$ be any nonempty set or class of modules. For any module M, define

$$H_{\mathcal{F}}(M) = \{N \leq M : M/N \in \mathcal{F}\} \text{ and}$$
$$H_{\mathcal{F}}(R) = \{I \leq R : R/I \in \mathcal{F}\}.$$

The proof below is omitted. Later, a more general proof is given for pre-natural classes and see (2.5.6) for this. We let $\mathrm{L}(R)$ be the lattice of right ideals of R.

2.3.9. The correspondence $\mathcal{K} \longmapsto H_{\mathcal{K}}(R)$ from $\mathcal{N}(R)$ to $\mathcal{P}(\mathrm{L}(R))$, the power set of $\mathrm{L}(R)$, is one-to-one. Hence $|\mathcal{N}(R)| \leq |\mathcal{P}(\mathrm{L}(R))| \leq 2^{|\mathcal{P}(R)|}$.

In the same way as a torsion theory is determined by a filter of right ideals of the ring, it will also turn out that a natural class (and later also pre-natural class) can be determined by sets of right ideals, as in the next definition. This definition first appeared in [136, 2.1]. Later we will expand it for pre-natural classes in (2.5.7).

2.3.10. DEFINITION. A nonempty set $\mathfrak{A}$ of right ideals of R is a **natural set** of right ideals if the following hold:

1. If $I, J \in \mathfrak{A}$, then $I \cap J \in \mathfrak{A}$.

2. If $I \in \mathfrak{A}$, then $a^{-1}I \in \mathfrak{A}$ for all $a \in R$.

3. If $I \notin \mathfrak{A}$, then there exists a right ideal J of R such that J properly contains I and $a^{-1}I \notin \mathfrak{A}$ for all $a \in J \backslash I$.

A nonempty set $\mathfrak{B}$ of right ideals of R is a **torsion free set** if $\mathfrak{B}$ is a natural set, and if $\mathfrak{B}$ is closed under arbitrary intersections.

As already observed in Page and Zhou [98], it was the sets of the form "$H_{\mathcal{K}}(R)$" which played a significant role in various module theoretic proofs. The next theorem due to [136, 2.2] shows why these proofs are really proofs about natural classes.

The proof of the next result is given in a more general context in (2.5.8).

2.3.11. THEOREM. For a set $\mathfrak{A}$ of right ideals of R, the following are equivalent:

1. $\mathfrak{A} = H_{\mathcal{K}}(R)$ for some $\mathcal{K} \in \mathcal{N}(R)$.

2. $\mathfrak{A}$ is a natural set.

2.3.12. COROLLARY. For a natural class $\mathcal{K}$ and any R-module M, $M \in \mathcal{K}$ iff $x^{\perp} \in H_{\mathcal{K}}(R)$ for all $x \in M$.

The next proposition was proved in [99, Prop.20].

2.3.13. PROPOSITION. The following hold for a natural class $\mathcal{K}$:

1. $\mathcal{K}$ is a hereditary torsion class iff, for any $I \subseteq J \leq R$, $I \in H_{\mathcal{K}}(R)$ implies $J \in H_{\mathcal{K}}(R)$.

2. $\mathcal{K}$ is a hereditary torsion free class iff $H_{\mathcal{K}}(R)$ is closed under arbitrary intersections.

PROOF. See [99, Prop.20]. □

As seen in (2.1.9), there is a bijective correspondence between hereditary torsion classes and hereditary filters. In [136, Prop.2.3] the following similar and complementary result was obtained.

2.3.14. COROLLARY. For any ring R, there is a bijective correspondence between hereditary torsion free classes of right R-modules and torsion free sets of right ideals of R.

For the readers' convenience we summarize in one place some of the ways how a natural class can be described.

2.3.15. THEOREM. The following are equivalent for a class $\mathcal{K}$ of modules:

1. $\mathcal{K}$ is a natural class.

2. $\mathcal{K} = c(\mathcal{F})$ for a class $\mathcal{F}$.

3. $\mathcal{K} = d(\mathcal{F})$ for a class $\mathcal{F}$.

4. $\mathcal{K} = c(X)$ for a module X.

5. $\mathcal{K} = d(Y)$ for a module Y.

6. $\mathcal{K} = d(\mathcal{K})$.

7. $\mathcal{K} = d(M_{\mathcal{K}})$.

8. $\mathcal{K} = c(M_{c(\mathcal{K})})$.

9. $H_{\mathcal{K}}(R)$ is a natural set.

Below, (3) was first observed in [99, Example (ii)]. See (3.1.18) for its proof.

2.3.16. EXAMPLES. All of the following are natural classes:

1. A hereditary torsion free class.

2. A stable hereditary torsion class.

3. $\mathcal{K} = \{N \in \text{Mod-}R : E(N) \text{ is singular}\}$.

4. $\mathcal{K} = \{N \in \text{Mod-}R : Soc(N) \leq_e N\}$.

PROOF. (1) and (2) are clear. See (3.1.18) for the proof of (3).

(4) If $\mathcal{D}$ is the class of all simple modules, then $\mathcal{K} = d(\mathcal{D})$, and by (2.3.4), $\mathcal{K}$ is a natural class. □

A natural class is, in general, not a hereditary torsion class or a hereditary torsion free class. The torsion $\mathbb{Z}$-modules is a natural class that is not closed under products, and hence is not a hereditary torsion free class. The torsion free $\mathbb{Z}$-modules is a natural class that is not closed under quotient modules, and hence is not a hereditary torsion class.

2.3.17. EXAMPLE. A natural class which is neither a hereditary torsion class nor a hereditary torsion free class is given as follows.

Let R be the **trivial extension** of the ring $\mathbb{Z}$ and the $\mathbb{Z}$-module $\oplus_{i=1}^{\infty}\mathbb{Z}_{p_i}$ where p_i is the ith prime number, i.e., $R = \left\{\begin{pmatrix} a & x \\ 0 & a \end{pmatrix} : a \in \mathbb{Z}, x \in \oplus_{i=1}^{\infty}\mathbb{Z}_{p_i}\right\}$, with addition and multiplication as matrix addition and matrix multiplication. Let

$$I_0 = \left\{\begin{pmatrix} 0 & x \\ 0 & 0 \end{pmatrix} : x \in \oplus_{i=1}^{\infty}\mathbb{Z}_{p_i}\right\},\ I = \left\{\begin{pmatrix} 0 & x \\ 0 & 0 \end{pmatrix} : x \in \mathbb{Z}_{p_1}\right\}, \text{ and}$$

$$J = \left\{\begin{pmatrix} 0 & x \\ 0 & 0 \end{pmatrix} : x \in \oplus_{i=2}^{\infty}\mathbb{Z}_{p_i}\right\}.$$

Let $\mathcal{K} = d(J)$. Then $\mathcal{K}$ is a natural class and $R/I \in \mathcal{K}$ but $R/I_0 \notin \mathcal{K}$. So $\mathcal{K}$ is not closed under quotient modules and hence is not a hereditary torsion class. Let $I_i = \left\{\begin{pmatrix} a & x \\ 0 & a \end{pmatrix} : a \in p_i\mathbb{Z}, x \in \oplus_{i=1}^{\infty}\mathbb{Z}_{p_i}\right\}$. Then, for $i \geq 2$, $R/I_i \in \mathcal{K}$ and $\cap_{i\geq 2}I_i = I_0$ and $R/I_0 \notin \mathcal{K}$. Thus $\mathcal{K}$ is not closed under products and hence is not a hereditary torsion free class.

2.4 M-Natural Classes

After [98] was written about the injectivity of direct sums of singular injective modules, it became apparent to these authors that the natural setting for investigating the injectivity of direct sums of injectives, was an M-natural class. Thus for the very first time M-natural classes were introduced in [99]. Later they were further studied in [135, 138], and many subsequent publications. Throughout, M is an arbitrary fixed R-module, and $\sigma[M]$ is as in section 2.2.

2.4.1. DEFINITION. A subclass of $\sigma[M]$ which is closed under isomorphic copies, submodules, arbitrary direct sums, and M-injective hulls is called an M-**natural class**. The collection of all M-natural classes is denoted by $\mathcal{N}(R, M)$.

The next lemma shows that M-natural classes are closed under essential extensions inside $\sigma[M]$.

2.4.2. LEMMA. Let $\mathcal{K}$ be an M-natural class and $X \leq_e N \in \sigma[M]$. If $X \in \mathcal{K}$, then $N \in \mathcal{K}$.

PROOF. Since $X \leq_e N$, $E(X) = E(N)$ and hence $E_M(X) = E_M(N)$. By (2.2.9), $N \subseteq E_M(N) \in \mathcal{K}$, and hence $N \in \mathcal{K}$. □

2.4.3. LEMMA. Let $\mathcal{F}$ be a subclass of $\sigma[M]$. Then

1. $d(\mathcal{F}) \cap \sigma[M]$ is the unique smallest M-natural class containing $\mathcal{F}$.
2. $\mathcal{F}$ is an M-natural class iff $\mathcal{F} = d(\mathcal{F}) \cap \sigma[M]$.

PROOF. (1) Since $d(\mathcal{F})$ is a natural class, $d(\mathcal{F}) \cap \sigma[M]$ is clearly closed under submodules, direct sums, and M-injective hulls, and hence is an M-natural class. Moreover, it is obvious that $\mathcal{F} \subseteq d(\mathcal{F}) \cap \sigma[M]$. Suppose that $\mathcal{F} \subseteq \mathcal{L}$ where $\mathcal{L}$ is an M-natural class. For $N \in d(\mathcal{F}) \cap \sigma[M]$, N contains as an essential submodule a direct sum $\oplus\{N_i : i \in I\}$ where all $N_i \in \mathcal{F}$. Then $\oplus\{N_i : i \in I\} \in \mathcal{L}$ since $\mathcal{L}$ is closed under direct sums. Thus by (2.4.2), $N \in \mathcal{L}$. Thus $d(\mathcal{F}) \cap \sigma[M] \subseteq \mathcal{L}$, so (1) follows.

(2) follows from (1). □

2.4.4. COROLLARY. A class $\mathcal{F}$ of modules is an M-natural class iff $\mathcal{F} = \mathcal{K} \cap \sigma[M]$ for some natural class $\mathcal{K}$.

PROOF. One direction is by (2.4.3). Suppose that $\mathcal{F} = \mathcal{K} \cap \sigma[M]$ where $\mathcal{K}$ is a natural class. It is easy to see that $\mathcal{F}$ is closed under submodules, direct sums, and M-injective hulls, and hence is an M-natural class. □

2.4.5. COROLLARY. Let $\mathcal{K}$ be an M-natural class and $X \subseteq N \in \sigma[M]$.

1. If $N \notin \mathcal{K}$, then there is a $0 \neq Y \leq N$ with $Y \in c(\mathcal{K})$.
2. If $X \in \mathcal{K}$ and $N/X \in \mathcal{K}$, then $N \in \mathcal{K}$.

PROOF. (1) Since $N \in \sigma[M]$ but $N \notin \mathcal{K} = d(\mathcal{K}) \cap \sigma[M]$, $N \notin d(\mathcal{K})$. This requires the existence of a $0 \neq Y \leq N$ with $Y \in c(\mathcal{K})$.

(2) Suppose that $N \notin \mathcal{K}$. By (1) there exists a $0 \neq Y \subseteq N$ such that $Y \in c(\mathcal{K})$. Since $X \in \mathcal{K}$, it must be that $X \cap Y = 0$. Thus, $Y \hookrightarrow N/X \in \mathcal{K}$, showing that $Y \in \mathcal{K}$, a contradiction. □

The next definition appeared in [135, Lemma 1.5].

2.4.6. DEFINITION. Given an M-natural class $\mathcal{K}$, a module U is $\mathcal{K}$-**cocritical** if $0 \neq U \in \mathcal{K}$, but $U/P \notin \mathcal{K}$ for any $0 \neq P \subset U$; a submodule V of U is called $\mathcal{K}$-**critical** if U/V is $\mathcal{K}$-cocritical.

This is the same as saying that $0 \neq U \in \mathcal{K}$, but every proper nonzero quotient U/P is in $c(\mathcal{K})$. For if not, then by Zorn's Lemma there exists an $A/P \leq U/P$ maximal with respect to $A/P \in \mathcal{K}$. Let B/P be the complement of A/P in U/P. Then $B/P \in c(\mathcal{K})$ and A/P embeds as an essential submodule in U/B. By (2.4.2), this implies that $U/B \in \mathcal{K}$ with $0 \neq B \leq U$. Hence $U/B \in \mathcal{K} \cap c(\mathcal{K}) = \{0\}$ and hence $U = B$, a contradiction.

2.4.7. LEMMA. For an M-natural class $\mathcal{K}$, suppose that U is a $\mathcal{K}$-cocritical module. Then the following hold:

1. Every nonzero submodule of U is $\mathcal{K}$-cocritical.
2. U is strongly uniform (see 2.2.16).
3. $\forall X \leq U, \forall Y \in \mathcal{K}$, every nonzero R-map $X \longrightarrow E(Y)$ is monic.

PROOF. (1) Let $0 \neq B \leq U$. Then $B \in \mathcal{K}$. Suppose that $B/V \in \mathcal{K}$ for some $0 \neq V \subset B$. Let C/V be a complement of B/V in U/V. Then B/V embeds in U/C as an essential submodule. By (2.4.2), $0 \neq U/C \in \mathcal{K}$, a contradiction.

(2) If not, then there exists an $0 \neq L \subseteq U$ such that U is not a rational extension of L. Hence by (2.2.16), there exist $m_1, 0 \neq m_2 \in U$ with $m_2(m_1^{-1}L) = 0$. Then $m_1 \notin L$ and $m_1R + L$ is $\mathcal{K}$-cocritical by (1). But

$$f : (m_1R + L)/L \longrightarrow m_2R, \; m_1r + L \mapsto m_2r \; (r \in R)$$

is a well defined nonzero homomorphism. Then, with $Ker(f) = U/L$, $0 \neq (m_1R + L)/U \cong m_2R \in \mathcal{K}$ contradicts that $m_1R + L$ is $\mathcal{K}$-cocritical.

(3) Let $0 \neq f : X \longrightarrow E(Y)$ for $X \leq U$ with $Y \in \mathcal{K}$. Define $0 \neq W = f^{-1}[Y \cap f(X)] \leq X$. Then $Ker(f) \subseteq W$, and $f|_W : W \longrightarrow Y$. If $Ker(f) \neq 0$, then $0 \neq W/Ker(f) \cong f(W) \leq Y$ shows that $W/Ker(f) \in \mathcal{K}$. But by (1), $0 \neq W \leq U$ is $\mathcal{K}$-cocritical, a contradiction. Hence $Ker(f) = 0$. □

Later, conditions will be given under which direct sums of M-injective modules belonging to an M-natural class $\mathcal{K}$ are M-injective. This will require that $\mathcal{K}$ contains enough $\mathcal{K}$-cocritical (uniform) modules. The condition $(*)$ in the next definition will assure this. This condition was introduced in [99] and [135].

2.4.8. DEFINITION. An M-natural class $\mathcal{K}$ satisfies the condition $(*)$ if for any cyclic submodule $0 \neq X \leq M$ and every ascending chain $\mathcal{C} = \{X_i\}$ with $X_i \subset X$ and $X_i \in H_{\mathcal{K}}(X)$ for each i, the union $\cup_i X_i \in H_{\mathcal{K}}(X)$. Since $X = xR$ is cyclic with $0 \neq x \in M$, and since all $X_i \neq X$, we see $\cup_i X_i \neq X$. Since every totally ordered set such as $\mathcal{C}$ contains a so called cofinal ordinal indexed subset $\{X_1 \subseteq X_2 \subseteq \cdots\} \subseteq \mathcal{C}$ such that, for every $Y \in \mathcal{C}$, $Y \subseteq X_i \in \mathcal{C}$ for some ordinal i, we may just as well assume that the above chain $\mathcal{C}$ is of the form $X_1 \subseteq X_2 \subseteq \cdots \subseteq X_i \subseteq \cdots < X$ and is indexed by ordinals $i \geq 1$.

For any i, $X_i = xI_i$, where $I_i = x^{-1}X_i$, and $X/X_i = xR/xI_i \cong R/I_i$ with $x^{\perp} \subseteq I_i$. Equivalently, $(*)$ holds if, for every $X = xR \leq M$ with $x \in M$ and every chain $x^{\perp} \subseteq I_1 \subseteq I_2 \subseteq \cdots \subseteq I_i \subseteq \cdots < R$ with $I_i \in H_{\mathcal{K}}(R)$, we have $\cup_i I_i \in H_{\mathcal{K}}(R)$. In particular, if R has ACC on $H_{\mathcal{K}}(R)$, then $\mathcal{K}$ satisfies $(*)$.

Note that the difference between $x^{\perp}$ and I_1 is that $R/x^{\perp} \hookrightarrow M$ but possibly $R/x^{\perp} \notin \mathcal{K}$, while perhaps $R/I_1 \not\hookrightarrow M$, but $R/I_1 \in \mathcal{K}$. For a natural class $\mathcal{K}$, we can omit "x" and "M" altogether. Let M be the direct sum of representatives of isomorphy classes of all cyclic R-modules. Then a natural class is an M-natural class and all cyclics embed in M.

By definition, a natural class $\mathcal{K}$ satisfies $(*)$ if for any chain of right ideals $I_1 \subseteq I_2 \subseteq \cdots \subseteq I_i \subseteq \cdots < R$ with $I_i \in H_{\mathcal{K}}(R)$, also $\cup_i I_i \in H_{\mathcal{K}}(R)$.

2.4.9. EXAMPLES. The following M-natural classes $\mathcal{K}$ satisfy the condition $(*)$:

(1) The M-natural class $\sigma[M]$.

(2) An M-natural class $\mathcal{K}$ such that $H_{\mathcal{K}}(X)$ satisfies the ascending chain condition for any cyclic submodule X of M.

(3) By (2.2.14) and (2.2.15), the class $\mathfrak{G}_M = \{N \in \sigma[M] : Z_M(N) \leq_e N\}$ is an M-natural class because it is closed under M-injective hulls. Since it is closed under quotient modules, it satisfies $(*)$.

(4) The same argument as in (3) shows that any hereditary torsion class in $\sigma[M]$ that is closed under M-injective hulls satisfies $(*)$.

(5) Any stable hereditary torsion class is a natural class that satisfies $(*)$. In particular, the class of all Goldie torsion modules satisfies $(*)$.

(6) The following example of a natural class $\mathcal{K}$ satisfying $(*)$ such that $\mathcal{K}$ is not closed under quotient modules and $H_{\mathcal{K}}(R)$ does not have ACC appeared in [99, Example 5(3)].

Let $R = \left\{\begin{pmatrix} a & x \\ 0 & a \end{pmatrix} : a \in \mathbb{Z}, x \in \mathbb{Z}_{2^\infty}\right\}$ be the trivial extension of $\mathbb{Z}$ and the $\mathbb{Z}$-module $\mathbb{Z}_{2^\infty}$. Then $J = \left\{\begin{pmatrix} a & x \\ 0 & a \end{pmatrix} : a \in 3\mathbb{Z}, x \in \mathbb{Z}_{2^\infty}\right\}$ is a maximal right ideal of R, and $V = R/J$ is a simple module. Then

$$\mathcal{K} = c(V) = \{N \in \text{Mod-}R : V \not\hookrightarrow N\}$$

is a natural class. Nonzero right ideals of R are of one of the following forms:

$$I_t = \left\{\begin{pmatrix} 0 & x \\ 0 & 0 \end{pmatrix} : x \in \mathbb{Z}_{2^t}\right\} \ (t = 1, \cdots, \infty),$$

$$I_\infty = \left\{\begin{pmatrix} 0 & x \\ 0 & 0 \end{pmatrix} : x \in \mathbb{Z}_{2^\infty}\right\}, \text{ and}$$

$$K_n = \left\{\begin{pmatrix} a & x \\ 0 & a \end{pmatrix} : a \in n\mathbb{Z}, x \in \mathbb{Z}_{2^\infty}\right\} \ (n = 1, 2, \cdots).$$

Then

$$H_\mathcal{K}(R) = \{I_t : t = 1, \cdots\} \cup \{I_\infty\} \cup \{K_n : n \in \mathbb{Z} \text{ with } (n, 3) = 1\}.$$

Thus $\mathcal{K}$ does not have the ACC, and since $R \in \mathcal{K}$, $\mathcal{K}$ is not closed under quotient modules. □

We next show how $(*)$ guarantees abundantly many $\mathcal{K}$-cocritical modules.

2.4.10. LEMMA. Let $\mathcal{K}$ be an M-natural class satisfying $(*)$. Then every nonzero cyclic submodule of M which also belongs to $\mathcal{K}$ has a $\mathcal{K}$-cocritical image.

PROOF. For a fixed cyclic module $0 \neq xR \leq M$ with $xR \in \mathcal{K}$, by $(*)$ the set $\{I < R_R : x^\perp \subseteq I, 0 \neq R/I \in \mathcal{K}\} \neq \emptyset$ is inductive and contains a maximal element J. Then xR has the $\mathcal{K}$-cocritical image $xR/xJ \cong (R/x^\perp)/(J/x^\perp) \cong R/J$. □

In general, an M-natural class need not contain any nonzero cyclic submodules of M. However, as seen in (2.4.8), for natural classes we can dispense with M and $xR \leq M$.

2.4.11. COROLLARY. Let $\mathcal{K}$ be a natural class satisfying (*). Then every nonzero cyclic module in $\mathcal{K}$ has a $\mathcal{K}$-cocritical image. □

2.4.12. PROPOSITION. Suppose that $\mathcal{K}$ is an M-natural class satisfying (*). Then every $0 \neq N \in \mathcal{K}$ has a $\mathcal{K}$-cocritical subfactor $B/A \in \mathcal{K}$ where $0 \leq A < B \leq N$.

PROOF. By (2.2.1), take $N \leq_e M^{(I)}/W$. Since $W \neq M^{(I)}$, without loss of generality there is an element in $M^{(I)}/W$ which we may take to be of the form $0 \neq x = (m, 0, \cdots, 0, \cdots) + W$ with exactly one nonzero component $m \in M$ only. There is an $r_0 \in R$ with $0 \neq xr_0 \in N$. Then

$$N \geq \frac{xr_0R + W}{W} \cong \frac{(mr_0, 0, \cdots, 0, \cdots)R}{(mr_0, 0, \cdots, 0, \cdots)R \cap W} \cong \frac{X}{X_1}$$

where the last isomorphism is obtained by taking the projection on the first component in the numerator and denominator. Thus, $X = xr_0R$, while

$$X_1 = \{mr_0b : b \in R, (mr_0b, 0, \cdots, 0, \cdots) \in W\} \in H_{\mathcal{K}}(X).$$

By $(*)$, the inductive set $\{Y : X_1 \leq Y < X, 0 \neq X/Y \in \mathcal{K}\}$ has a maximal element, say Y. Set $A = Y/Y_1 < X/X_1 = B \leq N$. Then $B/A \cong X/Y \in \mathcal{K}$ is $\mathcal{K}$-cocritical. □

2.5 Pre-Natural Classes

Pre-natural classes were first introduced in [138] and later exhaustively studied in [142]. One reason for their importance is that they contain almost all the other module classes which are of interest. The readers should recall (2.2.5) for the next definition.

2.5.1. DEFINITION. A nonempty class $\mathcal{K}$ of modules is a **pre-natural class** if it is closed under submodules, arbitrary direct sums, and $\mathrm{tr}(\mathcal{K}, EN) \in \mathcal{K}$ for every $N \in \mathcal{K}$. Let $\mathcal{N}_r^p(R)$ denote the collection of all pre-natural classes of right R-modules, and similarly $\mathcal{N}_l^p(R)$ for the left ones. As before for $\mathcal{N}(R)$, we frequently abbreviate $\mathcal{N}^p(R) = \mathcal{N}_r^p(R)$.

The following simple facts in the next two lemmas sometimes will be the deciding factors in many subsequent arguments.

2.5.2. LEMMA.

1. For any module M, every M-natural class is a pre-natural class.
2. For any pre-natural class $\mathcal{K}$, $\mathcal{K}$ is an $M_{\mathcal{K}}$-natural class.

PROOF. (1) If $\mathcal{K} \subseteq \sigma[M]$ is an M-natural class, then for any $N \in \mathcal{K}$, $\mathrm{tr}(\mathcal{K}, EN) \leq \mathrm{tr}(\sigma[M], EN) = \mathrm{tr}(M, EN) \in \mathcal{K}$. Since $\mathcal{K}$ is closed under submodules, $\mathrm{tr}(\mathcal{K}, EN) \in \mathcal{K}$, so $\mathcal{K}$ is a pre-natural class.

(2) In view of (2.5.2)(1), it suffices to show that for $N \in \mathcal{K}$, $\mathrm{tr}(M_{\mathcal{K}}, EN) \in \mathcal{K}$. Since $M_{\mathcal{K}} \in \mathcal{K}$, $\mathrm{tr}(M_{\mathcal{K}}, EN) \leq \mathrm{tr}(\mathcal{K}, EN) \in \mathcal{K}$. As above, $\mathrm{tr}(M_{\mathcal{K}}, EN) \in \mathcal{K}$, and $\mathcal{K}$ is an $M_{\mathcal{K}}$-natural class. □

We can now use the last lemma to conclude that pre-natural classes have many of the same properties that M-natural classes have.

2.5.3. COROLLARY. Let $\mathcal{K}$ be a pre-natural class, and suppose that $P \leq_e N \in \sigma[M_{\mathcal{K}}]$. If $P \in \mathcal{K}$, then $N \in \mathcal{K}$. □

2.5.4. LEMMA. Let $\mathcal{F}$ be a class of modules. Then

1. $d(M_{\mathcal{F}}) \cap \sigma[M_{\mathcal{F}}]$ is the smallest pre-natural class containing $\mathcal{F}$.
2. $\mathcal{F}$ is a pre-natural class iff $\mathcal{F} = d(M_{\mathcal{F}}) \cap \sigma[M_{\mathcal{F}}]$.

PROOF. We note that $d(M_{\mathcal{F}}) = d(\mathcal{F})$ by (2.3.6).

(1) By (2.4.3), $d(\mathcal{F}) \cap \sigma[M_{\mathcal{F}}]$ is an $M_{\mathcal{F}}$-natural class, and hence is a pre-natural class by (2.5.2). Moreover, it is clear that $\mathcal{F} \subseteq d(\mathcal{F}) \cap \sigma[M_{\mathcal{F}}]$. Suppose that $\mathcal{F} \subseteq \mathcal{L} \in \mathcal{N}^p(R)$. By (2.5.2), $\mathcal{L}$ is an $M_{\mathcal{L}}$-natural class, and hence $\mathcal{L} = \mathcal{K} \cap \sigma[M_{\mathcal{L}}]$ for some $\mathcal{K} \in \mathcal{N}(R)$ by (2.4.4). Since $\mathcal{F} \subseteq \mathcal{K} \cap \sigma[M_{\mathcal{L}}]$,

$d(\mathcal{F}) \subseteq \mathcal{K}$ and $\sigma[M_{\mathcal{F}}] \subseteq \sigma[M_{\mathcal{L}}]$, so $d(\mathcal{F}) \cap \sigma[M_{\mathcal{F}}] \subseteq \mathcal{K} \cap \sigma[M_{\mathcal{L}}] = \mathcal{L}$. Thus, (1) follows. And (2) follows by (1). □

2.5.5. LEMMA. A class of modules is a pre-natural class iff it is the intersection of a natural class and a hereditary pretorsion class.

PROOF. One direction is by (2.5.4). Suppose that $\mathcal{F} = \mathcal{K} \cap \mathcal{L}$ where $\mathcal{K}$ is a natural class and $\mathcal{L}$ a hereditary pretorsion class. By (2.2.4), $\mathcal{L} = \sigma[M]$ for some module M. Thus, $\mathcal{F} = \mathcal{K} \cap \sigma[M]$ is an M-natural class by (2.4.4), and hence is a pre-natural class by (2.5.2). □

2.5.6. THEOREM. The correspondence $\mathcal{K} \longmapsto H_{\mathcal{K}}(R)$ from $\mathcal{N}^p(R)$ to $\mathcal{P}(\mathrm{L}(R))$, the power set of $\mathrm{L}(R)$, is one-to-one. Consequently, $\mathcal{N}^p(R) = \bigcup\{\mathcal{N}(R, M) : M \in \text{Mod-}R\}$ is a set.

PROOF. For $\mathcal{K}, \mathcal{L} \in \mathcal{N}^p(R)$, if $H_{\mathcal{K}}(R) = H_{\mathcal{L}}(R)$, then $\mathcal{K}$ and $\mathcal{L}$ have the same cyclics. Let $0 \neq N \in \mathcal{K} \backslash \mathcal{L}$. Then for any $x \in N$, xR is in $\mathcal{K}$ and hence in $\mathcal{L}$. By (2.5.2), there exists a module M such that $\mathcal{L} \subseteq \sigma[M]$ and $\mathcal{L}$ is an M-natural class. Thus by (2.4.5), there exists a $0 \neq Y \leq N$ with $Y \in c(\mathcal{L})$. But, $yR \in \mathcal{L}$ for all $y \in Y$, a contradiction. Finally, $\mathcal{N}^p(R) = \bigcup\{\mathcal{N}(R, M) : M \in \text{Mod-}R\}$ by (2.5.2). □

Let $\mathfrak{A}$ be a nonempty set of right ideals, and set $G_{\mathfrak{A}} = \oplus\{R/I : I \in \mathfrak{A}\}$. If there exists a pre-natural class $\mathcal{K}$ such that $\mathfrak{A} = H_{\mathcal{K}}(R)$, then $M_{\mathcal{K}} \hookrightarrow G_{\mathfrak{A}}$, and in fact, $\sigma[M_{\mathcal{K}}] = \sigma[G_{\mathfrak{A}}]$. Suppose that $I \leq R, R/I \notin \mathcal{K} = d(\mathcal{K}) \cap \sigma[M_{\mathcal{K}}]$. So either $R/I \notin \sigma[G_{\mathfrak{A}}]$, or $R/I \notin d(\mathcal{K})$. The latter implies that there is a $0 \neq J/I \leq R/I$ with $J/I \in c(\mathcal{K})$. And the latter is equivalent to (3)(a) below. It is not difficult to see why the next definition has the form that it has.

2.5.7. DEFINITION. For a nonempty set $\mathfrak{A}$ of right ideals of R, set $G_{\mathfrak{A}} = \oplus\{R/I : I \in \mathfrak{A}\}$. Then $\mathfrak{A}$ is called a **pre-natural set** if the following conditions are satisfied:

1. If $I \in \mathfrak{A}$ and $J \in \mathfrak{A}$, then $I \cap J \in \mathfrak{A}$.
2. If $I \in \mathfrak{A}$, then $a^{-1}I \in \mathfrak{A}$ for all $a \in R$.
3. For $I \leq R$ with $I \notin \mathfrak{A}$, at least one of (a) and (b) below holds:
   (a) $\exists J \leq R$ such that $I \subset J$ and $a^{-1}I \notin \mathfrak{A}$ for all $a \in J \backslash I$.
   (b) $R/I \notin \sigma[G_{\mathfrak{A}}]$.

Recall that $\mathfrak{A}$ is called a natural set if (1), (2), and (3)(a) are satisfied.

The next theorem was proved in two stages, first for a natural class in [136] and later in [142, Prop.1.6] for a pre-natural class. We have already encountered a result very similar to the next theorem, namely Proposition (2.1.9).

2.5.8. THEOREM. The following hold for a set $\mathfrak{A}$ of right ideals of R:

1. $\mathfrak{A}$ is a natural set iff $\mathfrak{A} = H_{\mathcal{K}}(R)$ for some $\mathcal{K} \in \mathcal{N}(R)$.

2. $\mathfrak{A}$ is a pre-natural set iff $\mathfrak{A} = H_{\mathcal{K}}(R)$ for some $\mathcal{K} \in \mathcal{N}^p(R)$.

PROOF. "$\Longleftarrow$" of (1) and (2). Assume $\mathfrak{A} = H_{\mathcal{K}}(R)$. For right ideals I and J of R and any $a \in R$, from $R/(I \cap J) \hookrightarrow R/I \oplus R/J$ and $R/a^{-1}I \cong (aR+I)/I \subseteq R/I$ it follows that (1), (2) of (2.5.7) hold.

In case (1), if $I \notin H_{\mathcal{K}}(R)$, then $R/I \notin \mathcal{K} = d(\mathcal{K})$. The latter implies that there exists a $0 \neq J/I \leq R/I$ with $J/I \in c(\mathcal{K})$, or equivalently for any $a \in J \backslash I$, $R/a^{-1}I \cong (aR+I)/I \notin \mathcal{K}$, i.e., $a^{-1}I \notin H_{\mathcal{K}}(R) = \mathfrak{A}$. So (2.5.7)(3a) holds.

In case (2), if $I \notin H_{\mathcal{K}}(R)$, then $R/I \notin \mathcal{K} = d(\mathcal{K}) \cap \sigma[G_{\mathfrak{A}}]$. If $R/I \notin d(\mathcal{K})$, then (2.5.7)(3a) holds as above. Otherwise we have $R/I \notin \sigma[G_{\mathfrak{A}}]$ and (2.5.7)(3b) holds.

"$\Longrightarrow$" of (1) and (2). In either case, let $\mathcal{K} = \{V \in \text{Mod-}R : x^{\perp} \in \mathfrak{A}$ for all $x \in V\}$ and form $H_{\mathcal{K}}(R) = \{I \leq R : R/I \in \mathcal{K}\} = \{I \leq R : x^{-1}I \in \mathfrak{A}$ for all $x \in R\}$. It follows from (2.5.7)(2) that $\mathfrak{A} \subseteq H_{\mathcal{K}}(R)$. If $I \in H_{\mathcal{K}}(R)$, for $x = 1$, $I = x^{-1}I \in \mathfrak{A}$. So $H_{\mathcal{K}}(R) = \mathfrak{A}$.

Clearly, $\mathcal{K}$ is closed under submodules. Let $V = \oplus_i V_i$ with each $V_i \in \mathcal{K}$. For any $x \in V$, write $x = x_1 + x_2 + \cdots + x_n$ with $x_i \in V_i$. Then $x^{\perp} = \cap_{i=1}^n x_i^{\perp}$. Since each $x_i^{\perp} \in \mathfrak{A}$, $x^{\perp} \in \mathfrak{A}$ and hence $V \in \mathcal{K}$. Thus, $\mathcal{K}$ is closed under direct sums.

We now assume $\mathfrak{A}$ is a pre-natural set and show that $\mathcal{K}$ is an M-natural class where $M = G_{\mathfrak{A}}$. For $V \in \mathcal{K}$, since the sum map $f : \oplus\{R/x^{\perp} : x \in V\} \longrightarrow V$ is surjective, $V \cong \oplus\{R/x^{\perp} : x \in V\}/Ker(f) \in \sigma[M]$. Thus $\mathcal{K} \subseteq \sigma[G_{\mathfrak{A}}]$. If $\mathcal{K}$ is not an M-natural class, then for some $N \in \mathcal{K}$, $E_M(N) \notin \mathcal{K}$. Thus for some $0 \neq x \in E_M(N)$, $x^{\perp} \notin \mathfrak{A}$. Since $N \in \mathcal{K}$ and $\mathcal{K} \subseteq \sigma[M]$, $N \in \sigma[M]$, which insures that $R/x^{\perp} \cong xR \in E_M(N) \in \sigma[M]$. Thus (2.5.7)(3a) holds for $I = x^{\perp} \notin \mathfrak{A}$. Hence there exists a right ideal J of R such that $x^{\perp} \subset J$ and $(xa)^{\perp} = a^{-1}x^{\perp} \notin \mathfrak{A}$ for all $a \in J \backslash x^{\perp}$. Thus every nonzero submodule of xJ does not belong to $\mathcal{K}$. Next, $0 \neq J/x^{\perp} \cong xJ \leq E_M(N)$, and $0 \neq xJ \cap N \leq N \in \mathcal{K}$, a contradiction.

Assume $\mathfrak{A}$ is a natural set. To show that $\mathcal{K}$ is a natural class, let $N \in \mathcal{K}$ but $E(N) \notin \mathcal{K}$. Then for some $0 \neq x \in E(N)$, $x^{\perp} \notin \mathfrak{A}$. Using (2.5.7)(3a), we get a $J \leq R$ as in the previous paragraph with $0 \neq J/x^{\perp} \cong xJ \leq E(N)$, and the rest of the proof is the same as above. □.

2.5.9. REMARK. In the proof of "$\Longrightarrow$" of (2) above, reference to an $I \notin \mathfrak{A}$ with $R/I \notin \sigma[G_{\mathfrak{A}}]$ never occurred. This does not mean that the latter condition could be omitted from the definition of a pre-natural set. Condition (2.5.7)(3b) in the definition of a pre-natural set may be replaced by the following condition $(\bar{b})$.

$(\bar{b})$ For any $I \leq R$ with $R/I \in \sigma[G_{\mathfrak{A}}]$ but $I \notin \mathfrak{A}$, there exists an $I \subset J \leq R$ with $a^{-1}I \notin \mathfrak{A}$ for all $a \in J \backslash I$.

Natural classes and hereditary pretorsion classes are pre-natural classes. The following example, which appeared in [138], gives a pre-natural class which is neither a natural class nor a hereditary pretorsion class.

2.5.10. EXAMPLE. Let $R = \left\{ \begin{pmatrix} a & x \\ 0 & a \end{pmatrix} : a, x \in \mathbb{Z} \right\}$ be the trivial extension of $\mathbb{Z}$ and the $\mathbb{Z}$-module $\mathbb{Z}$. Let $I = \left\{ \begin{pmatrix} 0 & x \\ 0 & 0 \end{pmatrix} : x \in \mathbb{Z} \right\}$ and $M = I$. Set $\mathcal{K} = d(M) \cap \sigma[M]$. Then $\mathcal{K}$ is a pre-natural class by (2.4.4). Clearly, $\mathcal{K}$ is not closed under quotient modules and hence not a hereditary pretorsion class. Note that $M \in \mathcal{K}$. We consider $E(M)$. Since $M = I \leq_e R_R$, $E(M) = E(R)$. Suppose that $E(M) \in \mathcal{K}$. Then $R \in \mathcal{K}$ and hence $R \in \sigma[M]$. It follows that R is embeddable in a quotient of $M^{(J)}$ for an index set J. But $MI = 0$, implying $I = RI = 0$, a contradiction. So $E(M) \notin \mathcal{K}$. Thus $\mathcal{K}$ is not a natural class.

Chapter 3

Finiteness Conditions

Natural classes and pre-natural classes provide a nice setting for study of chain conditions of rings and modules. For a fixed pre-natural class $\mathcal{K}$ of right R-modules and a fixed right R-module M, the ascending chain condition and descending chain condition on $H_{\mathcal{K}}(M)$ are characterized in terms of certain decomposition properties of modules in $\mathcal{K}$. This gives a unifying treatment of chain conditions of rings and modules in torsion theory and in the category $\sigma[M]$, because of the fact that all hereditary torsion classes, hereditary torsion free classes, and $\sigma[M]$ are pre-natural classes. When $\mathcal{K}$ is specialized to some well-selected pre-natural classes that are not of the previous three kinds, then new results are obtained by this approach.

Section 3.1 focuses on the ascending chain conditions, and section 3.2 contains discussions on the descending chain conditions as well as applications to endomorphism rings. Section 3.3 gives a connection between covers of modules with the ascending chain conditions of rings.

3.1 Ascending Chain Conditions

For a fixed module M, and an M-natural class $\mathcal{K}$, a certain chain condition on submodules of M will guarantee that every direct sum of M-injective modules in $\mathcal{K}$ is M-injective. This result arose out of [98] and was proved in the above generality in [99] and [135]. It has many consequences.

In this section M is an arbitrary but fixed right R-module, and $\mathcal{K} \subseteq \sigma[M]$ is an M-natural class. We can write

$$\mathcal{K} = c(\mathcal{F}) \cap \sigma[M] = \{N \in \sigma[M] : \forall 0 \neq V \leq N, \forall W \in \mathcal{F}, V \not\hookrightarrow W\}$$

where $\mathcal{F}$ is a module class. Recall that for $N \in \text{Mod-}R, H_{\mathcal{K}}(N)$ is defined by $H_{\mathcal{K}}(N) = \{P \leq N : N/P \in \mathcal{K}\}$; in particular $H_{\mathcal{K}}(R) = \{I \leq R : R/I \in \mathcal{K}\}$. We now investigate the ascending chain condition on $H_{\mathcal{K}}(N)$ for various modules N.

3.1.1. LEMMA. For $N \in \sigma[M]$, suppose $B_1 \subseteq B_2 \subseteq \cdots \subseteq B_i \subseteq \cdots$ is a chain of submodules of N such that $B_{i+1}/B_i \in \mathcal{K}$ for all i. Let $B = \cup\{B_i : i = 1, 2, \cdots\}$. Then $B/B_i \in \mathcal{K}$ for all i.

PROOF. Let i be smallest such that $B/B_i \notin \mathcal{K}$. Note that $\mathcal{K} = c(\mathcal{F}) \cap \sigma[M]$ for a module class $\mathcal{F}$. Then there exists a $0 \neq X/B_i \subseteq B/B_i$ such that $X/B_i \hookrightarrow F$ for some $F \in \mathcal{F}$. Since $X = \cup\{X \cap B_j : i \leq j\}$, there exists a smallest j such that $j > i$ and $(X \cap B_j)/B_i \neq 0$. So $B_i = X \cap B_{j-1} = X \cap B_j \cap B_{j-1}$. Then

$$B_j/B_{j-1} \geq (X \cap B_j + B_{j-1})/B_{j-1} \cong (X \cap B_j)/B_i \leq X/B_i \hookrightarrow F.$$

All of the modules are not zero, and $F \in \mathcal{F} \subseteq c(\mathcal{K})$. This contradicts that $B_j/B_{j-1} \in \mathcal{K}$. Thus all $B/B_i \in \mathcal{K}$. □

3.1.2. LEMMA. Let $B_1 \subseteq B_2 \subseteq \cdots \subseteq B_i \subseteq \cdots \leq N \in \sigma[M]$ be a chain as in (3.1.1). Then there exists a $K \subseteq N$ such that $N/K \in \mathcal{K}$, $B_1 \subseteq K$, and $(B_{i+1} + K)/(B_i + K) \cong B_{i+1}/B_i$ for all i.

PROOF. If $N/B_1 \in \mathcal{K}$, then let $K = B_1$ and we are done. The following proof will work in general, irrespective of whether $N/B_1 \in \mathcal{K}$ or not. As before, set $B = \cup\{B_i : 1 \leq i\}$. Then there exists a submodule $B_1 \subseteq K \leq N$ such that K/B_1 is a complement of B/B_1 in N/B_1. Thus $K \cap B = B_1$ and hence $K \cap B_j = B_1$ for all $j \geq 1$. Note that

$$B_{i+1} \cap (B_i + K) = B_i + B_{i+1} \cap K = B_i + B_1 = B_i.$$

Consequently for all $i \geq 1$, we have

$$(B_{i+1} + K)/(B_i + K) \cong B_{i+1}/[B_{i+1} \cap (B_i + K)] = B_{i+1}/B_i.$$

Since K/B_1 is a complement of B/B_1 in N/B_1, B/B_1 embeds as an essential submodule in N/K. Thus, $E_M(B/B_1) \cong E_M(N/K)$. By (3.1.1), $B/B_1 \in \mathcal{K}$. Hence $E_M(B/B_1) \in \mathcal{K}$, showing that $N/K \in \mathcal{K}$. □

3.1.3. LEMMA. The following are equivalent for a module $N \in \sigma[M]$ and an M-natural class $\mathcal{K}$:

1. Every chain of submodules $B_1 \subseteq B_2 \subseteq \cdots \leq N$ with all $B_{i+1}/B_i \in \mathcal{K}$ terminates.

2. $H_{\mathcal{K}}(N)$ has the ascending chain condition (abbreviation ACC).

PROOF. (1) $\Longrightarrow$ (2). This follows because $\mathcal{K}$ is closed under submodules.
(2) $\Longrightarrow$ (1). Suppose there exists a strictly ascending chain

$$B_1 \subset B_2 \subset \cdots \subset B_n \subset \cdots$$

of submodules of N such that B_{i+1}/B_i is in $\mathcal{K}$ for every i. We show that this leads to a contradiction by constructing a strictly ascending chain

$$K_1 \subset K_2 \subset \cdots \quad \text{with every } N/K_i \in \mathcal{K}.$$

By (3.1.2), there is a $K_1 \subseteq N$ such that $N/K_1 \in \mathcal{K}$, $B_1 \subseteq K_1$, and

$$(B_{i+1} + K_1)/(B_i + K_1) \cong B_{i+1}/B_i \quad \text{for all } i.$$

Then

$$B_1 \subseteq K_1 \subset B_2 + K_1$$

and

$$B_2 + K_1 \subset B_3 + K_1 \subset \cdots$$

is a strictly ascending chain with $(B_{i+1} + K_1)/(B_i + K_1)$ in $\mathcal{K}$ for all $i \geq 1$. Suppose we have constructed $K_1, K_2, \cdots, K_n$ such that all N/K_i are in $\mathcal{K}$,

$$K_1 \subset K_2 \subset \cdots \subset K_n, \; K_n \subset B_{n+1} + K_n$$

and

$$B_{n+1} + K_n \subset B_{n+2} + K_n \subset \cdots$$

is a strictly ascending chain with

$$(B_{i+1} + K_n)/(B_i + K_n) \in \mathcal{K} \text{ for all } i \geq n.$$

Applying (3.1.2) to the chain

$$B_{n+1} + K_n \subset B_{n+2} + K_n \subset \cdots,$$

we have a $K_{n+1} \subseteq N$ such that

$$N/K_{n+1} \in \mathcal{K}, \; B_{n+1} + K_n \subseteq K_{n+1} \subset B_{n+2} + K_{n+1}$$

and

$$B_{n+2} + K_{n+1} \subset B_{n+3} + K_{n+1} \subset \cdots$$

is a strictly ascending chain with

$$(B_{i+1} + K_{n+1})/(B_i + K_{n+1}) \in \mathcal{K} \text{ for all } i \geq n+1.$$

The induction principle implies that there exists a sequence $\{K_i : i \in \mathbb{N}\}$ such that

$$N/K_i \in \mathcal{K} \quad \text{for all } i \quad \text{and } K_1 \subset K_2 \subset \cdots K_n \subset \cdots$$

is a strictly ascending chain. The lemma is proved. □

3.1.4. COROLLARY. Let $\mathcal{K}$ be an M-natural class and $Y \leq N \in \sigma[M]$. Then $H_{\mathcal{K}}(N)$ has ACC iff both $H_{\mathcal{K}}(Y)$ and $H_{\mathcal{K}}(N/Y)$ have ACC.

PROOF. "$\Longrightarrow$". The ACC on $H_{\mathcal{K}}(N)$ clearly implies the ACC on $H_{\mathcal{K}}(Y)$ by (3.1.3). Any chain

$$B_1/Y \subseteq B_2/Y \subseteq \cdots \subseteq B_i/Y \subseteq \cdots$$

in $H_{\mathcal{K}}(N/Y)$ gives rise to a chain

$$B_1 \subseteq B_2 \subseteq \cdots \subseteq B_i \subseteq \cdots$$

in $H_{\mathcal{K}}(N)$, and hence must be finite.

"$\Longleftarrow$". Suppose that

$$X_1 \subseteq X_2 \subseteq \cdots \leq N$$

is a chain in $H_{\mathcal{K}}(N)$. Thus,

$$X_1 \cap Y \subseteq X_2 \cap Y \subseteq \cdots \leq Y$$

and

$$(X_1 + Y)/Y \subseteq (X_2 + Y)/Y \subseteq \cdots \leq N/Y.$$

Since $Y/(X_i \cap Y) \cong (X_i + Y)/X_i \leq N/X_i$, we have that $Y/(X_i \cap Y) \in \mathcal{K}$ for all i. By the ACC on $H_{\mathcal{K}}(Y)$, there exists an m such that $X_{m+s} \cap Y = X_m \cap Y$ for all $s \geq 0$. Then for any $j \geq m$, we have that

$$\begin{aligned}
[(X_{j+1} + Y)/Y]/[(X_j + Y)/Y] &\cong (X_{j+1} + Y)/(X_j + Y) \\
&\cong X_{j+1}/[X_j + (X_{j+1} \cap Y)] \\
&= X_{j+1}/X_j \in \mathcal{K}.
\end{aligned}$$

Because $H_{\mathcal{K}}(N/Y)$ has ACC, by using (3.1.3), there exists an n ($n \geq m$) such that $X_{n+k} + Y = X_n + Y$ for all $k \geq 0$. It follows that, for all k,

$$\begin{aligned}
X_{n+k} &= X_{n+k} \cap (X_n + Y) \\
&= X_n + (X_{n+k} \cap Y) \\
&= X_n + (X_n \cap Y) = X_n.
\end{aligned}$$

□

The next theorem is due to [99, Theorem 10].

3.1.5. THEOREM. The following are equivalent for an M-natural class $\mathcal{K}$:

1. Every direct sum of M-injective modules in $\mathcal{K}$ is M-injective.

2. For every cyclic (or finitely generated) submodule A of M, any chain of submodules $B_1 \subseteq B_2 \subseteq \cdots \leq A$ with all $B_{i+1}/B_i \in \mathcal{K}$ terminates.

3. For any cyclic (or finitely generated) submodule $A \subseteq M$, $H_{\mathcal{K}}(A)$ has ACC.

4. R satisfies ACC on $H_{\mathcal{K}}(R)$.

PROOF. (1) $\Longrightarrow$ (2). Suppose that (2) fails. Then there exists a finitely generated submodule A of M and a strictly ascending chain $B_1 \subset B_2 \subset \cdots \leq A$ such that all $B_{i+1}/B_i \in \mathcal{K}$. Set $F = \oplus_i E_M(B_{i+1}/B_i)$ and $B = \cup_i B_i$. Let p_i be the canonical map from B_{i+1} onto B_{i+1}/B_i and ℓ_i be the inclusion of B_{i+1}/B_i to $E_M(B_{i+1}/B_i)$. Because $E_M(B_{i+1}/B_i)$ is M-injective, the map $\ell_i p_i : B_{i+1} \longrightarrow E_M(B_{i+1}/B_i)$ extends to $f_i : M \longrightarrow E_M(B_{i+1}/B_i)$. Define $f : B \longrightarrow F$ by $\pi_i f(b) = f_i(b)$, where π_i is the projection of F onto $E_M(B_{i+1}/B_i)$. Then f is well-defined. Since F is M-injective by (1), f extends to $g : A \longrightarrow F$. Then $f(B) \subseteq g(A) \subseteq \oplus_{i=1}^m E_M(B_{i+1}/B_i)$ for some m since A is finitely generated. Hence $\pi_i f = 0$ for all $i > m$. For $b \in B_{m+1}$, $0 = \pi_{m+1} f(b) = f_{m+1}(b) = b + B_m$. This implies that $B_{m+1} = B_m$, a contradiction.

(2) $\Longrightarrow$ (1). By [87, Theorem 1.7, p.3], it suffices to show that every countable direct sum of M-injective modules in $\mathcal{K}$ is M-injective. So let $N = \oplus_{i=1}^\infty N_i$, where each $N_i \in \mathcal{K}$ is M-injective. The generalized Baer's criterion ([87, Prop.1.4, p.2]) will be verified for N; that is that for every $x \in M$, N is xR-injective. So let $B \leq xR$ and $f : B \longrightarrow N$ a homomorphism. Set $B_k = \{b \in B : f(b) \in \oplus_{i=1}^k N_i\}$. Thus $B_1 \subseteq B_2 \subseteq \cdots \subseteq B_i \subseteq \cdots \leq B$ and, for all $k \geq 1$,

$$B_{k+1}/B_k \overset{\phi}{\hookrightarrow} (\oplus_{i=1}^{k+1} N_i)/(\oplus_{i=1}^k N_i) \cong N_{k+1} \in \mathcal{K}$$
$$\text{by } \phi(b + B_k) = f(b) + \oplus_{i=1}^k N_i.$$

Thus $B_{k+1}/B_k \in \mathcal{K}$ for all $k \geq 1$. By (2), there exists an m such that $B_{m+i} = B_m$ for all i. Consequently $f(B) \subseteq \oplus_{i=1}^m N_i$ where the latter is M-injective. Hence f extends to $g : xR \longrightarrow N_1 \oplus \cdots \oplus N_m \subseteq N$. Hence N is M-injective.

(2) $\Longleftrightarrow$ (3). It follows from (3.1.3).

(4) $\Longrightarrow$ (3). Every n-generated submodule A of M is an image of R^n. By (3.1.4), (4) implies the ACC on $H_{\mathcal{K}}(R^n)$ and further the ACC on $H_{\mathcal{K}}(A)$.

(3) $\Longrightarrow$ (4). For any $I \in H_{\mathcal{K}}(R)$, R/I is embeddable in a quotient of Y, where Y is a finite direct sum of cyclic submodules of M. In view of (3.1.4), (3) implies the ACC on $H_{\mathcal{K}}(Y)$. Hence $H_{\mathcal{K}}(R/I)$ has ACC again by (3.1.4). Thus, we proved that $H_{\mathcal{K}}(R/I)$ has ACC for every $I \in H_{\mathcal{K}}(R)$. It follows that $H_{\mathcal{K}}(R)$ has ACC. □

3.1.6. Recall [8, p.141] that a **maximal direct summand** of a module N is any submodule K such that $N = K \oplus P$ for some indecomposable submodule $P \leq N$. A direct sum decomposition $N = \oplus\{C_i : i \in I\}$ of a module N is said to **complement direct summands (complement maximal direct summands)** if for any direct summand (any maximal direct summand) K of N, $N = K \oplus (\oplus\{C_i : i \in J\})$ for some $J \subseteq I$.

3.1.7. LEMMA. For a module decomposition $D = \oplus\{D_\alpha : \alpha \in \Gamma\}$, let Λ be a subset of Γ such that (i) for α and β in Λ, $D_\alpha \cong D_\beta$ iff $\alpha = \beta$, and (ii) for any $\alpha \in \Gamma$, $D_\alpha \cong D_\gamma$ for some $\gamma \in \Lambda$. Set $C = \oplus\{D_\alpha : \alpha \in \Lambda\}$ and let M be a module.

1. If $C^{(\mathbb{N})}$ is M-injective, then D is M-injective.

2. If all D_α $(\alpha \in \Gamma)$ are uniform M-injective and $E_M(C^{(\mathbb{N})})$ has a decomposition that complements maximal direct summands, then $C^{(\mathbb{N})}$ is M-injective.

PROOF. Partition $\Gamma = \cup\{\Gamma_t : t \in T\}$ in such a way that, for $\alpha_1 \in \Gamma_{t_1}$ and $\alpha_2 \in \Gamma_{t_2}$, $D_{\alpha_1} \cong D_{\alpha_2}$ iff $t_1 = t_2$. Therefore $D = \oplus_{t\in T}\{D_\alpha : \alpha \in \Gamma_t\}$. For each t, choose one $\alpha_t \in \Gamma_t$ and let $F = \{\alpha_t : t \in T\}$. Then $C \cong \oplus_{\alpha \in F} D_\alpha$.

(1) Let Γ be the disjoint union of $\{\Gamma_t : t \in T\}$. By [87, Theorem 1.7], $C^{(\mathbb{N})}$ being M-injective implies that $C^{(\Gamma)}$ is M-injective. We have

$$\begin{aligned} D = \oplus_{t\in T}\{D_\alpha : \alpha \in \Gamma_t\} &\cong \oplus_{t\in T} {D_{\alpha_t}}^{(\Gamma_t)} \\ &\subseteq^{\oplus} \oplus_{t\in T} {D_{\alpha_t}}^{(\Gamma)} \cong (\oplus_{t\in T} D_{\alpha_t})^{(\Gamma)} \cong C^{(\Gamma)}. \end{aligned}$$

It follows that D is M-injective.

(2) Set $E = E_M(C^{(\mathbb{N})})$. By (2), E has a decomposition $E = \oplus\{E_\alpha : \alpha \in A\}$ that complements maximal direct summands of E. For each $\alpha \in F$, set $A(\alpha) = \{\beta \in A : E_\beta \cong D_\alpha\}$. Now for each $n > 0$ and each $\alpha \in F$, the module ${D_\alpha}^{(n)}$ is isomorphic to a summand of E. At this point, [8, Lemma 12.2] has to be invoked to conclude that the cardinality of $A(\alpha)$ satisfies $|A(\alpha)| \geq n$. Hence $|A(\alpha)| \geq |\mathbb{N}|$. Let $U = \cup_{\alpha\in F} A(\alpha)$. Then $C^{(\mathbb{N})}$ is isomorphic to a summand of $\oplus_U E_\alpha \subseteq^{\oplus} E$. Therefore $C^{(\mathbb{N})}$ is M-injective. □

3.1.8. THEOREM. The following are equivalent for an M-natural class $\mathcal{K}$:

1. Every direct sum of M-injective modules in $\mathcal{K}$ is M-injective.

2. $N^{(\mathbb{N})}$ is M-injective for every M-injective module $N \in \mathcal{K}$.

3. $D_0^{(\mathbb{N})}$ is M-injective, where $D_0 = \oplus\{E_M(R/I) : I \in H_{\mathcal{K}}(R)\}$.

PROOF. (1) $\Longrightarrow$ (2) is obvious.

(2) $\Longrightarrow$ (3). Set $E = E_M(D_0) \in \mathcal{K}$. By (2), $E^{(\mathbb{N})}$ is M-injective. Thus $E^{(H_{\mathcal{K}}(R))}$ is M-injective by [87, Theorem 1.7]. For each $I \in H_{\mathcal{K}}(R)$, $E_M(R/I)$ is a direct summand of E, and hence

$$D_0 = \oplus\{E_M(R/I) : I \in H_{\mathcal{K}}(R)\} \subseteq^{\oplus} E^{(H_{\mathcal{K}}(R))}.$$

So D_0 is M-injective. Thus $D_0^{(\mathbb{N})}$ is M-injective by (2).

(3) $\Longrightarrow$ (1). First we note that if $N = \oplus_\alpha E_\alpha$ such that each $E_\alpha \cong E_M(R/I)$ for some $I \in H_{\mathcal{K}}(R)$, then N is M-injective by (3) and (3.1.7). Now to show (1), it suffices to show that every M-injective module $P \in \mathcal{K}$ is a direct sum of modules of form $E_M(R/I)$ with $I \in H_{\mathcal{K}}(R)$. Since $xR \cong R/x^\perp$ for any $x \in P$, $E_M(xR)$ is isomorphic to $E_M(R/I)$ for some $I \in H_{\mathcal{K}}(R)$. Set $\mathcal{F} = \{A : A \text{ is an independent set of nonzero cyclic submodules of } P\}$. We

define a partial order $\leq$ on $\mathcal{F}$ by $A_1 \leq A_2$ iff $A_1 \subseteq A_2$. Then $(\mathcal{F}, \leq)$ is an inductive set. By Zorn's Lemma, there exists a maximal element, say A, in $\mathcal{F}$. Then $\sum_{X \in A} E_M(X) = \oplus_{X \in A} E_M(X)$ is M-injective as seen above. Thus $P = [\oplus_{X \in A} E_M(X)] \oplus Y$ for some Y. But the maximality of A implies that $Y = 0$. Therefore $P = \oplus_{X \in A} E_M(X)$, with each $E_M(X)$ isomorphic to $E_M(R/I)$ for some $I \in H_{\mathcal{K}}(R)$. □

3.1.9. Consider the following conditions on a module N:
($C1$) Every submodule of N is essential in a direct summand of N.
($C2$) For $A, B \leq N$, if $A \cong B \subseteq^{\oplus} N$, then also $A \subseteq^{\oplus} N$.
($C3$) For $A \subseteq^{\oplus} N$ and $B \subseteq^{\oplus} N$, if $A \cap B = 0$, then also $A \oplus B \subseteq^{\oplus} N$.

The module N is **extending** if it satisfies ($C1$), and **quasi-continuous** if it satisfies ($C1$) and ($C3$). Any direct summand of an extending module is again extending [87, Prop.2.7, p.20]. If a module N is quasi-injective, then it is quasi-continuous by [87, p.18].

3.1.10. A family $\{N_\lambda : \lambda \in \Lambda\}$ of submodules of a module N is said to be a **local summand** of N if the sum $\sum_{\lambda \in \Lambda} N_\lambda$ is direct, and every finite direct sum of the N_λ's is a summand of N. In this case, the sum $\sum_{\lambda \in \Lambda} N_\lambda$ is also called a local summand of N. If every local summand of a module N is a summand, then N is a direct sum of indecomposable modules by [87, Theorem 2.17].

3.1.11. THEOREM. Let $\mathcal{K}$ be an M-natural class and $N \in \mathcal{K}$ be an extending module. If $H_{\mathcal{K}}(A)$ has ACC for every cyclic submodule A of M, then every local summand of N is a summand.

PROOF. Let $\Gamma = \{X_\lambda : \lambda \in \Lambda\}$ be a local summand of N and $X = \sum_{\lambda \in \Lambda} X_\lambda$. Since N is an extending module, $X \leq_e Y \subseteq^{\oplus} N$ for some Y. Then $Y \in \mathcal{K}$ is an extending module. So without loss of generality, we can assume that

$$X \leq_e Y = N \leq_e E_M(N) = \Sigma\{f(M) : f \in \text{Hom}(M, EN)\}.$$

We need to show that $X = N$. Suppose that $X \neq N$. There exists a smallest positive integer t such that, for some $x \in N \setminus X$ and some $f_i \in \text{Hom}(M, EN)$ $(i = 1, \cdots, t)$, $x \in f_1(M) + \cdots + f_t(M)$. Choose $a_i \in M$ $(i = 1, \cdots, t)$ such that $\Sigma_{i=1}^t f_i(a_i) \in N \setminus X$.

For $g_i \in \text{Hom}(M, EN)$ $(i = 1, \cdots, t)$, we denote by $\oplus_{i=1}^t g_i$ the sum map $M^{(t)} \longrightarrow E(N)$ which maps $(x_1, \cdots, x_t) \in M^{(t)}$ to $g_1(x_1) + \cdots + g_t(x_t)$, and denote by $\oplus_{i=1}^t g_i|$ the restriction of $\oplus_{i=1}^t g_i$ on $a_1 R \oplus \cdots \oplus a_t R \leq M^{(t)}$. Now let

$$\Omega = \{Ker(\oplus_{i=1}^t g_i|) : [g_1(a_1 R) + \cdots + g_t(a_t R)] \cap (N \setminus X) \neq \emptyset, \quad g_i \in \text{Hom}(M, EN)\,(i = 1, \cdots, t)\}.$$

By the choice of t and the a_i's, we see that Ω is a non-empty subset of $H_{\mathcal{K}}(a_1 R \oplus \cdots \oplus a_t R)$. Since each $H_{\mathcal{K}}(a_i R)$ has ACC by our assumption, $H_{\mathcal{K}}(a_1 R \oplus \cdots \oplus a_t R)$ has ACC by (3.1.4), and thus there exist $y \in N \setminus X$ and

$g_i \in \mathrm{Hom}(M, E_M(N))$ $(i = 1, \cdots, t)$, such that $y \in g_1(a_1R) + \cdots + g_t(a_tR)$ and $Ker(\oplus_{i=1}^t g_i|)$ is a maximal element in Ω. Since $X \leq_e N$, $0 \neq yr \in X$ for some $r \in R$. Let $yr \in \oplus_{i=1}^n X_{\lambda_i}$. By assumption, $N = (\oplus_{i=1}^n X_{\lambda_i}) \oplus Z$ for some $Z \subseteq N$. Then $E(N) = E(\oplus_{i=1}^n X_{\lambda_i}) \oplus E(Z)$. Let p_Z be the projection of $E(N)$ onto $E(Z)$. Write $y = y_1 + y_2$, where $y_1 \in \oplus_{i=1}^n X_{\lambda_i}$, $y_2 \in Z$. Clearly, $y_2 \notin X$ and $yr = y_1r$. Let $h_i = p_Z g_i$ $(i = 1, \cdots, t)$. Then

$$y_2 = p_Z(y) \in \Sigma_{i=1}^t h_i(a_iR) \subseteq E(Z).$$

Therefore, $Ker(\oplus_{i=1}^t h_i|) \in \Omega$. It is easy to check that

$$Ker(\oplus_{i=1}^t g_i|) \subseteq Ker(\oplus_{i=1}^t h_i|).$$

Choose $r_i \in R$ $(i = 1, \cdots, t)$ such that $g_1(a_1r_1) + \cdots + g_t(a_tr_t) = y$. Then $\Sigma_{i=1}^t g_i(a_ir_ir) = yr \neq 0$, but $\Sigma_{i=1}^t h_i(a_ir_ir) = p_Z(yr) = 0$. Thus, we have

$$(a_1r_1r, \cdots, a_tr_tr) \in Ker(\oplus_{i=1}^t h_i|) \setminus Ker(\oplus_{i=1}^t g_i|).$$

Hence

$$Ker(\oplus_{i=1}^t g_i|) \subset Ker(\oplus_{i=1}^t h_i|),$$

contradicting the maximality of $Ker(\oplus_{i=1}^t g_i|)$. Hence $X = N$, and every local summand of N is a summand. □

The proof of a subsequent theorem here will require the following result from [87, Theorems 2.17]. Note that an indecomposable extending module is uniform.

3.1.12. The following are equivalent for a quasi-continuous module N:

1. Every local summand of N is a summand.

2. N is a direct sum of indecomposable (uniform) modules.

3. N has a direct sum decomposition that complements direct summands.

Finally we have essentially all the ingredients needed to explain the next theorem due to [99, Theorem 16] and [135, Theorem 2.4]. A module M is said to be c-**generated** if there is an indexed set $\{x_t : t \in C\}$ that spans M with $c = |C|$. A module N is called Σ-**quasi-injective** if $N^{(\Lambda)}$ is quasi-injective for every index set Λ.

3.1.13. THEOREM. The following are equivalent for an M-natural class $\mathcal{K}$:

1. Every direct sum of M-injective modules in $\mathcal{K}$ is M-injective.

2. Every M-injective module in $\mathcal{K}$ is a direct sum of uniform modules.

3. Every extending module in $\mathcal{K}$ is a direct sum of uniform modules.

4. Every M-injective module in $\mathcal{K}$ has a direct sum decomposition that complements direct summands.

5. Every module in $\mathcal{K}$ contains a maximal M-injective submodule.

6. Every quasi-injective module in $\mathcal{K}$ is Σ-quasi-injective.

7. There exists a cardinal c such that every M-injective module in $\mathcal{K}$ is a direct sum of c-generated modules.

PROOF. Before beginning, note that an M-injective module in $\sigma[M]$ is quasi-injective by (2.2.10)(5). As explained in (3.1.9), a quasi-injective module is quasi-continuous and hence extending. Thus (3) implies (2).

(2) $\Longleftrightarrow$ (4). Since a quasi-injective module is quasi-continuous, the claim follows from (3.1.12).

(1) $\Longrightarrow$ (3). Let $N \in \mathcal{K}$ be an extending module. By (3.1.5), (1) implies that $H_{\mathcal{K}}(A)$ has ACC for any cyclic submodule A of M. The latter and (3.1.11) show that every local summand of N is a summand. Hence, as explained in (3.1.10), N is a direct sum of indecomposable modules. But every extending indecomposable module is a uniform module, so N is a direct sum of uniform modules.

(4) $\Longrightarrow$ (1). Let $D = \oplus\{D_\alpha : \alpha \in \Gamma\}$ where all $D_\alpha \in \mathcal{K}$ are M-injective. Because of the equivalence (2) $\Longleftrightarrow$ (4), each D_α is a direct sum of M-injective uniform modules. Without loss of generality, we may assume that each D_α is an M-injective uniform module in $\mathcal{K}$. Then (4) ensures that (2) of (3.1.7) holds, and hence D is M-injective by (3.1.7).

(1) $\Longrightarrow$ (5). For any $N \in \mathcal{K}$, 0 is an injective submodule of N. By Zorn's Lemma, there exists a maximal independent set of M-injective submodules of N, say $\mathcal{X}$. Set $X = \Sigma_{P\in\mathcal{X}}P = \oplus_{P\in\mathcal{X}}P$. Then X is M-injective by (1). Suppose that $Y \leq N$ is M-injective with $X \subseteq Y$. Then $Y = X \oplus Z$ for some $Z \leq Y$ by (2.2.10)(1). Thus $Z = 0$, otherwise $\mathcal{X} \cup \{Z\}$ is an independent set of M-injective submodules of N, contradicting the maximality of $\mathcal{X}$. It follows that X is a maximal M-injective submodule of N.

(5) $\Longrightarrow$ (1). Let $E = \oplus_i E_i$ be a direct sum of M-injective modules with all $E_i \in \mathcal{K}$. By (5), E has a maximal M-injective submodule, say E_0. If $E_0 \cap E_i = 0$, then $E_0 \oplus E_i$ is M-injective. Therefore, the maximality of E_0 implies that $E_0 \cap E_i \neq 0$ for all i. Moreover, $E_0 \cap E_i$ is essential in E_i. For, if $A \subseteq E_i$ and $A \cap E_0 = 0$, then $E_M(A) \subseteq E_M(E_i) = E_i$ and hence $E_M(A) \cap E_0 = 0$. It follows that $E_M(A) \oplus E_0$ is M-injective. The maximality of E_0 implies that $E_M(A) = 0$, so $A = 0$. Consequently, $E_0 \cap E_i \leq_e E_i$ for all i, and so $E_0 \leq_e E$. But $E_0 \subseteq^{\oplus} E$ by (2.2.10)(1), so $E = E_0$ is M-injective.

(1) $\Longrightarrow$ (6). Let $N \in \mathcal{K}$ be quasi-injective. Then, for every index set I, $E_M(N)^{(I)}$ is M-injective by (1), and hence is $E_M(N)$-injective. By [87, 1.7], for any choice of $m_i \in E_M(N)$ $(i = 1, 2, \cdots)$ and $a \in E_M(N)$ such that $a^\perp \subseteq \cap_{i=1}^\infty m_i^\perp$, the ascending sequence $\cap_{i\geq n} m_i^\perp$ terminates. Thus, for any choice of $m_i \in N$ $(i = 1, 2, \cdots)$ and $a \in N$ such that $a^\perp \subseteq \cap_{i=1}^\infty m_i^\perp$, the

ascending sequence $\cap_{i\geq n} m_i^{\perp}$ terminates. Since N is quasi-injective, it follows from [87, 1.7] that $N^{(I)}$ is N-injective. So $N^{(I)}$ is quasi-injective by [87, 1.5].

(6) $\Longrightarrow$ (1). It suffices to show that $H_{\mathcal{K}}(R)$ has ACC by (3.1.5). So let $I_1 \subseteq I_2 \subseteq \cdots$ be an ascending chain of right ideals of R with all $I_i \in H_{\mathcal{K}}(R)$. Set $N = \oplus_i E_M(R/I_i)$. Then $N \in \mathcal{K}$. By (2.2.10)(1), for each i, $E_M(N) = E_M(R/I_i) \oplus X_i$ for some $X_i \leq E_M(N)$. Then

$$E_M(N)^{(\mathbb{N})} \cong \oplus_i[E_M(R/I_i) \oplus X_i] = [\oplus_i E_M(R/I_i)] \oplus (\oplus_i X_i).$$

Since $E_M(N)^{(\mathbb{N})}$ is quasi-injective by (6), N is quasi-injective. Let $I = \cup_i I_i$. For each i, we have a map $f_i : I/I_i \longrightarrow E_M(R/I_i)$ defined by $f_i(a+I_i) = a+I_i$ (for $a \in I$). Define $f : I/I_1 \longrightarrow N$ by $\pi_i(f(a+I_1)) = f_i(a+I_i)$, where π_i is the projection of N onto $E_M(R/I_i)$. Then f is well-defined, and thus f extends to $g : N \longrightarrow N$ since N is quasi-injective. It follows that $f(I/I_1) \subseteq g(R/I_1) \subseteq N$. Let $x = g(1+I_1) \in N$, where 1 is the identity of R. Then $f(I/I_1) \subseteq xR \subseteq \Sigma_{i=1}^k E_M(R/I_i)$ for some k, implying $I_{k+1} = I_{k+2} = \cdots = I$.

(2) $\Longrightarrow$ (7). Since the set $\{E_M(R/I) : I \in H_{\mathcal{K}}(R)\}$ contains an isomorphic copy of each uniform M-injective module in $\mathcal{K}$, there exists a cardinal c (e.g., $c = |\oplus\{E_M(R/I) : I \in H_{\mathcal{K}}(R)\}|$) such that every uniform M-injective module in $\mathcal{K}$ is c-generated. Consequently (2) implies that every M-injective module in $\mathcal{K}$ is a direct sum of c-generated modules.

(7) $\Longrightarrow$ (1). It suffices to show that $N^{(\mathbb{N})}$ is M-injective whenever $N \in \mathcal{K}$ is M-injective by (3.1.8). But the necessary argument has been given by Faith and Walker (see [49] or the proof of [8, 25.8, p.293]). □

3.1.14. REMARK. The class $\mathcal{K} = \{N \in \sigma[M] : N \text{ is non } M\text{-singular}\}$ is an M-natural class. Applying (3.1.5) and (3.1.13) to this class yields the characterizations of modules M for which direct sums of non M-singular M-injective modules are M-injective.

3.1.15. COROLLARY. The following are equivalent for a natural class $\mathcal{K}$:

1. R satisfies ACC on $H_{\mathcal{K}}(R)$.
2. Every direct sum of injective modules in $\mathcal{K}$ is injective.
3. Every injective module in $\mathcal{K}$ is a direct sum of uniform modules.
4. Every extending module in $\mathcal{K}$ is a direct sum of uniform modules.
5. Every injective module in $\mathcal{K}$ has a decomposition that complements direct summands.
6. Every module in $\mathcal{K}$ contains a maximal injective submodule.
7. Every quasi-injective module in $\mathcal{K}$ is Σ-quasi-injective.
8. There exists a cardinal c such that every injective module in $\mathcal{K}$ is a direct sum of c-generated modules.

3.1.16. REMARK. Let $\tau = (\mathcal{T}_\tau, \mathcal{F}_\tau)$ be a hereditary torsion theory. Applying (3.1.15) to $\mathcal{F}_\tau$ yields the characterizations of rings R for which direct sums of injective τ-torsion free R-modules are injective (see Golan [59]). In particular, we obtain Teply's characterizations of the rings R for which direct sums of non-singular injective modules are injective when τ is the Goldie torsion theory (see Teply [115]).

Next, we characterize the rings R for which direct sums of singular injective R-modules are injective. The starting point is the following lemma.

3.1.17. LEMMA. For a module M, $Z(EM) = EM$ iff, for any $0 \neq X \leq M$, X is not embeddable in R.

PROOF. Suppose EM is singular. If $0 \neq N \leq M$ and $f : N \hookrightarrow R$ is an R-monomorphism, then $E(N) \cong E(f(N))$ is singular and $E(R) = E(f(N)) \oplus X$ for some X. Write $1 = q + w$, $q \in E(f(N))$ and $w \in X$. Then $q^\perp$ is essential in R, implying that $q^\perp \cap f(N) \neq 0$. Hence

$$\begin{aligned} 0 \neq q^\perp \cap f(N) &= (q + w)(q^\perp \cap f(N)) \\ &= w(q^\perp \cap f(N)) \subseteq f(N) \cap X. \end{aligned}$$

This is a contradiction.

Conversely, if $E(M)$ is not singular, then $x^\perp$ is not essential in R for some $0 \neq x \in E(M)$. There exists a nonzero right ideal I of R such that $x^\perp \cap I = 0$. Thus $xI \cong I \hookrightarrow R$. □

3.1.18. COROLLARY. For a ring R, the class

$$\mathcal{S}(R) = \{N \in \text{Mod-}R : Z(EN) = EN\}$$

is a natural class.

PROOF. It suffices to show that, for $W_\alpha \in \mathcal{S}(R)$ $(\alpha \in \Gamma)$, $E(\oplus_{\alpha \in \Gamma} W_\alpha) \in \mathcal{S}(R)$. If not, then there exists a $0 \neq x \in E(\oplus_{\alpha \in \Gamma} W_\alpha)$ and an embedding $xR \hookrightarrow R$ by (3.1.17). By the projection argument (2.3.3), for some $r \in R$, $0 \neq xrR \cong wR \leq W_\alpha$ for some $\alpha \in \Gamma$. Thus $0 \neq wR \hookrightarrow xR \hookrightarrow R$ shows that $Z(EW_\alpha) \neq EW_\alpha$, a contradiction. □

For $\mathcal{K} = \mathcal{S}(R)$, we now reinterpret (3.1.15) and obtain the following result in [98].

3.1.19. THEOREM. Let $\mathcal{S} = \mathcal{S}(R)$. The following are equivalent:

1. R satisfies ACC on $H_{\mathcal{S}}(R) = \{I \leq R : Z(E(R/I)) = E(R/I)\}$.
2. Every direct sum of singular injective modules is injective.
3. Every singular injective module is a direct sum of uniform modules.
4. Every singular extending module is a direct sum of uniform modules.

5. Every singular injective module has a decomposition that complements direct summands.

6. There exists a cardinal c such that every singular injective module is a direct sum of c-generated modules.

We next give a ring R such that $\mathcal{S}(R)$ is closed neither under quotient modules nor under arbitrary products. So $\mathcal{S}(R)$ is neither a pretorsion class nor a torsion free class. This explains that Theorem (3.1.19) can not be obtained in Torsion Theory setting or in the setting of $\sigma[M]$. We need the following facts for the example.

3.1.20. PROPOSITION. Let R be a ring and $\mathcal{S} = \mathcal{S}(R)$.

1. $H_{\mathcal{S}}(R)$ contains all dense right ideals of R.

2. Let H_1 be a non-empty subset of $H_{\mathcal{S}}(R)$ which contains all dense right ideals of R. Then H_1 forms a Gabriel topology iff H_1 is the set of all dense right ideals of R.

PROOF. Note that I is a dense right ideal of R iff $\mathrm{Hom}_R(J/I, R) = 0$ for all right ideals J with $I \subseteq J \subseteq R$ (see 2.1.12).

(1) If I is a dense right ideal of R, then clearly no submodules of R/I embed in R. So $I \in H_{\mathcal{S}}(R)$ by (3.1.17).

(2) The set of all dense right ideals of R is a Gabriel topology by (2.1.12). Now suppose H_1 is a Gabriel topology and $I \in H_1$. If $0 \neq f \in \mathrm{Hom}(J/I, R)$, where $J \subseteq R$, then $Ker(f) = K/I$ for some $K \subseteq J$, and J/K embeds in R. Thus, $E(R/K)$ is not singular by (3.1.17). Hence $K \notin H_1$, contracting the fact that H_1 is a Gabriel topology. So $\mathrm{Hom}_R(J/I, R) = 0$ for all right ideals J with $I \subseteq J \subseteq R$. Thus, I is a dense right ideal. □

3.1.21. PROPOSITION. Given a ring R, let $S_0 = \cap\{I : R/I \in \mathcal{S}(R)\}$ and $Z_0 = \{r \in R : rI = 0 \text{ for some } I \subseteq R \text{ with } R/I \in \mathcal{S}(R)\}$. Then the following hold:

1. Z_0 and S_0 both are ideals of R.

2. $\mathcal{S}(R)$ is closed under quotient modules iff $Z_0 = 0$.

3. $\mathcal{S}(R)$ is closed under arbitrary products iff S_0 is essential in R_R.

PROOF. Let $\mathcal{S} = \mathcal{S}(R)$.

(1) Suppose $r_1, r_2 \in Z_0$ and $a \in R$. Then there exist $I_1, I_2 \in H_{\mathcal{S}}(R)$ such that $r_1 I_1 = 0$, $r_2 I_2 = 0$. Hence $(ar_1)I_1 = 0$, $(r_1 + r_2)(I_1 \cap I_2) = 0$, and $(r_1 a)(a^{-1}I_1) = 0$. Since $\mathcal{S}(R)$ is a natural class by (3.1.18), $I_1 \cap I_2$ and $a^{-1}I_1$ both are in $H_{\mathcal{S}}(R)$ by (2.3.11). It follows that $ar_1, r_1 - r_2$, and $r_1 a$ all are in Z_0. Hence Z_0 is an ideal.

Suppose S_0 is not an ideal. Then $aS_0 \not\subseteq I$ for some $a \in R$ and some $I \in H_{\mathcal{S}}(R)$. Then $ab \notin I$ for some $b \in S_0$. This implies that

$$0 \neq (abR + I)/I \cong R/(ab)^{-1}I = R/[b^{-1}(a^{-1}I)]$$

and thus $b^{-1}(a^{-1}I) \neq R$. But, on the other hand, we have

$$R/a^{-1}I \cong (aR + I)/I \subseteq R/I \in \mathcal{S}.$$

So $R/a^{-1}I \in \mathcal{S}$, since $\mathcal{S}$ is closed under submodules. Therefore, $b \in S_0 \subseteq a^{-1}I$ and we have $b^{-1}(a^{-1}I) = R$. This is a contradiction. So S_0 is an ideal.

(2) First we note that $\mathcal{S}$ is closed under quotient modules iff $\mathcal{S}$ is a hereditary torsion class iff $H_{\mathcal{S}}(R)$ is a Gabriel topology iff, by (3.1.20), $H_{\mathcal{S}}(R)$ is the set of all dense right ideals of R.

Suppose $0 \neq x \in Z_0$. Then $I \subseteq x^{\perp} \subset R$ for some $I \in H_{\mathcal{S}}(R)$. Since $R/x^{\perp} \cong xR$, we have $\mathrm{Hom}(R/x^{\perp}, R) \neq 0$. Hence $x^{\perp}$ is not a dense right ideal of R. It follows that I is not a dense right ideal.

For the converse, suppose $Z_0 = 0$. If $I \in H_{\mathcal{S}}(R)$, then $a^{-1}I \in H_{\mathcal{S}}(R)$ for any $a \in R$ by (3.1.18) and (2.3.11). So ${}^{\perp}(a^{-1}I) = 0$ since $Z_0 = 0$. Hence I is a dense right ideal of R by (2.1.12).

(3) Suppose S_0 is an essential right ideal of R. Let $M = \Pi_t M_t$ with each $M_t \in \mathcal{S}$. Then $E(M) = \Pi_t E(M_t)$. For any $x \in E(M)$, write $x = (x_t)$ with $x_t \in E(M_t)$ for all t. We have $x_t^{\perp} \in H_{\mathcal{S}}(R)$ since $E(R/x_t^{\perp}) \cong E(x_tR) \subseteq E(M_t)$ and $E(M_t)$ is singular. Then $S_0 \subseteq x_t^{\perp}$ for all t, implying $xS_0 = 0$. It follows that $E(R)$ is singular. Therefore, $M \in \mathcal{S}$.

Now suppose $\mathcal{S}$ is closed under arbitrary products. Then $E(R/S_0)$ embeds in $\Pi\{E(R/I) : I \in H_{\mathcal{S}}(R)\} \in \mathcal{S}$. So $E(R/S_0) \in \mathcal{S}$ and hence $S_0 \in H_{\mathcal{S}}(R)$. It follows by (3.1.17) that S_0 is an essential right ideal of R. □

3.1.22. EXAMPLE. There exists a ring R such that $\mathcal{S}(R)$ is closed neither under quotient modules nor under arbitrary products.

Let $R_1 = \mathbb{Z}$ and $R_2 = \left\{\begin{pmatrix} a & x \\ 0 & a \end{pmatrix} : a \in \mathbb{Z}, x \in \mathbb{Z}_{2^\infty}\right\}$. And let $R = R_1 \times R_2$ be the direct product of rings. Let $A = \left\{\begin{pmatrix} 0 & x \\ 0 & 0 \end{pmatrix} : x \in \mathbb{Z}_{2^\infty}\right\}$. Then A is an ideal of R_2, and so $I = R_1 \times A$ is an ideal of R.

Claim 1. $R/I \in \mathcal{S}(R)$. First, $(R/I)_R \cong (R_2/A)_R$ which has no nonzero submodules embeddable in $(R_1)_R$. Since $A_{\mathbb{Z}}$ is torsion and $(R_2/A)_{\mathbb{Z}}$ is torsion free, $A_{\mathbb{Z}}$ and $(R_2/A)_{\mathbb{Z}}$ have no nonzero isomorphic submodules. Thus, A_R and $(R_2/A)_R$ have no nonzero isomorphic submodules. Since $A_R \leq_e (R_2)_R$, it follows that $(R_2)_R$ and $(R_2/A)_R$ have no nonzero isomorphic submodules. So, by (2.3.3), no nonzero submodules of $(R/I)_R$ embed in R_R. Hence $R/I \in \mathcal{S}(R)$ by (3.1.17). Because ${}^{\perp}I \neq 0$, $\mathcal{S}(R)$ is not closed under quotient modules by (3.2.21).

For any prime number p, let $I_p = p\mathbb{Z} \times R_2$. Then all I_p are ideal of R.

Claim 2. $R/I_p \in \mathcal{S}(R)$ for any prime $p \neq 2$. In fact, $(R/I_p)_R \cong (\mathbb{Z}/p\mathbb{Z})_R$ has no nonzero submodules embeddable in $(R_1)_R$ or in $(R_2)_R$. So no nonzero submodules of R/I_p embed in R_R by (2.3.3). Thus, $R/I_p \in \mathcal{S}(R)$ by (3.1.17). Now $\cap\{I_p : p \neq 2 \text{ a prime}\} = 0 \times R_2 \subseteq R$. So $S_0 \subseteq 0 \times R_2$, and hence S_0 is not an essential right ideal of R. By (3.1.21), $\mathcal{S}(R)$ is not closed under arbitrary products. □

It is known that every direct sum of injective modules is injective (= right Noetherian) iff every direct sum of uniform injective modules is injective iff every injective module has a decomposition that complements maximal direct summand. This result can be extended to the setting of an M-natural class which contains "enough" uniform modules. For a cardinal c, a module N is called **c-limited** if every direct sum of nonzero submodules of N contains at most c direct summands. For the condition $(*)$, see (2.4.8). The following result is proved in [135].

3.1.23. THEOREM. The following are equivalent for an M-natural class $\mathcal{K}$ with $(*)$:

1. Every direct sum of M-injective modules in $\mathcal{K}$ is M-injective.

2. Every direct sum of uniform M-injective modules in $\mathcal{K}$ is M-injective.

3. Every direct sum of M-injective hulls of $\mathcal{K}$-cocritical modules is M-injective.

4. Every M-injective module in $\mathcal{K}$ has a decomposition that complements maximal direct summands.

5. There exists a cardinal c such that every direct sum of M-injective modules in $\mathcal{K}$ is a direct sum of an M-injective module and a c-limited module.

PROOF. The implications $(1) \Longrightarrow (2) \Longrightarrow (3)$ and $(1) \Longrightarrow (5)$ are obvious.

$(3) \Longrightarrow (1)$. Suppose that $A \leq M$ is a finitely generated submodule and $X_1 \subset X_2 \subset \cdots \leq A$ is a strictly ascending chain with all $X_i \in H_{\mathcal{K}}(A)$. Then, by (2.4.10), there exist right ideals K_i, J_{i+1} of R such that

$$X_i \subseteq K_i \subset J_{i+1} \subseteq X_{i+1}$$

and J_{i+1}/K_i is $\mathcal{K}$-cocritical. Let $f_i : J_{i+1} \longrightarrow E_M(J_{i+1}/K_i)$ be the canonical map. Then f_i extends to $g_i : A \longrightarrow E_M(J_{i+1}/K_i)$. Set $X = \cup X_i$ and let $E = \oplus_i E_M(J_{i+1}/K_i)$. Define $f : X \longrightarrow E = \oplus_i E_M(J_{i+1}/K_i)$ by $\pi_i f(a) = g_i(a)$, where π_i is the projection of E onto $E_M(J_{i+1}/K_i)$. Clearly f is well-defined. By (3), E is M-injective, and hence f extends to $g : A \longrightarrow E$. This implies that, for some n, $f(X) \subseteq g(A) \subseteq \oplus_{j<n} E_M(J_{j+1}/K_j)$. For any $a \in J_{n+1}$, we have $0 = \pi_n f(a) = g_n(a) = a + K_n$, implying that $a \in K_n$. So, $J_{n+1} = K_n$, a contradiction. This shows that $H_{\mathcal{K}}(A)$ has ACC. So (1) follows by (3.1.5).

$(1) \Longrightarrow (4)$ is by (3.1.13).

(4) $\Longrightarrow$ (3). Let $D = \oplus_A D_\alpha$ where each $D_\alpha = E_M(N_\alpha)$ for a $\mathcal{K}$-cocritical module N_α. Since all D_α are uniform, (2) of (3.1.7) holds. Thus, by (3.1.7), D is M-injective.

(5) $\Longrightarrow$ (3). Suppose that (5) holds. Let $\{S_x : x \in I\}$ denote a collection of representatives of the isomorphism classes of cyclic $\mathcal{K}$-cocritical modules and let $C = \oplus_{x\in I} S_x$. Let J be an index set with $|J| \geq c$ and, for each $y \in J$, let $C_y = C$. We define $B = \oplus_{y\in J} C_y$. Let N_α ($\alpha \in A$) be $\mathcal{K}$-cocritical modules and $E_\alpha = E_M(N_\alpha)$ for each α. Set $E = \oplus_\alpha E_\alpha$. It suffices to show that E is M-injective. Choose an index set K with $|K| > |E(B)|$. For each $\beta \in K$ let $D_\beta = E$ and consider $D = \oplus_{\beta\in K} D_\beta$. By (5), $D = X \oplus Y$ for an M-injective module X and a c-limited module Y. Let $N = \oplus_\alpha N_\alpha$ be the submodule of E. Then N is essential in E. Let $Z = \oplus_{\beta\in K} Z_\beta$ with each $Z_\beta = N$. Then $Z \leq_e D$ and so $Z \cap Y \leq_e Y$. By Zorn's Lemma, there exists a maximal independent set of cyclic $\mathcal{K}$-cocritical submodules of $Z \cap Y$, say $\mathcal{F}$. Write $F = \Sigma_{P\in\mathcal{F}} P = \oplus_{P\in\mathcal{F}} P$. We claim that F is essential in $Z \cap Y$. For, if $F \cap Q = 0$ for some nonzero submodule Q of $Z \cap Y$, then we can find a smallest t such that, for some $a \in Q$, $a = a_1 + a_2 + \cdots + a_t$ with $0 \neq a_i \in N_{\alpha_i}$ $(i = 1, 2, \cdots, t)$. The choice of t assures us that $a_1 r \longmapsto ar$ is a well-defined nonzero epimorphism from $a_1 R$ to aR. By (2.4.7), $aR \in \mathcal{K}$ and $a_1 R$ is $\mathcal{K}$-cocritical, and hence $aR \cong a_1 R$ is $\mathcal{K}$-cocritical. Then $\mathcal{F} \cup \{aR\}$ is an independent set of cyclic $\mathcal{K}$-cocritical submodules of $Z \cap Y$, contradicting the maximality of $\mathcal{F}$. It follows that F is essential in $Z \cap Y$, implying that F is essential in Y. Since Y is a c-limited module, F is the direct sum of at most c cyclic $\mathcal{K}$-cocritical submodules. Thus, we have an embedding $F \hookrightarrow B$ which can be extended to an embedding $Y \hookrightarrow E(B)$ since $F \leq_e Y$. Then we have that $|Y| \leq |E(B)| < |K|$. For each $b \in Y$, there exists a finite subset K_b of K such that $b \in \oplus_{\beta\in K_b} D_\beta$. We let $K_1 = \cup_{b\in Y} K_b$ and $K_2 = K \setminus K_1$. Clearly, $|K_1| \leq |E(B)|$ and hence K_2 is not empty since $|E(B)| < |K|$. Now let $G = \oplus_{\beta\in K_1} D_\beta$ and $H = \oplus_{\beta\in K_2} D_\beta$. Then $D = G \oplus H = X \oplus Y$ and $Y \subseteq G$. It follows that $G = (G \cap X) \oplus Y$ and so $D = (G \cap X) \oplus Y \oplus H = X \oplus Y$. This implies that $X \cong (G \cap X) \oplus H$. Since X is an M-injective module, H is M-injective, and thus E is M-injective since $E \cong D_\beta$ (for some $\beta \in K_2$) is a summand of H. □

The next example shows that the condition ($*$) in (3.1.23) cannot be removed.

3.1.24. EXAMPLE. Let R be a domain such that $R^2 \cong R^3$ as right R-modules. Such a ring R exists by J.D. O'Neill [93]. Clearly, R_R has infinite Goldie dimension. Since every nonzero cyclic right ideal of R is isomorphic to R_R, every nonzero right ideal of R has infinite Goldie dimension. It follows that every nonzero right ideal of R is not uniform. Let $\mathcal{K} = d(R_R)$. That is, $M \in \mathcal{K}$ iff, for any $0 \neq N \subseteq M$, N has a nonzero submodule embeddable in R. Then $\mathcal{K}$ is a natural class. Now for any $0 \neq M \in \mathcal{K}$, M has a nonzero submodule embeddable in R, so M is not uniform. This shows that there are no uniform modules in $\mathcal{K}$.

For the M-natural class $\mathcal{K}$ above, (2), (3), and (4) of (3.1.23) hold since there are no uniform modules in $\mathcal{K}$. But (1) of (3.1.23) fails to hold, because (1) holds iff every injective module in $\mathcal{K}$ is a direct sum of uniform modules.

The next corollary is obtained by applying (3.1.5), (3.1.13), and (3.1.23) to the M-natural class $\sigma[M]$. A **locally Noetherian module** is a module whose finitely generated submodules are Noetherian.

3.1.25. COROLLARY. The following are equivalent for an R-module M:

1. M is a locally Noetherian module.
2. Every direct sum of M-injective modules in $\sigma[M]$ is M-injective.
3. Every direct sum of uniform M-injective modules in $\sigma[M]$ is M-injective.
4. Every direct sum of M-injective hulls of simple modules in $\sigma[M]$ is M-injective.
5. Every M-injective module in $\sigma[M]$ is a direct sum of uniform modules.
6. Every extending module in $\sigma[M]$ is a direct sum of uniform modules.
7. Every M-injective module in $\sigma[M]$ has a decomposition that complements direct summands.
8. Every M-injective module in $\sigma[M]$ has a decomposition that complements maximal direct summands.
9. Every module in $\sigma[M]$ contains a maximal M-injective submodule.
10. Every quasi-injective module in $\sigma[M]$ is Σ-quasi-injective.
11. There exists a cardinal c such that every M-injective module in $\sigma[M]$ is a direct sum of c-generated modules.
12. There exists a cardinal c such that every direct sum of M-injective modules in $\sigma[M]$ is a direct sum of an M-injective module and a c-limited module.

3.1.26. REMARK. For a stable hereditary torsion theory $\tau = (\mathcal{T}_\tau, \mathcal{F}_\tau)$, $\mathcal{T}_\tau$ is a natural class satisfying $(*)$. Thus, applying (3.1.15) and (3.1.23) (with $M = R$) to $\mathcal{T}_\tau$ yields the various characterizations of the rings R for which direct sums of τ-torsion injective modules are injective.

3.1.27. REMARK. In view of Remark (3.1.14) and Corollary (3.1.18), it is interesting to characterize the modules M for which direct sums of M-singular M-injective modules are M-injective. This relates to the open question whether the class $\{N \in \sigma[M] : E_M(N) \text{ is } M\text{-singular}\}$ is an M-natural class (see [31]).

3.1.28. REFERENCES. Anderson and Fuller [8]; Dauns [31]; Faith and Walker [49]; Mohamed and Müller [87]; Golan [59]; Page and Zhou [98,99,100]; Teply [115]; Wisbauer [131]; Zhou [135].

3.2 Descending Chain Conditions

For a module class $\mathcal{K}$, recall that for $N \in$ Mod-$R, H_{\mathcal{K}}(N)$ is defined by $H_{\mathcal{K}}(N) = \{P \leq N : N/P \in \mathcal{K}\}$. In particular, $H_{\mathcal{K}}(R) = \{I \leq R : R/I \in \mathcal{K}\}$. In this section, we investigate the DCC (the descending chain condition) on $H_{\mathcal{K}}(N)$ and the relation between DCC and ACC on $H_{\mathcal{K}}(N)$ when $\mathcal{K}$ is a pre-natural class. As an application, a nil-imply-nilpotent theorem is proved for the endomorphism ring of a module $N \in \mathcal{K}$ satisfying ACC or DCC on $H_{\mathcal{K}}(N)$.

Throughout this section, $\tau_{\mathcal{K}} = (\mathcal{T}_{\mathcal{K}}, \mathcal{F}_{\mathcal{K}})$ is *the torsion theory cogenerated by* a module class $\mathcal{K}$, that is,

$$\mathcal{T}_{\mathcal{K}} = \{T \in \text{Mod-}R : \text{Hom}(T, C) = 0 \text{ for all } C \in \mathcal{K}\},$$
$$\mathcal{F}_{\mathcal{K}} = \{F \in \text{Mod-}R : \text{Hom}(T, F) = 0 \text{ for all } T \in \mathcal{T}_{\mathcal{K}}\}.$$

By (2.1.5)-(2.1.7), $\mathcal{T}_{\mathcal{K}}$ is a torsion class and $\mathcal{F}_{\mathcal{K}}$ is a torsion free class, and every module N has a largest submodule, denoted by $\tau_{\mathcal{K}}(N)$, in $\mathcal{T}_{\mathcal{K}}$. It should be noted that $\tau_{\mathcal{K}}$ is not necessarily a hereditary torsion theory, even for a hereditary pretorsion class $\mathcal{K}$. For example, let $R = \mathbb{Z}$ and $M = \mathbb{Z}/2\mathbb{Z}$. Set $\mathcal{K} = \sigma[M]$. Then $\mathcal{K}$ is a hereditary pretorsion class. Since M is a simple module, every module in $\sigma[M]$ is semisimple. Thus, $\mathbb{Z}_{2^\infty} \in \mathcal{T}_{\mathcal{K}}$. But $M \hookrightarrow \mathbb{Z}_{2^\infty}$. So $\mathcal{T}_{\mathcal{K}}$ is not closed under submodules.

3.2.1. LEMMA. Let $\mathcal{K}$ be an M-natural class and $P \subseteq N \in \sigma[M]$. If $N \in \mathcal{T}_{\mathcal{K}}$ then $P \in \mathcal{T}_{\mathcal{K}}$.

PROOF. If $P \notin \mathcal{T}_{\mathcal{K}}$, then $P/X \in \mathcal{K}$ for some $X \subset P$. There exists a submodule Q/X of N/X which is maximal with respect to the property that $(P/X) \cap (Q/X) = 0$. Then P/X is embeddable as an essential submodule in N/Q. Since $N/Q \in \sigma[M]$, it follows from (2.4.2) that $N/Q \in \mathcal{K}$, contradicting the fact that $N \in \mathcal{T}_{\mathcal{K}}$. □

3.2.2. LEMMA. Let P be a submodule of N and E be a quasi-injective module such that $N/P \overset{\theta}{\hookrightarrow} E$. If $W = \{\alpha \in \text{Hom}_R(N, E) : P \subseteq Ker(\alpha)\}$ and $S = \text{End}(E_R)$, then $W = S(\theta v)$ is a cyclic submodule of the left S-module $\text{Hom}_R(N, E)$ where $v : N \longrightarrow N/P$ is the natural homomorphism, and $P = \cap\{Ker(\alpha) : \alpha \in W\}$.

PROOF. It is clear that $S(\theta v) \subseteq W$. Let $\alpha \in W$. Since $\alpha(P) = 0$, $\alpha = \beta v$ for some $\beta \in \text{Hom}_R(N/P, E)$. Since E is quasi-injective, there exists a $\gamma \in S$ such that $\beta = \gamma\theta$. Thus, $\alpha = \gamma(\theta v) \in S(\theta v)$, so $W = S(\theta v)$. Now it is clear that $P = \cap\{Ker(\alpha) : \alpha \in W\}$. □

3.2.3. LEMMA. Let E be a quasi-injective module with $S = \text{End}(E_R)$. If $W = S\alpha_1 + \cdots + S\alpha_n$ is a finitely generated submodule of the left S-module $\text{Hom}_R(N, E)$, then $W = \{\beta \in \text{Hom}_R(N, E) : \cap_{\alpha \in W} Ker(\alpha) \subseteq Ker(\beta)\}$.

PROOF. For $\beta \in \text{Hom}_R(N, E)$ with $\cap_{\alpha \in W} Ker(\alpha) \subseteq Ker(\beta)$, $\cap_{i=1}^n Ker(\alpha_i) \subseteq Ker(\beta)$. Since E is quasi-injective, the diagram with exact row

$$\begin{array}{l} 0 \longrightarrow N/Ker(\beta) \stackrel{(\alpha_i)}{\longrightarrow} E^n \\ \qquad\qquad \downarrow \beta \\ \qquad\qquad E \end{array}$$

can be extended commutatively by some $\Sigma s_i : E^n \to E$, $s_i \in S$. Thus $\beta = \Sigma_{i=1}^n s_i \alpha_i \in W$, and so $W = \{\beta \in \text{Hom}_R(N, E) : \cap_{\alpha \in W} Ker(\alpha) \subseteq Ker(\beta)\}$. □

A module E is called a **cogenerator** of $\mathcal{K}$ if $E \in \mathcal{K}$ and every module in $\mathcal{K}$ can be cogenerated by E. For an M-natural class $\mathcal{K}$ and a family $\{X_\alpha : \alpha \in \Lambda\} \subseteq \mathcal{K}$, if $X = (\oplus_\alpha X_\alpha) \oplus M_{\mathcal{K}}$, then $E_M(X)$ is an M-injective cogenerator of $\mathcal{K}$ and contains each X_α.

3.2.4. PROPOSITION. The following are equivalent for an M-natural class $\mathcal{K}$ and a module $N \in \sigma[M]$:

1. $H_{\mathcal{K}}(N)$ has ACC.
2. For any $X \subseteq N$, $X/Y \in \mathcal{T}_{\mathcal{K}}$ for some finitely generated submodule Y of X.
3. $H_{\mathcal{F}_{\mathcal{K}}}(N)$ has ACC.
4. For any chain $N_1 \subseteq N_2 \subseteq \cdots$ of submodules of N, there exists a positive integer k such that $N_{i+1}/N_i \in \mathcal{T}_{\mathcal{K}}$ for all $i \geq k$.
5. Every chain $N_1 \subseteq N_2 \subseteq \cdots$ of submodules of N with each $N_{i+1}/N_i \in \mathcal{K}$ terminates.
6. For any M-injective cogenerator E of $\mathcal{K}$, let $S = \text{End}(E_R)$ and let $N^* = \text{Hom}_R(N, E)$. Then N^* satisfies DCC on finitely generated left S-submodules.

PROOF. (1) $\Longleftrightarrow$ (5) is by (3.1.3).

(5) $\Longrightarrow$ (2). Suppose that there exists a submodule X of N such that $X/Y \notin \mathcal{T}_{\mathcal{K}}$ for any finitely generated submodule Y of X. Choose $0 \neq x_1 \in X$. Then $X/(x_1R) \notin \mathcal{T}_{\mathcal{K}}$ and so $X/N_1 \in \mathcal{K}$ for some submodule N_1 with $x_1R \subseteq N_1 \subset X$. Choose $x_2 \in X \setminus N_1$. By the assumption on X, $X/(x_1R + x_2R) \notin \mathcal{T}_{\mathcal{K}}$. Therefore, we have $X/N_2 \in \mathcal{K}$ for some N_2 with $x_1R + x_2R \subseteq N_2 \subset X$. By a simple induction, we can choose a sequence $\{x_i : i = 1, 2, \cdots\}$ and a family $\{N_i : i = 1, 2, \cdots\}$ of submodules of X such that $X/N_i \in \mathcal{K}$, $x_1R + \cdots + x_iR \subseteq N_i$ and $x_{i+1} \notin N_i$ for all i. Let $Y = \oplus_{i=1}^\infty X/N_i$ and $K = \Sigma_{i=1}^\infty x_iR$. Define $f : K \longrightarrow Y$ by $\pi_i f(x) = x + N_i$, where $x \in K$ and π_i is the projection of Y onto X/N_i. Then f is a well-defined homomorphism. Let $K_i = \{x \in K : f(x) \in \oplus_{j=1}^{j=i} X/N_j\}$. Then $K_1 \subseteq K_2 \subseteq \cdots$ is a chain of submodules of K and

$$K_{i+1}/K_i \stackrel{\phi}{\hookrightarrow} (\oplus_{j=1}^{i+1} X/N_j)/(\oplus_{j=1}^{i} X/N_j) \cong X/N_{i+1} \in \mathcal{K} \text{ via}$$
$$\phi(x + K_i) = f(x) + \oplus_{j=1}^{i} X/N_j.$$

Thus $K_{i+1}/K_i \in \mathcal{K}$ for all i. By (5), we have $K_{m+j} = K_m$ for some m and all j. Thus, $f(K) \subseteq \oplus_{j=1}^{m} X/N_j$, implying $0 = \pi_{m+1}f(x_{m+2}) = x_{m+2} + N_{m+1}$, i.e., $x_{m+2} \in N_{m+1}$, a contradiction.

(2) $\Longrightarrow$ (4). Let $N_1 \subseteq N_2 \subseteq \cdots$ be a chain of submodules of N and $X = \cup_i N_i$. Then, there exists a finitely generated submodule Y of X such that $X/Y \in \mathcal{T}_{\mathcal{K}}$ by (2). It follows that $Y \subseteq N_k$ for some k and hence $X/N_i \in \mathcal{T}_{\mathcal{K}}$ for all $i \geq k$. If $N_{i+1}/N_i \notin \mathcal{T}_{\mathcal{K}}$ for some $i \geq k$, then $N_{i+1}/P \in \mathcal{K}$ for some P with $N_i \subseteq P \subset N_{i+1}$. Take a submodule Q/P of X/P maximal with respect to $(Q/P) \cap (N_{i+1}/P) = 0$. Then N_{i+1}/P is embeddable as an essential submodule in X/Q. Noting that $X/Q \in \sigma[M]$ and $N_{i+1}/P \in \mathcal{K}$, we have $X/Q \in \mathcal{K}$ by (2.4.2). But $X/Q \in \mathcal{T}_{\mathcal{K}}$ since X/Q is a quotient of X/N_i. So $X/Q = 0$, implying $N_{i+1}/P = 0$, a contradiction.

(4) $\Longrightarrow$ (3). Let $N_1 \subseteq N_2 \subseteq \cdots$ be a chain of submodules of N with $N/N_i \in \mathcal{F}_{\mathcal{K}}$ for all i. By (4), there exists a $k > 0$ such that $N_{i+1}/N_i \in \mathcal{T}_{\mathcal{K}}$ for all $i \geq k$. But since $\mathcal{F}_{\mathcal{K}}$ is a torsion free class, $N_{i+1}/N_i \in \mathcal{F}_{\mathcal{K}}$ for all $i \geq 1$. It follows that $N_k = N_{k+1} = \cdots$.

(3) $\Longrightarrow$ (6). Let $W_1 \supseteq W_2 \supseteq \cdots$ be a descending chain of finitely generated S-submodules of $\mathrm{Hom}_R(N, E)$. Then $\cap_{\alpha \in W_1} Ker(\alpha) \subseteq \cap_{\alpha \in W_2} Ker(\alpha) \subseteq \cdots$ is an ascending chain of modules in $H_{\mathcal{F}_{\mathcal{K}}}(N)$. By (3), there exists a $k > 0$ such that $\cap_{\alpha \in W_k} Ker(\alpha) = \cap_{\alpha \in W_{k+1}} Ker(\alpha) = \cdots$. Thus, by (3.2.3), $W_k = W_{k+1} = \cdots$.

(6) $\Longrightarrow$ (1). Let $N_1 \subseteq N_2 \subseteq \cdots$ be a chain of submodules of N with $N/N_i \in \mathcal{K}$ for all i. There exists an M-injective cogenerator E of $\mathcal{K}$ such that for each i there is an embedding $N/N_i \overset{\theta_i}{\hookrightarrow} E$ (see the remark before 3.2.4). Note that E is quasi-injective by (2.2.10). Let $S = \mathrm{End}(E_R)$ and, for each i, let $W_i = \{\alpha \in \mathrm{Hom}_R(N, E) : N_i \subseteq Ker(\alpha)\}$. Then, by (3.2.2), $W_1 \supseteq W_2 \supseteq \cdots$ is a chain of cyclic submodules of the left S-module $\mathrm{Hom}_R(N, E)$. Therefore, by (6), there exists a $k > 0$ such that $W_k = W_{k+1} = \cdots$. It follows that $N_k = N_{k+1} = \cdots$, since $N_i = \cap\{Ker(\alpha) : \alpha \in W_i\}$ for all i by (3.2.2). □

The assumption $N \in \sigma[M]$ in (3.2.4) cannot be removed because of the following example.

3.2.5. EXAMPLE. Let $R = \mathbb{Z}, N = \mathbb{Z}_{2^\infty}$, and $N_j = \{x \in \mathbb{Z}_{2^\infty} : 2^j x = 0\}$ for all $j > 0$. Set $\mathcal{K} = \sigma[M]$ where $M = N_1$. Then $H_{\mathcal{K}}(N) = \{N\}$ and so N has ACC on $H_{\mathcal{K}}(N)$. Moreover, $N_1 \subset N_2 \subset \cdots$ is a strictly ascending chain of submodules of N such that $N_{i+1}/N_i \cong N_1 \in \mathcal{K}$ for all $i > 0$.

3.2.6. PROPOSITION. The following are equivalent for an M-natural class $\mathcal{K}$ and a module $N \in \sigma[M]$:

1. $H_{\mathcal{K}}(N)$ has DCC.
2. For any non-empty set $\{N_\alpha : \alpha \in \Lambda\} \subseteq H_{\mathcal{K}}(N)$, $\cap_{\alpha \in \Lambda} N_\alpha = \cap_{\alpha \in F} N_\alpha$ for a finite subset F of Λ.
3. $H_{\mathcal{F}_{\mathcal{K}}}(N)$ has DCC.

4. For any chain $N_1 \supseteq N_2 \supseteq \cdots$ of submodules of N, there exists a positive integer k such that $N_i/N_{i+1} \in \mathcal{T}_\mathcal{K}$ for all $i \geq k$.

5. Every chain $N_1 \supseteq N_2 \supseteq \cdots$ of submodules of N with each $N_i/N_{i+1} \in \mathcal{K}$ terminates.

6. For any M-injective cogenerator E of $\mathcal{K}$, if $S = \mathrm{End}(E_R)$ and $N^* = \mathrm{Hom}_R(N, E)$, then N^* is a Noetherian left S-module.

PROOF. (1) $\Longrightarrow$ (3). Suppose that there exists a strictly descending chain $N_1 \supset N_2 \supset \cdots$ of submodules of N such that $N/N_i \in \mathcal{F}_\mathcal{K}$ for all i. Consider the set

$$\mathcal{A}_1 = \{X \subseteq N : N_1 \subseteq X \subseteq N, N/X \in \mathcal{K}\}.$$

By (1), there exists a minimal element, say X_1, in $\mathcal{A}_1$. The same reason implies that there exists a minimal element, say X_2, in

$$\mathcal{A}_2 = \{X \subseteq N : N_2 \subseteq X \subseteq X_1, N/X \in \mathcal{K}\}.$$

By a simple induction, we have a descending chain $X_1 \supseteq X_2 \supseteq \cdots$ of submodules of N such that, for each i, $N/X_i \in \mathcal{K}$ and X_{i+1} is a minimal element in

$$\{X \subseteq N : N_{i+1} \subseteq X \subseteq X_i, N/X \in \mathcal{K}\}.$$

By (1), there exists a natural number m such that $X_{m+j} = X_m$ for all $j \geq 0$. Without loss of generality, we may assume that $m = 1$ and $N_1 \subset X_1$. Since $N/N_2 \in \mathcal{F}_\mathcal{K}$, we have $X_1/N_2 \notin \mathcal{T}_\mathcal{K}$. Then $X_1/P \in \mathcal{K}$ for some P with $N_2 \subseteq P \subset X_1$. Consider the exact sequence $0 \longrightarrow X_1/P \longrightarrow N/P \longrightarrow N/X_1 \longrightarrow 0$. Since $N/P \in \sigma[M]$, we have $N/P \in \mathcal{K}$ by (2.4.5). But $N_2 \subseteq P \subset X_1 = X_2$, contradicting the minimality of X_2.

(3) $\Longrightarrow$ (4). Let $N_1 \supseteq N_2 \supseteq \cdots$ be a chain of submodules of N and, for each i, let $P_i/N_i = \tau_\mathcal{K}(N/N_i)$. Then we have $P_1 \supseteq P_2 \supseteq \cdots$ and $N/P_i \in \mathcal{F}_\mathcal{K}$ for all i. By (3), there exists a number k such that $P_k = P_{k+1} = \cdots$. Then for $i \geq k$, $N_i/N_{i+1} \subseteq P_i/N_{i+1} = P_{i+1}/N_{i+1} \in \mathcal{T}_\mathcal{K}$. By (3.2.1), we have $N_i/N_{i+1} \in \mathcal{T}_\mathcal{K}$ for all $i \geq k$.

(4) $\Longrightarrow$ (2). Suppose that (2) does not hold. Then there exists a sequence $\{\alpha_i : i = 1, 2, \cdots\} \subseteq \Lambda$ such that $N_{\alpha_1} \supset N_{\alpha_1} \cap N_{\alpha_2} \supset \cdots$ is a strictly descending chain. By (4), there exists a $k > 0$ such that

$$\cap_{i=1}^{n} N_{\alpha_i} / \cap_{i=1}^{n+1} N_{\alpha_i} \in \mathcal{T}_\mathcal{K} \quad \text{for all } n \geq k.$$

But

$$\cap_{i=1}^{n} N_{\alpha_i} / \cap_{i=1}^{n+1} N_{\alpha_i} \cong [\cap_{i=1}^{n} N_{\alpha_i} + N_{\alpha_{n+1}}]/N_{\alpha_{n+1}} \in \mathcal{K},$$

so $\cap_{i=1}^{n} N_{\alpha_i} = \cap_{i=1}^{n+1} N_{\alpha_i}$, a contradiction.

The implications (2) $\Longrightarrow$ (1), (5) $\Longrightarrow$ (1), and (4) $\Longrightarrow$ (5) are clear.

(3) $\Longrightarrow$ (6). Assume that (3) holds and let $W_1 \subseteq W_2 \subseteq \cdots$ be an ascending chain of finitely generated S-submodules of $\mathrm{Hom}_R(N, E)$. Then

$$\cap_{\alpha \in W_1} Ker(\alpha) \supseteq \cap_{\alpha \in W_2} Ker(\alpha) \supseteq \cdots$$

is a descending chain of modules in $H_{\mathcal{F}_\mathcal{K}}(N)$. By (3), there exists a $k > 0$ such that

$$\cap_{\alpha \in W_k} Ker(\alpha) = \cap_{\alpha \in W_{k+1}} Ker(\alpha) = \cdots .$$

Thus, by (3.2.3), $W_k = W_{k+1} = \cdots$.

(6) $\Longrightarrow$ (1). Let $N_1 \supseteq N_2 \supseteq \cdots$ be a chain of submodules of N with $N/N_i \in \mathcal{K}$ for all i. There exists an M-injective cogenerator E of $\mathcal{K}$ such that for each i there is an embedding $N/N_i \overset{\theta_i}{\hookrightarrow} E$ (see the remark before 3.2.4). Note that E is quasi-injective by (2.2.10). Let $S = \mathrm{End}(E_R)$ and, for each i, let $W_i = \{\alpha \in \mathrm{Hom}_R(N, E) : N_i \subseteq Ker(\alpha)\}$. Then, by (3.2.2), $W_1 \subseteq W_2 \subseteq \cdots$ is a chain of cyclic submodules of the left S-module $\mathrm{Hom}_R(N, E)$. Therefore, by (6), there exists a $k > 0$ such that $W_k = W_{k+1} = \cdots$. It follows that $N_k = N_{k+1} = \cdots$, since $N_i = \cap\{Ker(\alpha) : \alpha \in W_i\}$ for all i by (3.2.2). □

The next example shows that we cannot remove the condition $N \in \sigma[M]$ from (3.2.6).

3.2.7. EXAMPLE. Let $R = \mathbb{Z}$ and $N_i = 2^i\mathbb{Z}$ for $i \geq 0$. We let $N = N_0$, $M = N_0/N_1$, and $\mathcal{K} = \sigma[M]$. Then $H_\mathcal{K}(N) = \{N_1, N\}$ and hence N has DCC on $H_\mathcal{K}(N)$. But we have $N = N_0 \supset N_1 \supset N_2 \supset \cdots$ such that $N_i/N_{i+1} \in \mathcal{K}$ for all $i \geq 0$.

3.2.8. COROLLARY. Let $\mathcal{K}$ be an M-natural class and $X \subseteq N \in \sigma[M]$. Then $H_\mathcal{K}(N)$ has DCC iff both $H_\mathcal{K}(X)$ and $H_\mathcal{K}(N/X)$ have DCC.

PROOF. Because of the equivalence (1) $\Longleftrightarrow$ (5) of (3.2.6), the proof is similar to that of (3.1.4). □

3.2.9. DEFINITION. Let $\mathcal{K}$ be a pre-natural class and let N be a module. By a $\mathcal{K}$-**composition series** for N we mean a chain

$$N_0 \subset N_1 \subset N_2 \subset \cdots \subset N_m = N$$

of submodules of N such that $N_0 = \cap\{X : X \in H_\mathcal{K}(N)\}$, $N/N_0 \in \mathcal{K}$, and N_j/N_{j-1} is $\mathcal{K}$-cocritical for each $0 < j \leq m$.

3.2.10. LEMMA. Let $\mathcal{K}$ be a pre-natural class and let the module N have DCC on $H_\mathcal{K}(N)$.

1. If $N \in \mathcal{K}$, then every nonzero submodule A of N has a $\mathcal{K}$-cocritical submodule B with $A/B \in \mathcal{K}$.

2. Every nonzero factor module in $\mathcal{K}$ of N has a $\mathcal{K}$-cocritical submodule. In particular, if R has DCC on $H_\mathcal{K}(R)$, then every nonzero module in $\mathcal{K}$ has a $\mathcal{K}$-cocritical submodule.

PROOF. (1) Let $0 \neq A \subseteq N$. Then A has DCC on $H_{\mathcal{K}}(A)$ by (3.2.8); so A has a submodule B which is a minimal element of

$$\{0 \neq X \subseteq A : A/X \in \mathcal{K}\}.$$

Suppose that B is not $\mathcal{K}$-cocritical. Then $B/Y \in \mathcal{K}$ for some Y with $0 \subset Y \subset B$. Let us assume $\mathcal{K}$ is an M-natural class for some module M. Consider the exact sequence $0 \longrightarrow B/Y \longrightarrow A/Y \longrightarrow A/B \longrightarrow 0$. Note that $A/Y \in \sigma[M]$ since $A \subseteq N \in \mathcal{K}$. Therefore, by (2.4.5), $A/Y \in \mathcal{K}$, contradicting the minimality of B.

(2) For $P \subseteq N$, DCC on $H_{\mathcal{K}}(N)$ clearly implies DCC on $H_{\mathcal{K}}(N/P)$. So if $0 \neq N/P \in \mathcal{K}$, then N/P has a $\mathcal{K}$-cocritical submodule by (1). □

3.2.11. THEOREM. Let $\mathcal{K}$ be a pre-natural class. Then a module N has ACC and DCC on $H_{\mathcal{K}}(N)$ iff N has a $\mathcal{K}$-composition series.

PROOF."$\Longrightarrow$". The DCC on $H_{\mathcal{K}}(N)$ implies that $N_0 := \cap\{X : X \in H_{\mathcal{K}}(N)\}$ is an intersection of a finite number of elements in $H_{\mathcal{K}}(N)$. So $N/N_0 \in \mathcal{K}$, and we have DCC on $H_{\mathcal{K}}(N/N_0)$. By (3.2.10), there exists a submodule N_1 of N such that N_1/N_0 is $\mathcal{K}$-cocritical and $N/N_1 \in \mathcal{K}$. Inductively, we have a chain $N_0 \subset N_1 \subset N_2 \subset \cdots \subset N_m$ of submodules of N such that $N/N_{i+1} \in \mathcal{K}$ and N_{i+1}/N_i is $\mathcal{K}$-cocritical for each $i = 0, \cdots, m-1$. Since N has ACC on $H_{\mathcal{K}}(N)$, the inductive process stops at m steps for some $m > 0$. So the chain $N_0 \subset N_1 \subset N_2 \subset \cdots \subset N_m = N$ is a $\mathcal{K}$-composition series.

"$\Longleftarrow$". If necessary, we may replace N by N/N_0. So, without loss of generality, we may assume that $N_0 = 0$ and $0 = N_0 \subset N_1 \subset \cdots \subset N_m = N$ is a $\mathcal{K}$-composition series for N. Then $H_{\mathcal{K}}(N_{i+1}/N_i)$ has ACC and DCC for $i = 0, 1, \cdots, m-1$. Note that $N \in \mathcal{K}$. Therefore, by (3.1.4) and (3.2.8), $H_{\mathcal{K}}(N)$ has ACC and DCC. □

3.2.12. COROLLARY. The following are equivalent for an M-natural class $\mathcal{K}$ and $N \in \sigma[M]$:

1. N has a $\mathcal{K}$-composition series.

2. For any M-injective cogenerator E of $\mathcal{K}$, $\mathrm{Hom}_R(N, E)$ has a composition series as a left module over $\mathrm{End}(N_R)$.

PROOF. This follows from (3.2.4), (3.2.6), and (3.2.11). □

3.2.13. LEMMA. Let $\mathcal{K}$ be a pre-natural class and N a module for which 0 is an intersection of finitely many $\mathcal{K}$-critical submodules of N. Then there exist submodules $N_i \subseteq N$ $(i = 1, \cdots, n)$ such that each N_i is $\mathcal{K}$-cocritical, $\Sigma_{i=1}^n N_i = \oplus_{i=1}^n N_i \leq_e N$, and $N/(\oplus_{i=1}^n N_i) \in \mathcal{T}_{\mathcal{K}}$.

PROOF. If N is $\mathcal{K}$-cocritical, we are done. So we can assume that N is not $\mathcal{K}$-cocritical. Thus 0 is an intersection of finitely many nonzero $\mathcal{K}$-critical submodules of N. So N embeds in a direct sum of finitely many $\mathcal{K}$-cocritical modules. Note that every $\mathcal{K}$-cocritical module X has ACC and DCC on $H_{\mathcal{K}}(X)$ by (3.2.11). So $N \in \mathcal{K}$ has DCC and ACC on $H_{\mathcal{K}}(N)$ by (3.1.4) and (3.2.8). By (3.2.10), there exists $N_1 \subseteq N$ such that N_1 is $\mathcal{K}$-cocritical. Let $0 \neq x \in N_1$. Then there exists a $\mathcal{K}$-critical submodule C_1 of N such that $x \notin C_1$ by hypothesis. Thus, $0 \neq N_1/(N_1 \cap C_1) \cong (N_1 + C_1)/C_1 \in \mathcal{K}$, so $N_1 \cap C_1 = 0$ since N_1 is $\mathcal{K}$-cocritical. Assume that $N_1, N_2, \cdots, N_t$ and $C_1, C_2, \cdots, C_t$ have been chosen such that N_i is $\mathcal{K}$-cocritical, C_i is $\mathcal{K}$-critical, $N_i \subseteq \cap_{j \leq i-1} C_j$, and $N_i \cap C_i = 0$ for each $i \leq t$. If $\cap_{j=1}^t C_j \neq 0$, then there exists an $N_{t+1} \subseteq \cap_{j=1}^t C_j$ such that N_{t+1} is $\mathcal{K}$-cocritical by (3.2.10). As in the case N_1, there exists a $\mathcal{K}$-critical submodule C_{t+1} of N such that $N_{t+1} \cap C_{t+1} = 0$. Since N has DCC on $H_{\mathcal{K}}(N)$, there exists a $k > 0$ such that $\cap_{j=1}^k C_j = 0$; so the inductive process stops after k steps. It now can be seen by induction that

$$(N_1 + N_2 + \cdots + N_i) \cap N_{i+1} \subseteq (N_1 + \cdots + N_i) \cap (C_1 \cap \cdots \cap C_i) = 0$$

for all $i = 1, \cdots, k-1$. So $\Sigma_{i=1}^k N_i$ is direct.

Next we show that $N/(\oplus_{i=1}^k N_i) \in \mathcal{T}_{\mathcal{K}}$. It suffices to show by induction that $N/(\Sigma_{i=1}^j N_i + \cap_{i=1}^j C_j) \in \mathcal{T}_{\mathcal{K}}$ for all $j = 1, 2, \cdots, k$. Since N/C_1 is $\mathcal{K}$-cocritical and $(N_1 + C_1)/C_1 \neq 0$, $N/(N_1 + C_1) \in \mathcal{T}_{\mathcal{K}}$. Assume that

$$N/(\Sigma_{i=1}^{k-1} N_i + \cap_{i=1}^{k-1} C_i) \in \mathcal{T}_{\mathcal{K}}.$$

Since N/C_k is $\mathcal{K}$-cocritical and

$$0 \neq (N_k + C_k)/C_k \subseteq (\cap_{i=1}^{k-1} C_i + C_k)/C_k,$$

we have $(\cap_{i=1}^{k-1} C_i + C_k)/C_k$ is $\mathcal{K}$-cocritical and so

$$\cap_{i=1}^{k-1} C_i/(N_k + \cap_{i=1}^k C_i) \cong (\cap_{i=1}^{k-1} C_i + C_k)/(N_k + C_k) \in \mathcal{T}_{\mathcal{K}}.$$

Since $\mathcal{T}_{\mathcal{K}}$ is closed under quotient modules and

$$(\Sigma_{i=1}^{k-1} N_i + \cap_{i=1}^{k-1} C_i)/(\Sigma_{i=1}^k N_i + \cap_{i=1}^k C_i)$$

is an epic image of $\cap_{i=1}^{k-1} C_i/(N_k + \cap_{i=1}^k C_i)$, we have

$$(\Sigma_{i=1}^{k-1} N_i + \cap_{i=1}^{k-1} C_i)/(\Sigma_{i=1}^k N_i + \cap_{i=1}^k C_i) \in \mathcal{T}_{\mathcal{K}}.$$

Since $\mathcal{T}_{\mathcal{K}}$ is closed under extensions, we obtain that $N/(\Sigma_{i=1}^k N_i + \cap_{i=1}^k C_i) \in \mathcal{T}_{\mathcal{K}}$ from the induction hypothesis and from the following exact sequence

$$0 \longrightarrow \frac{\Sigma_{i=1}^{k-1} N_i + \cap_{i=1}^{k-1} C_i}{\Sigma_{i=1}^k N_i + \cap_{i=1}^k C_i} \longrightarrow \frac{N}{\Sigma_{i=1}^k N_i + \cap_{i=1}^k C_i} \longrightarrow \frac{N}{\Sigma_{i=1}^{k-1} N_i + \cap_{i=1}^{k-1} C_i} \longrightarrow 0.$$

If $\oplus_{i=1}^k N_i \leq_e N$, it is done. If not, since N has DCC and ACC on $H_{\mathcal{K}}(N)$, there exist finitely many submodules $N_{k+1}, \cdots, N_n$ of N with N_j $\mathcal{K}$-cocritical for all $j = k+1, \cdots, n$ such that $\Sigma_{i=1}^n N_i = \oplus_{i=1}^n N_i \leq_e N$. Thus, $N/(\oplus_{i=1}^n N_i) \in \mathcal{T}_\mathcal{K}$. □

3.2.14. THEOREM. Let R have DCC on $H_\mathcal{K}(R)$ where $\mathcal{K}$ is a pre-natural class. If a module N has DCC on $H_\mathcal{K}(N)$ then N has ACC on $H_\mathcal{K}(N)$. In particular, R has ACC on $H_\mathcal{K}(R)$.

PROOF. Suppose that N does not have ACC on $H_\mathcal{K}(N)$. Let

$$N_0 = \cap\{X : X \in H_\mathcal{K}(N)\}.$$

The DCC on $H_\mathcal{K}(N)$ implies that N_0 is an intersection of finitely many elements of $H_\mathcal{K}(N)$. So $N/N_0 \in \mathcal{K}$. Let N_1 be a minimal element of $\{X \subseteq N : N_0 \subset X \text{ and } N/X \in \mathcal{K}\}$; the DCC on $H_\mathcal{K}(N)$ guarantees the existence of such a submodule. Moreover, N_1/N_0 is $\mathcal{K}$-cocritical by the minimality of N_1. Since N does not have a $\mathcal{K}$-composition series by (3.2.12), we have $N_1 \subset N$. So as above (with replacing N_0 by N_1), there exists a minimal element, say N_2, of $\{X \subseteq N : N_1 \subset X \text{ and } N/X \in \mathcal{K}\}$ and N_2/N_1 is $\mathcal{K}$-cocritical. A simple induction shows that there exists an infinite strictly ascending chain $N_0 \subset N_1 \subset N_2 \subset \cdots$ of submodules of N such that N_j/N_{j-1} are $\mathcal{K}$-cocritical and $N/N_{j-1} \in \mathcal{K}$ for all $j \geq 1$.

Now let

$$C = \cap\{x^\perp : x \in W, W \text{ is a } \mathcal{K}\text{-cocritical module}\}.$$

Since R has DCC on $H_\mathcal{K}(R)$, C is an intersection of finitely many $\mathcal{K}$-critical right ideals. Set

$$h(0) = 0,$$
$$h(t+1) = max\{j > 0 : xC \subseteq N_{h(t)} \text{ for some } x \in N_j \setminus N_{j-1}\}.$$

Note that $h(t) + 1 \leq h(t+1)$ since, by the $\mathcal{K}$-cocriticalness of $N_{h(t)+1}/N_{h(t)}$, $N_{h(t)+1}C \subseteq N_{h(t)}$. Inductively, assume $h(t)$ exists; we show next that $h(t+1)$ exists.

Suppose that $h(t+1)$ does not exist. Then for an infinite set Ω of indices j greater than $h(t)$ we can choose elements $x_j \in N_j \setminus N_{j-1}$ satisfying $x_jC \subseteq N_{h(t)}$. By (3.2.13), there exist right ideals $A_1, \cdots, A_k$ of R such that

$$\Sigma_{i=1}^k (A_i/C) = \oplus_{i=1}^k (A_i/C) \leq_e (R/C)_R, \ (R/C)/[\oplus_{i=1}^k (A_i/C)] \in \mathcal{T}_\mathcal{K},$$

and each A_i/C is $\mathcal{K}$-cocritical. Note that if $x_j(\Sigma_{i=1}^k A_i) \subseteq N_{h(t)}$, then $0 \neq (x_jR + N_{h(t)})/N_{h(t)} \in \mathcal{T}_\mathcal{K}$, which will contradict the fact that $N/N_{h(t)} \in \mathcal{K}$. Hence for at least one of the A_i, $x_jA_i \not\subseteq N_{h(t)}$. We continue to show that, for some such A_i with $x_jA_i \not\subseteq N_{h(t)}$, $(x_jA_i + N_{h(t)}) \cap N_{j-1} = N_{h(t)}$. If not, then $(x_jA_i + N_{h(t)}) \cap N_{j-1} \supset N_{h(t)}$ whenever $x_jA_i \not\subseteq N_{h(t)}$. Since A_i/C is $\mathcal{K}$-cocritical and $(x_jA_i + N_{h(t)})/N_{h(t)} \in \mathcal{K}$, the epimorphism

$$A_i/C \longrightarrow (x_jA_i + N_{h(t)})/N_{h(t)}$$

is an isomorphism. So $(x_jA_i + N_{h(t)})/N_{h(t)}$ is $\mathcal{K}$-cocritical and hence we have

$$\frac{x_jA_i + N_{h(t)}}{(x_jA_i + N_{h(t)}) \cap N_{j-1}} \cong \frac{(x_jA_i + N_{h(t)})/N_{h(t)}}{[(x_jA_i + N_{h(t)}) \cap N_{j-1}]/N_{h(t)}} \in \mathcal{T}_{\mathcal{K}}.$$

But $(x_jA_i + N_{h(t)})/[(x_jA_i + N_{h(t)}) \cap N_{j-1}] \cong (x_jA_i + N_{j-1})/N_{j-1} \in \mathcal{K}$, so we conclude that $x_jA_i \subseteq N_{j-1}$ whenever $x_jA_i \not\subseteq N_{h(t)}$. For each of the remaining A_i, we have $x_jA_i \subseteq N_{h(t)} \subseteq N_{j-1}$. Hence $x_j(\Sigma_{i=1}^k A_i) \subseteq N_{j-1}$, which leads to $(x_jR + N_{j-1})/N_{j-1} \in \mathcal{T}_{\mathcal{K}} \cap \mathcal{K}$ since $R/(\Sigma_{i=1}^k A_i) \in \mathcal{T}_{\mathcal{K}}$, i.e., $x_j \in N_{j-1}$, a contradiction.

So we have established that, for each $j \in \Omega$, there exists a right ideal A_j of R and an $x_j \in N_j \backslash N_{j-1}$ such that $x_jA_j \not\subseteq N_{h(t)}$, $(x_jA_j + N_{h(t)})/N_{h(t)}$ is $\mathcal{K}$-cocritical, and $(x_jA_j + N_{h(t)}) \cap N_{j-1} = N_{h(t)}$. One easily checks that $Y := \Sigma_{j\in\Omega}[(x_jA_j + N_{h(t)})/N_{h(t)}] \subseteq N/N_{h(t)}$ is direct. By (3.2.8), the DCC on $H_{\mathcal{K}}(N)$ implies the DCC on $H_{\mathcal{K}}(Y)$ which, since $Y \in \mathcal{K}$, in turn implies that Ω must be a finite set. This contradiction establishes our claim that $h(t+1)$ exists.

Therefore, $\{h(t)\}_{t=1}^{\infty}$ is a strictly increasing infinite sequence. Moreover, since $H_{\mathcal{K}}(R)$ has DCC, there exists an $n > 0$ such that $C^i/C^{i+1} \in \mathcal{T}_{\mathcal{K}}$ for all $i \geq n$ by (3.2.6)(4). Let $x \in N_{h(n)+1} \backslash N_{h(n)}$. Then $xC \not\subseteq N_{h(n-1)}$, so $xc_1 \notin N_{h(n-1)}$ for some $c_1 \in C$. But $(xc_1)C \not\subseteq N_{h(n-2)}$. Thus, we inductively obtain $c_1, c_2, \cdots, c_{n-1} \in C$ such that $(xc_1 \cdots c_i)C \not\subseteq N_{h(n-i-1)}$ for $i = 1, \cdots, n-1$. In particular, $(xc_1 \cdots c_{n-1})C \not\subseteq N_{h(0)} = N_0$. Hence $xC^n \not\subseteq N_0$. However, since $x \in N_{h(n)+1}$, we have $xC^{h(n)+1} \subseteq N_0$. So there exists an integer $l \geq n$ such that $xC^l \not\subseteq N_0$, but $xC^{l+1} \subseteq N_0$.

Consider the R-epimorphism

$$\alpha : R/C^{l+1} \longrightarrow (xR + N_0)/N_0, \; r + C^{l+1} \longmapsto xr + N_0.$$

Then $0 \neq \alpha(C^l/C^{l+1}) \subseteq N/N_0 \in \mathcal{K}$. But, since $l \geq n$, $C^l/C^{l+1} \in \mathcal{T}_{\mathcal{K}}$ and hence $\alpha(C^l/C^{l+1}) \in \mathcal{T}_{\mathcal{K}}$. Thus, $\mathcal{K} \cap \mathcal{T}_{\mathcal{K}} \neq 0$. This contradiction shows that N has ACC on $H_{\mathcal{K}}(N)$. □

3.2.15. COROLLARY. For a pre-natural class $\mathcal{K}$, R satisfies DCC on $H_{\mathcal{K}}(R)$ iff R has a $\mathcal{K}$-composition series.

PROOF. This follows by (3.2.11) and (3.2.14). □

3.2.16. COROLLARY. The following are equivalent for an M-natural class $\mathcal{K}$:

1. R has DCC on $H_{\mathcal{K}}(R)$.
2. (a) Every M-injective module in $\mathcal{K}$ is a direct sum of M-injective hulls of $\mathcal{K}$-cocritical modules.

   (b) $H_{\mathcal{K}}(R)$ is closed under arbitrary intersections.
3. (a) R has ACC on $H_{\mathcal{K}}(R)$.

(b) Every nonzero module in $\mathcal{K}$ has a $\mathcal{K}$-cocritical submodule.

(c) $H_{\mathcal{K}}(R)$ is closed under arbitrary intersections.

PROOF. (1) $\Longrightarrow$ (2). By (3.2.14), R has ACC on $H_{\mathcal{K}}(R)$; so every M-injective module in $\mathcal{K}$ is a direct sum of M-injective uniform modules by (3.1.5) and (3.1.13). By (3.2.10)(2), every nonzero module in $\mathcal{K}$ has a $\mathcal{K}$-cocritical submodule. Thus, every M-injective uniform module in $\mathcal{K}$ is an M-injective hull of some $\mathcal{K}$-cocritical module, so $(2a)$ follows.

(2) $\Longrightarrow$ (3). Since every $\mathcal{K}$-cocritical module is uniform (see 2.4.7), $(2a)$ certainly implies that every M-injective module in $\mathcal{K}$ is a direct sum of uniform modules. So R has ACC on $H_{\mathcal{K}}(R)$ by (3.1.5) and (3.1.13). Moreover, for $0 \neq N \in \mathcal{K}$, $E_M(N) = \oplus_t E_M(N_t)$ where each N_t is a $\mathcal{K}$-cocritical module by $(2a)$. Thus, by (2.3.3), N contains a nonzero submodule X which embeds in some N_t, so X is $\mathcal{K}$-cocritical by (2.4.7).

(3) $\Longrightarrow$ (1). Let $I_1 \supseteq I_2 \supseteq \cdots$ be a chain of right ideals of R with $R/I_i \in \mathcal{K}$ for all i and let $I = \cap_{i \geq 1} I_i$. Then $R/I \in \mathcal{K}$ by $(3c)$, so $E_M(R/I) \in \mathcal{K}$. By (3.1.5) and (3.1.13), the hypothesis implies that $E_M(R/I) = \oplus_{j=1}^{n} E_M(N_j)$ where each N_j is a $\mathcal{K}$-cocritical module. Let $N = \oplus_{j=1}^{n} N_j$. Then N has DCC on $H_{\mathcal{K}}(N)$ by (3.2.11), and hence N' has DCC on $H_{\mathcal{K}}(N')$ by (3.2.8) where $N' = (R/I) \cap N$. There is a descending chain

$$N' = (R/I) \cap N \supseteq (I_1/I) \cap N \supseteq \cdots$$

with

$$[(R/I) \cap N]/[(I_j/I) \cap N] \hookrightarrow (R/I)/(I_j/I) \cong R/I_j \in \mathcal{K}$$

for all $j \geq 1$. The DCC on $H_{\mathcal{K}}(N')$ implies that

$$(I_k/I) \cap N = (I_{k+1}/I) \cap N = \cdots$$

for some integer $k > 0$; so

$$\begin{aligned}(I_k/I) \cap N &= \cap_{j \geq k}[(I_j/I) \cap N] \\ &= [\cap_{j \geq k}(I_j/I)] \cap N \\ &= [(\cap_{j \geq k} I_j)/I] \cap N = 0.\end{aligned}$$

Thus $I_k = I$ since N is essential in $E_M(R/I)$, proving the DCC on $H_{\mathcal{K}}(R)$. □

Condition $(2b)$ or $(3c)$ cannot be removed from (3.2.16) as the following example shows.

3.2.17. EXAMPLE. Let $R = \left\{ \begin{pmatrix} a & x \\ 0 & a \end{pmatrix} : a \in \mathbb{Z}, x \in \mathbb{Z}_2 \right\}$ be the trivial extension of $\mathbb{Z}$ and the $\mathbb{Z}$-module $\mathbb{Z}_2$. Let $X = \left\{ \begin{pmatrix} 0 & x \\ 0 & 0 \end{pmatrix} : x \in \mathbb{Z}_2 \right\}$ and $\mathcal{K} = d(X)$, i.e., $\mathcal{K} = \{N \in \text{Mod-}R : X$ is embeddable in every nonzero submodule of $N\}$. Then $\mathcal{K}$ is a natural class. As a simple module, $X \in \mathcal{K}$ is $\mathcal{K}$-cocritical. It follows that every nonzero module in $\mathcal{K}$ contains a $\mathcal{K}$-cocritical submodule. Obviously R is a right Noetherian ring and hence $H_{\mathcal{K}}(R)$ has ACC. It is also easily noticed that $\left\{\begin{pmatrix} a & x \\ 0 & a \end{pmatrix} : a \in 2^n\mathbb{Z}, x \in \mathbb{Z}_2\right\}$ is in $H_{\mathcal{K}}(R)$ for any natural number n. Therefore, R does not have DCC on $H_{\mathcal{K}}(R)$.

If $\mathcal{K} = \sigma[M]$ is a hereditary pretorsion class, then ($2b$) or ($3c$) of (3.2.16) can be replaced by the condition that $\cap\{I : I \in H_{\mathcal{K}}(R)\} \in H_{\mathcal{K}}(R)$.

3.2.18. COROLLARY. The following are equivalent for a hereditary pretorsion class $\mathcal{K} = \sigma[M]$:

1. R has DCC on $H_{\mathcal{K}}(R)$.
2. (a) Every M-injective module in $\mathcal{K}$ is a direct sum of M-injective hulls of simple modules.
   (b) $\cap\{I : I \in H_{\mathcal{K}}(R)\} \in H_{\mathcal{K}}(R)$.
3. (a) R has ACC on $H_{\mathcal{K}}(R)$.
   (b) Every nonzero module in $\mathcal{K}$ has a simple submodule.
   (c) $\cap\{I : I \in H_{\mathcal{K}}(R)\} \in H_{\mathcal{K}}(R)$.

For a hereditary torsion theory $\tau = (\mathcal{T}, \mathcal{F})$, a submodule X of N is called τ-**closed** if $N/X \in \mathcal{F}$; the module N is called τ-**cocritical** if $0 \neq N \in \mathcal{F}$ and $N/X \in \mathcal{T}$ for any $0 \neq X \subseteq N$; and a τ-**composition series** for N means a finite chain $\tau(N) = N_0 \subset N_1 \subset \cdots N_m = N$ of submodules of N such that N_{i+1}/N_i is τ-cocritical for $i = 0, 1, \cdots, m-1$. Thus in our terminology, N is τ-cocritical iff N is $\mathcal{F}$-cocritical and τ-composition series for N are precisely $\mathcal{F}$-composition series for N. Applying the results to the hereditary torsion free class $\mathcal{F}$, one obtains several results in torsion theory; among them is the well-known generalization, due to Miller and Teply, of the Hopkins–Levitzki Theorem.

3.2.19. COROLLARY. [86, Miller-Teply] Let R have DCC on τ-closed right ideals for a hereditary torsion theory τ. If a module N has DCC on τ-closed submodules, then N has ACC on τ-closed submodules. In particular, R has ACC on τ-closed right ideals.

There is a ring R and a pre-natural class $\mathcal{K}$ such that $H_{\mathcal{K}}(R)$ has DCC, but $\mathcal{K}$ is neither a hereditary torsion free class nor a hereditary pretorsion class.

3.2.20. EXAMPLE. Let $R = \left\{\begin{pmatrix} a & x \\ 0 & a \end{pmatrix} : a, x \in \mathbb{Z}\right\}$ be the trivial extension of $\mathbb{Z}$ and the $\mathbb{Z}$-module $\mathbb{Z}$. Let $I = \left\{\begin{pmatrix} 0 & x \\ 0 & 0 \end{pmatrix} : x \in \mathbb{Z}\right\}$ and $M = I$. Set $\mathcal{K} = d(M) \cap \sigma[M]$. Then $\mathcal{K}$ is a pre-natural class, but it is neither a hereditary pretorsion class nor a natural class by (2.5.10). However, it can easily be seen that $H_{\mathcal{K}}(R) = \{I, R\}$, so R has DCC on $H_{\mathcal{K}}(R)$.

3.2.21. COROLLARY. The following are equivalent for a hereditary torsion theory τ:

1. R has DCC on τ-closed right ideals.
2. Every τ-torsion free injective module is a direct sum of injective hulls of τ-cocritical modules.
3. (a) R has ACC on τ-closed right ideals.
   (b) Every nonzero τ-torsion free module has a τ-cocritical submodule.

3.2.22. COROLLARY. Let τ be a hereditary torsion theory. Then a module N has ACC and DCC on τ-closed submodules iff N has a τ-composition series.

A result of Small says that for a Noetherian module M, every nil subring of the endomorphism ring $\text{End}(M)$ is nilpotent. Fisher [52] showed that the same result holds for an Artinian module. These results were then extended to any τ-torsion free module with ACC (or DCC) on τ-closed submodules for a hereditary torsion theory τ by Gómez Pardo [64] and Wu and Golan [133]. In the last part of this section, we present a generalization of those results.

3.2.23. LEMMA. Let $\mathcal{K}$ be a pre-natural class and $N \in \mathcal{K}$ with $S = \text{End}(N)$.

1. If $H_{\mathcal{K}}(N)$ has ACC, then S satisfies DCC on left annihilators.
2. If $H_{\mathcal{K}}(N)$ has DCC, then S satisfies ACC on left annihilators.

PROOF. We give the proof of (1) only. Let

$$^{\perp}S_1 \supseteq {}^{\perp}S_2 \supseteq \cdots$$

be a chain of left annihilators of S, where $^{\perp}S_i = \{r \in S : ra = 0 \text{ for all } a \in S_i\}$. We may assume $S_1 \subseteq S_2 \subseteq \cdots$. For each i, write $(S_iN)^c/(S_iN) = \tau_{\mathcal{K}}(N/S_iN)$. Then there exists a chain $(S_1N)^c \subseteq (S_2N)^c \subseteq \cdots$ of submodules of N, such that $N/(S_iN)^c \in \mathcal{F}_{\mathcal{K}}$ for each i. Therefore, by (3.2.4), we have

$$(*) \qquad (S_mN)^c = (S_{m+1}N)^c = \cdots \quad \text{for some } m.$$

Now if $f \in {}^{\perp}S_i$, then $fS_i = 0$ and so $f(S_iN) = 0$. Thus, f induces a homomorphism $\bar{f} : N/S_iN \longrightarrow N$ such that

$$\bar{f}((S_iN)^c/S_iN) = \bar{f}(\tau_{\mathcal{K}}(N/S_iN)) \in \mathcal{T}_{\mathcal{K}} \cap K.$$

Therefore, $\bar{f}((S_iN)^c/S_iN) = 0$, implying $f((S_iN)^c) = 0$.

Conversely, $f((S_iN)^c) = 0$ implies $f(S_iN) = 0$, i.e., $(fS_i)N = 0$, and hence $f \in {}^{\perp}S_i$ since N is a faithful S-module. Therefore,

$$^{\perp}S_i = \{f \in S : f((S_iN)^c) = 0\}.$$

By $(*)$, we have ${}^{\perp}S_m = {}^{\perp}S_{m+1} = \cdots$. □

Recall a concept used by Shock [108]. For modules $Y \subseteq X$, X is called a rational extension of Y if every homomorphism f from a submodule of X to X such that $Ker(f) \supseteq Y$ must be the zero map. And Y is called a **rationally closed** submodule of X if Y has no proper rational extension in X. Note that the rationally closed submodules of X are the $\mathcal{X}(X)$-closed submodules, where $\mathcal{X}(X)$ is the largest hereditary torsion theory relative to which X is torsion free. Therefore, Y is a rationally closed submodule of X iff X/Y is cogenerated by $E(X)$. A non-empty subset A of a ring R is said to be **left** T**-nilpotent** if, for any sequence $\{a_1, a_2, \cdots, a_n, \cdots\} \subseteq A$, there exists $n \geq 1$ such that $a_1 \cdots a_n = 0$.

3.2.24. THEOREM. Let $\mathcal{K}$ be a pre-natural class and $N \in \mathcal{K}$ with endomorphism ring $S = \text{End}(N)$. If N satisfies ACC (or DCC) on $H_{\mathcal{K}}(N)$ then every nil subring of S is nilpotent.

PROOF. We may assume that $\mathcal{K} \subseteq \sigma[M]$ is an M-natural class for a module M.

Suppose that N has ACC on $H_{\mathcal{K}}(N)$. If X is a rationally closed submodule of N, then N/X can be cogenerated by $E(N)$ and so we may write $N/X \subseteq E(N)^I$. Noting that, for each $\alpha \in I$, the α-component of N/X in the product is contained in $E_N(N)$, we have $N/X \hookrightarrow E_N(N)^I \in \mathcal{F}_{\mathcal{K}}$. Therefore, all rationally closed submodules of N are contained in $H_{\mathcal{F}_{\mathcal{K}}}(N)$. By (3.2.4), we deduce that N satisfies ACC on the rationally closed submodules. It follows from [108, Theorem 10] that every nil subring of S is nilpotent.

Suppose that N has DCC on $H_{\mathcal{K}}(N)$. By (3.2.23), S satisfies the ACC on left annihilators. For a nil subring A of S, to show that A is nilpotent, it suffices to show that A is left T-nilpotent by [51, Proposition 1.5]. Suppose A is not left T-nilpotent. Then

$$G = \{f_1 \in A : \exists \{f_2, f_3, \cdots\} \subseteq A \text{ such that } f_1 \cdots f_n \neq 0 \, \forall n\}$$

is not empty. For each $P \subseteq N$, we define P^c by $P^c/P = \tau(N/P)$, where $\tau = \tau_{\mathcal{K}}$, and write $c(f) = (f(N))^c$ for any $f \in S$. Note that, for any $f, g \in S$, $c(f) \in \mathcal{F}_{\mathcal{K}}$ and $c(fg) = [f(c(g))]^c$ (as $(fg)(N) \subseteq f(g(N)^c) \subseteq (fg(N))^c$). Since N satisfies DCC on $H_{\mathcal{K}}(N)$, N has DCC on $H_{\mathcal{F}_{\mathcal{K}}}(N)$ by (3.2.6). Therefore, by a simple induction, there exists a sequence $\{f_1, f_2, \cdots\}$ such that, for each $n \geq 1$, $c(f_n)$ is a minimal element of $\{c(f) : f \in G \text{ and } f_1 \cdots f_{n-1}f \in G\}$. Let $g_n = f_1 \cdots f_n$. For $n \geq m$, we have

$$g_{m-1}(f_m \cdots f_n) = g_n \in G \ \text{ and } (f_m \cdots f_n)(N) \subseteq f_m(N).$$

This implies that $c(f_m \cdots f_n) \subseteq c(f_m)$ and hence $c(f_m \cdots f_n) = c(f_m)$ by the choice of f_m. In particular, we have $c(g_n) = c(f_1) = c(g_{n+1})$ for $n \geq 1$. Then

$$g_n(N)/g_{n+1}(N) \subseteq c(g_n)/g_{n+1}(N) = c(g_{n+1})/g_{n+1}(N) \in \mathcal{T}_{\mathcal{K}}$$

and hence, by (3.2.1), $g_n(N)/g_{n+1}(N) \in \mathcal{T}_{\mathcal{K}}$. Note that g_n induces a homomorphism

$$\begin{aligned} \bar{g}_n : N/f_{n+1}(N) &\longrightarrow g_n(N)/g_{n+1}(N) \\ x + f_{n+1}(N) &\longmapsto g_n(x) + g_{n+1}(N) \end{aligned}$$

and the kernel of $\bar{g}_n$ is

$$Ker(\bar{g}_n) = [Ker(g_n) + f_{n+1}(N)]/f_{n+1}(N).$$

Therefore, we obtain

$$(*_1) \qquad N/[Ker(g_n) + f_{n+1}(N)] \cong g_n(N)/g_{n+1}(N) \in \mathcal{T}_{\mathcal{K}}, \ \forall n \geq 1.$$

We next show that $g_n f_m = 0$ for all $n \geq m$. Suppose that $g_n f_m \neq 0$ for some $n \geq m$. Then for each $l \geq m$, we have

$$c(g_n f_m \cdots f_l) = [g_n(c(f_m \cdots f_l))]^c = [g_n(c(f_m))]^c = c(g_n f_m).$$

Therefore, $g_{m-1} f_m \cdots f_n f_m \cdots f_l = g_n f_m \cdots f_l \neq 0$ for all $l \geq m$, and hence $c(f_m \cdots f_n f_m) = c(f_m)$ by the choice of f_m. Since $f = f_m \cdots f_n \in A$ is nilpotent, we have

$$c(f_m) = c(f f_m) = c(f^2 f_m) = \cdots = 0,$$

implying $f_m = 0$, a contradiction. Therefore, we have

$$(*_2) \qquad \Sigma_{i=2}^n f_i(N) \subseteq Ker(g_n) \ \forall n \geq 2.$$

For each $n \geq 1$, we let

$$D_n = \cap_{i=1}^n Ker(g_i) + \Sigma_{j=2}^{n+1} f_j(N).$$

By $(*_1)$, we see $N/D_1 \in \mathcal{T}_{\mathcal{K}}$. Suppose $N/D_{n-1} \in \mathcal{T}_{\mathcal{K}}$ for some $n-1 \geq 1$. We show that $N/D_n \in \mathcal{T}_{\mathcal{K}}$. Notice

$$\begin{aligned} &[Ker(g_n) + D_{n-1}]/D_{n-1} \\ &\cong Ker(g_n)/(Ker(g_n) \cap D_{n-1}) \\ &\twoheadrightarrow Ker(g_n)/[Ker(g_n) \cap D_{n-1} + Ker(g_n) \cap f_{n+1}(N)] \end{aligned}$$

and, by $(*_2)$, we have

$$\begin{aligned} &[Ker(g_n) + D_n]/D_n \\ &= [Ker(g_n) + Ker(g_n) \cap D_{n-1} + f_{n+1}(N)]/[Ker(g_n) \cap D_{n-1} + f_{n+1}(N)] \\ &\cong Ker(g_n)/\{Ker(g_n) \cap [Ker(g_n) \cap D_{n-1} + f_{n+1}(N)]\} \\ &= Ker(g_n)/[Ker(g_n) \cap D_{n-1} + Ker(g_n) \cap f_{n+1}(N)]. \end{aligned}$$

Since $N/D_{n-1} \in \mathcal{T}_{\mathcal{K}}$, we have $[Ker(g_n) + D_{n-1}]/D_{n-1} \in \mathcal{T}_{\mathcal{K}}$ by (3.2.1) and hence $[Ker(g_n) + D_n]/D_n \in \mathcal{T}_{\mathcal{K}}$ from above. Note that $\mathcal{T}_{\mathcal{K}}$ is closed under extensions. Thus the fact that $N/D_n \in \mathcal{T}_{\mathcal{K}}$ follows from the following exact sequence:

$$0 \longrightarrow [Ker(g_n) + D_n]/D_n \longrightarrow N/D_n \xrightarrow{\bar{g}_n} g_n(N)/g_{n+1}(N) \longrightarrow 0,$$

where $\bar{g}_n(x + D_n) = g_n(x) + g_{n+1}(N)$ for $x \in N$. Therefore, we have proved

$$(*_3) \qquad N/[\cap_{i=1}^{n} Ker(g_i) + \Sigma_{j=2}^{n+1} f_j(N)] \in \mathcal{T}_{\mathcal{K}},\ \forall n \geq 1.$$

Since $\cap_{i=1}^{n} Ker(g_i)$ are in $H_{\mathcal{F}_{\mathcal{K}}}(N)$ for all n, there exists a number l such that $\cap_{i=1}^{n} Ker(g_i) = \cap_{i=1}^{n+1} Ker(g_i)$ for all $n \geq l$ and hence

$$Ker(g_{l+1}) \supseteq \cap_{i=1}^{n} Ker(g_i) \text{ for all } n \geq l.$$

Then

$$Ker(g_{l+1}) \supseteq \cap_{i=1}^{l} Ker(g_i) + \Sigma_{i=2}^{l+1} f_i(N)$$

by $(*_2)$. It follows from $(*_3)$ that $N/Ker(g_{l+1}) \in \mathcal{T}_{\mathcal{K}}$, but $N/Ker(g_{l+1}) \in \mathcal{F}_{\mathcal{K}}$. Therefore, $N = Ker(g_{l+1})$, i.e., $g_{l+1} = 0$, a contradiction. □

3.2.25. REFERENCES. Albu and Nástásescu [5]; Fisher [51,52]; Golan [59]; Gómez Pardo [64]; Miller and Teply [86]; Shock [107]; Wu and Golan [133]; Zhou [135,138].

3.3 Covers and Ascending Chain Conditions

In this section, covers of modules are discussed and connected to certain ascending chain conditions of rings. As before, $\mathcal{F}$ and $\mathcal{K}$ indicate classes of R-modules.

3.3.1. DEFINITION. [47, Enochs] Let N be a module, $F \in \mathcal{F}$, and $f : F \longrightarrow N$ be a homomorphism. The homomorphism $f : F \longrightarrow N$ is called an $\mathcal{F}$-**precover** of N if $\operatorname{Hom}(G, F) \longrightarrow \operatorname{Hom}(G, N) \longrightarrow 0$ is exact for all $G \in \mathcal{F}$. If furthermore every $g : F \longrightarrow F$ such that $fg = f$ is an automorphism of F, then $f : F \longrightarrow N$ is called an $\mathcal{F}$-**cover** of N.

Note that the $\mathcal{F}$-cover of a module, if it exists, is unique up to isomorphism. The next lemma is well-known and easy to prove.

3.3.2. LEMMA. Let $f : F \longrightarrow N$ be a homomorphism and $X = \oplus_i X_i$. Then

$$\operatorname{Hom}(X, F) \stackrel{\operatorname{Hom}(X,f)}{\longrightarrow} \operatorname{Hom}(X, N) \longrightarrow 0$$

is exact if and only if

$$\operatorname{Hom}(X_i, F) \stackrel{\operatorname{Hom}(X_i,f)}{\longrightarrow} \operatorname{Hom}(X_i, N) \longrightarrow 0$$

is exact for every i. □

For any class $\mathcal{F}$ of R-modules and for any R-module N, we let

$$i_N(\mathcal{F}) = \{X \in \mathcal{F} : X \ \text{ is } N\text{-injective}\}.$$

3.3.3. THEOREM. Let $\mathcal{K}$ be an M-natural class. Then the following are equivalent:

1. Every R-module has an $i_M(\mathcal{K})$-precover.
2. Every module in $\mathcal{K}$ has an $i_M(\mathcal{K})$-precover.
3. R has ACC on $H_{\mathcal{K}}(R)$.

PROOF. (1) $\Longrightarrow$ (2) is clear.

(2) $\Longrightarrow$ (3). To show that R has ACC on $H_{\mathcal{K}}(R)$, it suffices to show that for any family $\{E_i : i \in I\}$ of M-injective modules in $\mathcal{K}$, the direct sum $\oplus_i E_i$ is again M-injective by (3.1.5). Since $\oplus_i E_i \in \mathcal{K}$, by (2), let $f : E \longrightarrow \oplus_i E_i$ be an $i_M(\mathcal{K})$-precover, and let $l_i : E_i \longrightarrow \oplus_i E_i$ be the canonical inclusion for each $i \in I$. Then there exists $f_i : E_i \longrightarrow E$ such that $ff_i = l_i$ for every $i \in I$. Thus, the composition map $\oplus_i E_i \stackrel{\oplus f_i}{\longrightarrow} E \stackrel{f}{\longrightarrow} \oplus_i E_i$ is the identity, showing that $\oplus_i E_i$ is a direct summand of E. So $\oplus_i E_i$ is M-injective.

(3) $\Longrightarrow$ (1). Suppose that R has ACC on $H_{\mathcal{K}}(R)$. Thus, by (3.1.5) and (3.1.13), every M-injective module in $\mathcal{K}$ is a direct sum of uniform M-injective modules. Let

$$\mathcal{Y} = \{E_M(R/I) : I \subseteq R, R/I \in \mathcal{K} \text{ is uniform}\}.$$

Then every uniform M-injective module in $\mathcal{K}$ is isomorphic to some module in $\mathcal{Y}$. Given an R-module N, for each $Y \in \mathcal{Y}$, let $Y^* = Y^{(\mathrm{Hom}(Y,N))}$ and let

$$\phi_Y : Y^* \longrightarrow N \text{ defined by } (y_f)_{f \in \mathrm{Hom}(Y,N)} \longmapsto \Sigma f(y_f).$$

We claim that $\oplus\phi_Y : \oplus Y^* \longrightarrow N$ is an $i_M(\mathcal{K})$-precover. By (3.1.5), $\oplus Y^* \in i_M(\mathcal{K})$. It is clear that

$$\mathrm{Hom}(Y, Y^*) \overset{\mathrm{Hom}(Y,\phi_Y)}{\longrightarrow} \mathrm{Hom}(Y, N) \longrightarrow 0$$

is exact for every $Y \in \mathcal{Y}$. It follows that

$$\mathrm{Hom}(Y, \oplus Y^*) \overset{\mathrm{Hom}(Y,\oplus\phi_Y)}{\longrightarrow} \mathrm{Hom}(Y, N) \longrightarrow 0$$

is exact for every $Y \in \mathcal{Y}$. Since every M-injective module in $\mathcal{K}$ is a direct sum of modules each isomorphic to an element in $\mathcal{Y}$, it follows from (3.3.2) that

$$\mathrm{Hom}(X, \oplus Y^*) \overset{\mathrm{Hom}(X,\oplus\phi_Y)}{\longrightarrow} \mathrm{Hom}(X, N) \longrightarrow 0$$

is exact for every M-injective module $X \in \mathcal{K}$. So $\oplus\phi_Y : \oplus Y^* \longrightarrow N$ is an $i_M(\mathcal{K})$-precover. $\square$

3.3.4. REMARKS. (3.3.3) is previously known to be true in the following cases: (1) $\mathcal{K} =$ Mod-R (Enochs [47] and Teply [117]); (2) $\mathcal{K}$ is a hereditary torsion free class (Ahsan and Enochs [2]); (3) $\mathcal{K}$ is the class of all Goldie torsion modules (Ahsan and Enochs [3]); $\mathcal{K}$ is the class of all non-singular modules (Ahsan and Enochs [2]).

A partially ordered set Λ is called **directed** if for every pair $a, b \in \Lambda$ there exists $c \in \Lambda$ such that $a \leq c$ and $b \leq c$. A **direct system** of R-modules over a directed partially ordered set Λ, denoted by $(M_a; f_a^b, \Lambda)$, is a family of R-modules $\{M_a : a \in \Lambda\}$ together with a family of morphisms $\{f_a^b : M_a \longrightarrow M_b, a \leq b, a, b \in \Lambda\}$ satisfying $f_a^a = 1$ and $f_b^c f_a^b = f_a^c$ whenever $a \leq b \leq c$. The **direct limit** of a direct system $(M_a; f_a^b, \Lambda)$ of modules is an R-module A together with a family of morphisms $\{f_a : M_a \longrightarrow A, a \in \Lambda\}$ satisfying the following two conditions:

1. $f_a = f_b f_a^b$ for all $a \leq b$ in Λ.
2. For any R-module B and any family of morphisms $\{g_a : M_a \longrightarrow B, a \in \Lambda\}$ with $g_a = g_b f_a^b$ for all $a \leq b$ in Λ, there exists a unique $g : A \longrightarrow B$ such that $g_a = g f_a$ for all $a \in \Lambda$.

The direct limit of a direct system of modules always exists and is unique up to isomorphism. We write $A = \underrightarrow{lim} M_a$ and $(f_a, \underrightarrow{lim} M_a)$ for the direct limit of the direct system of modules $(M_a; f_a^b, \Lambda)$. Note that $\underrightarrow{lim} M_a = \cup_\Lambda f_a(M_a)$ and $Ker(f_a) = \cup_{b \geq a} Ker(f_a^b)$ for all $a \in \Lambda$.

3.3.5. LEMMA. Let $\mathcal{K}$ be an M-natural class and $N \in \sigma[M]$.

1. If $N = \cup\{N_i : i \in I\}$ with all $N_i \in \mathcal{K}$, then $N \in \mathcal{K}$.

2. Suppose that R has ACC on $H_{\mathcal{K}}(R)$. If $\mathcal{X}$ is a directed set of submodules of N such that $N/X \in \mathcal{K}$ for each $X \in \mathcal{X}$, then $N/(\cup_{\mathcal{X}} X) \in \mathcal{K}$.

PROOF. By (2.3.5) and (2.4.4), we can write $\mathcal{K} = c(\mathcal{F}) \cap \sigma[M]$ for some class $\mathcal{F}$ of modules.

(1) If $N \notin \mathcal{K}$, then $N \notin c(\mathcal{F})$. Thus, there exists $0 \neq x \in N$ such that $xR \in \mathcal{F}$. But $x \in N_i$ for some $i \in I$, so $N_i \notin \mathcal{K}$, a contradiction.

(2) If $N/(\cup_{\mathcal{X}} X) \notin \mathcal{K}$, then there is a nonzero cyclic submodule $Y/(\cup_{\mathcal{X}} X)$ of $N/(\cup_{\mathcal{X}} X)$ such that $Y/(\cup_{\mathcal{X}} X) \in \mathcal{F}$. Write $Y/(\cup_{\mathcal{X}} X) = (yR + \cup_{\mathcal{X}} X)/(\cup_{\mathcal{X}} X)$.

Let $\mathcal{Y} = \{yR \cap X : X \in \mathcal{X}\}$. Clearly $\mathcal{Y}$ is a directed set of submodules of yR. For any $yR \cap X \in \mathcal{Y}$,

$$yR/(yR \cap X) \cong (yR + X)/X \subseteq N/X \in \mathcal{K}.$$

So $yR/(yR \cap X) \in \mathcal{K}$. But

$$\begin{aligned} yR/(\cup_{\mathcal{Y}} Z) &= yR/[\cup_{X \in \mathcal{X}}(yR \cap X)] \\ &= yR/[yR \cap (\cup_{\mathcal{X}} X)] \\ &\cong (yR + \cup_{\mathcal{X}} X)/(\cup_{\mathcal{X}} X) \notin \mathcal{K}. \end{aligned}$$

So, replacing N by xR and $\mathcal{X}$ by $\mathcal{Y}$, we can assume that N is cyclic. Thus $N \hookrightarrow P/Q$ where P is a cyclic submodule of $M^{(n)}$ for some $n > 0$ and $Q \leq P$. Since R has ACC on $H_{\mathcal{K}}(R)$, the ACC holds on $H_{\mathcal{K}}(N)$ by (3.1.4) and (3.1.5), so there exists a maximal element, say X', in $\mathcal{X}$. But since $\mathcal{X}$ is a directed set, $X' = \cup_{\mathcal{X}} X$. Therefore $N/(\cup_{\mathcal{X}} X) = N/X' \in \mathcal{K}$. □

3.3.6. THEOREM. Let $\mathcal{K}$ be an M-natural class. Then every direct limit of M-injective modules in $\mathcal{K}$ is M-injective iff R has ACC on $H_{\mathcal{K}}(R)$.

PROOF. "$\Longrightarrow$". The hypothesis clearly implies that every direct sum of M-injective modules in $\mathcal{K}$ is M-injective, so R has ACC on $H_{\mathcal{K}}(R)$ by (3.1.5).

"$\Longleftarrow$". Let $\{E = \varinjlim E_a, f_a\}$ be the direct limit of the directed system $\{E_a; f_a^b, \Lambda\}$ of M-injective modules in $\mathcal{K}$, and $f : N \longrightarrow E$ a homomorphism, where N is a submodule of a cyclic submodule C of M. It suffices to show that there is a homomorphism $C \longrightarrow E$ that extends f by [87, Proposition 1.4].

First, we show $E \in \mathcal{K}$. Since $E = \cup_{a \in \Lambda} Im(f_a)$, to show that $E \in \mathcal{K}$, it suffices to show each $Im(f_a) \in \mathcal{K}$ by (3.3.5). For each $a \in \Lambda$, we have that $Ker(f_a) = \cup_{a \leq b} Ker(f_a^b)$ and $\{Ker(f_a^b) : a \leq b, b \in \Lambda\}$ is a directed set of submodules of the module E_a. Note that $E_a/Ker(f_a^b) \cong Im(f_a^b) \subseteq E_b$ for any $b \in \Lambda$ with $a \leq b$. Thus $E_a/Ker(f_a^b) \in \mathcal{K}$ for any $b \in \Lambda$ with $a \leq b$. Thus, by (3.3.5), we have

$$Im(f_a) \cong E_a/Ker(f_a) = E_a/[\cup_{a \leq b} Ker(f_a^b)] \in \mathcal{K}$$

for every $a \in \Lambda$. So $E \in \mathcal{K}$.

We need to show that there is a homomorphism $g : C \longrightarrow E$ that extends f. Let

$$\mathcal{T}_{\mathcal{K}} = \{T \in \text{Mod-}R : \text{Hom}(T, C) = 0 \text{ for all } C \in \mathcal{K}\}.$$

Since $H_{\mathcal{K}}(R)$ has ACC, $H_{\mathcal{K}}(C)$ has ACC by (3.1.5). Therefore, by (3.2.4), there is a finitely generated submodule $N_1 \subseteq N$ such that $N/N_1 \in \mathcal{T}_{\mathcal{K}}$. Let $f_1 = f|_{N_1} : N_1 \longrightarrow E$. If there exists a homomorphism $g_1 : C \longrightarrow E$ that extends f_1, then $N_1 \subseteq Ker(f - g_1)$, and thus $f - g_1$ induces a homomorphism $\overline{f - g_1} : N/N_1 \longrightarrow E$. But, since $N/N_1 \in \mathcal{T}_{\mathcal{K}}$ and $E \in \mathcal{K}$, we have $\overline{f - g_1} = 0$. This shows that $(f - g_1)|_N = 0$, i.e., g_1 extends f. So, without loss of generality, we can assume N is a finitely generated submodule of C.

So $f(N)$ is a finitely generated submodule of $E = \cup_{a\in\Lambda} Im(f_a)$, and hence $f(N) \subseteq Im(f_a)$ for some $a \in \Lambda$. Forming a pullback, one obtains the commutative diagram with exact rows:

$$\begin{array}{ccccccccc}
0 & \longrightarrow & X & \longrightarrow & Y & \stackrel{\varphi}{\longrightarrow} & N & \longrightarrow & 0 \\
 & & \| & & \downarrow{\scriptstyle\psi} & & \downarrow{\scriptstyle f} & & \\
0 & \longrightarrow & X & \longrightarrow & E_a & \stackrel{f_a}{\longrightarrow} & Im(f_a) & \longrightarrow & 0.
\end{array}$$

Since N is finitely generated, there exists a finitely generated submodule Y_1 of Y such that $\varphi_1 = \varphi|_{Y_1} : Y_1 \longrightarrow N$ is epic. Since $Y \cong N \oplus Ker(f_a) \in \sigma[M]$, we have $\psi(Ker\varphi_1) \in \sigma[M]$. So the ACC on $H_{\mathcal{K}}(R)$ implies that ACC on $H_{\mathcal{K}}(\psi(Ker\varphi_1))$. By (3.2.4), there is a finitely generated submodule $Z \subseteq \psi(Ker\varphi_1)$ such that $\psi(Ker\varphi_1)/Z \in \mathcal{T}_{\mathcal{K}}$. But $Z \subseteq \psi(Ker\varphi_1) \subseteq Ker(f_a)$, so $Z \subseteq Ker(f_a^b)$ for some $b \geq a$. Thus,

$$\frac{\psi(Ker\varphi_1)}{Z} \subseteq \frac{E_a}{Z} \twoheadrightarrow \frac{Ea}{Kerf_a^b} \cong Imf_a^b \subseteq E_b,$$

here "$\twoheadrightarrow$" means an epimorphism. It must be that $\psi(Ker\varphi_1)/Z = 0$. So $\psi(Ker\varphi_1) = Z$, and hence $(f_a^b\psi)(Ker\varphi_1) = 0$. Thus there exists $\alpha : N \longrightarrow E_b$ that makes the diagram commutative:

$$\begin{array}{ccccccc}
Y_1 & \stackrel{\subseteq}{\longrightarrow} & Y & \stackrel{\psi}{\longrightarrow} & E_a & \stackrel{f_a^b}{\longrightarrow} & E_b \\
{\scriptstyle\varphi_1}\downarrow & & & & & & \| \\
N & & & \stackrel{\alpha}{\longrightarrow} & & & E_b.
\end{array}$$

Since E_b is M-injective, there exists a morphism $\beta : C \longrightarrow E_b$ that extends α. Let $g = f_b\beta : C \longrightarrow E$. For any $x \in N$, write $x = \varphi_1(y)$ with $y \in Y_1$. Then

$$g(x) = f_b\beta(x) = f_b\alpha(x) = f_b\alpha\varphi_1(y) = f_bf_a^b\psi(y) = f_a\psi(y) = f(x).$$

So g extends f. □

3.3.7. LEMMA. [134, Theorem 2.2.8] Let $\mathcal{F}$ be a class of R-modules that is closed under isomorphic copies and let M be an R-module. Suppose that $\mathcal{F}$ is closed under direct limits. If M has an $\mathcal{F}$-precover, then M has an $\mathcal{F}$-cover. □

3.3.8. THEOREM. Let $\mathcal{K}$ be an M-natural class. Then the following are equivalent:

1. Every R-module has an $i_M(\mathcal{K})$-cover.
2. Every R-module in $\mathcal{K}$ has an $i_M(\mathcal{K})$-cover.
3. R has ACC on $H_{\mathcal{K}}(R)$.

PROOF. (1) $\Longrightarrow$ (2) is obvious. (2) $\Longrightarrow$ (3) follows from (3.3.3).

(3) $\Longrightarrow$ (1). Suppose that (3) holds. Then every R-module has an $i_M(\mathcal{K})$-precover. By (3.3.6), the class $i_M(\mathcal{K})$ is closed under direct limits. Therefore, by (3.3.7), every R-module has an $i_M(\mathcal{K})$-cover. □

3.3.9. COROLLARY. Let M be an R-module. The following are equivalent:

1. M is locally Noetherian.
2. Every module in $\sigma[M]$ has an $i_M(\sigma[M])$-cover.
3. Every R-module has an $i_M(\sigma[M])$-cover. □

3.3.10. COROLLARY. [100, Theorems 11 and 18] Let $\mathcal{K}$ be a natural class. The following are equivalent:

1. R has the ACC on $H_{\mathcal{K}}(R)$.
2. Every module in $\mathcal{K}$ has an $i_R(\mathcal{K})$-cover.
3. Every R-module has an $i_R(\mathcal{K})$-cover. □

3.3.11. REMARKS. (3.3.10) is previously known to be true in the following cases: (1) $\mathcal{K}$ = Mod-R (Enochs [47]); (2) $\mathcal{K}$ is the class of the non-singular modules (Ahsan-Enochs [2]); (3) $\mathcal{K}$ is the class of the Goldie torsion modules (Ahsan-Enochs [3]).

3.3.12. REFERENCES. Ahsan and Enochs [2,3]; Enochs [47]; Page and Zhou [100]; Teply [117].

Chapter 4

Type Theory of Modules: Dimension

The first section defines type submodules and describes their basic properties. In order to be able to use type submodules effectively, one has to be able to identify which submodules are type submodules, and to be familiar with a few module operations through which type submodules are obtained.

Then a special kind of type submodules, those which are also atomic, are used to define the type dimension of any module. Some of the more useful properties of the type dimension are explained. Section 4.2 develops computational formulas for calculating type dimension, and concludes with formulas for the type dimension for several classes of modules and rings, such as formal triangular matrix rings and Laurent polynomial modules and rings.

Several classes of rings defined using type dimension, including those rings with type ascending and descending chain conditions, are investigated in section 4.3 as immediate applications of the type dimension.

The concept of the type dimension first appeared in [140], where many of its properties were proved. Then it was subsequently used throughout [141]. By then, its usefulness was apparent.

4.1 Type Submodules and Type Dimensions

Type submodule and type dimension are two fundamental concepts in the type theory of modules. There are two ways in which type submodules may be approached as explained in (4.1.2). One is that type submodules are precisely those submodules belonging to a natural class $\mathcal{K}$ which are maximal with respect to that property. This approach was used in [32]. The other approach from [140] is that type submodules of a module are those which have no proper "parallel" extensions in the module. The behavior of type submodules under various module operations, such as quotients, submodules, and direct sums is investigated. The atomic modules generalize uniform modules. Similarly, the type dimension of a module, which is based on atomic submodules, is analogous to the uniform dimension. Type submodules are complement submodules, and there is a partial resemblance between the two. For example, if X is a type submodule of N and N is a type submodule of M, then X is a type submodule of M; the same holds for complement submodules. Two symmetric properties of modules, orthogonal and parallel, are defined below.

4.1.1. DEFINITION. Two modules N and P are **orthogonal**, written as $N \perp P$, if they do not have nonzero isomorphic submodules. Modules N_1 and N_2 are **parallel**, denoted as $N_1 \| N_2$, if there does not exist a $0 \neq V_2 \leq N_2$ with $N_1 \perp V_2$, and also there does not exist a $0 \neq V_1 \leq N_1$ such that $V_1 \perp N_2$.

An equivalent definition of $N_1 \| N_2$ is that for any $0 \neq V_1 \leq N_1$, there exist $0 \neq aR \leq V_1$ and $0 \neq bR \leq N_2$ with $aR \cong bR$, and dually for any $0 \neq V_2 \leq N_2$, there exist $0 \neq aR \leq N_1$, $0 \neq bR \leq V_2$ with $aR \cong bR$.

4.1.2. DEFINITION. A submodule P of a module N is called a **type submodule**, denoted as $P \leq_t N$, if the following equivalent conditions hold:

1. If $P \subseteq Y \subseteq N$ with $P \| Y$, then $P = Y$.

2. If $P \subset Y \subseteq N$, then $P \perp X$ for some $0 \neq X \subseteq Y$.

3. P is a complement submodule of N such that $P \oplus D \leq_e N$ and $P \perp D$ for some $D \subseteq N$.

4. There exists a natural class $\mathcal{K}$ such that, among all the submodules of N, P is maximal with respect to $P \in \mathcal{K}$.

PROOF. (1) $\Longleftrightarrow$ (2) is clear.

(1) $\Longrightarrow$ (3). (1) clearly implies that P has no proper essential extensions in N; so P is a complement submodule. There exists a $D \subseteq N$ such that $P \oplus D \leq_e N$. If P is not orthogonal to D, then there exists an $0 \neq X \subseteq D$ such that $X \hookrightarrow P$, so $P \| (P \oplus X)$, contradicting (1). So $P \perp D$.

(3) $\Longrightarrow$ (4). Set $\mathcal{K} = d(P)$. Suppose that $P \subset Y \subseteq N$ with $Y \in \mathcal{K}$. Since P is a complement submodule, there is a $0 \neq V \subseteq Y$ such that $P \cap V = 0$; so V embeds in N/P. Let D be as in (3). Then D embeds as an essential submodule in N/P because P is a complement submodule of N. Thus, there exists a $0 \neq W \subseteq V$ such that $W \hookrightarrow D$. But from $Y \in \mathcal{K}$, we have $W \in d(P)$; so W has a nonzero submodule embeddable in P. This contradicts that $P \perp D$.

(4) $\Longrightarrow$ (2). Let $\mathcal{K}$ be as in (4). Then P is a complement submodule of N since $\mathcal{K}$ is closed under essential extensions. Suppose $P \subset Y \subseteq N$. Then $P \cap X = 0$ for some $0 \neq X \subseteq Y$. If P and X are not orthogonal, then X has a nonzero submodule V embeddable in P, so $P \oplus V \in \mathcal{K}$, contradicting the maximality of P. It follows that $P \perp X$. □

A submodule P satisfying (4.1.2)(4) is sometimes called a type submodule of **type** $\mathcal{K}$ of N. Note that if N is a type submodule of the module M of type $\mathcal{K}$ and if C is any complement of N in M, then $N \perp C$ and $C \in c(\mathcal{K})$; moreover, for any R-homomorphism $f : N \longrightarrow C$, we have $Ker(f) \leq_e N$, and consequently $f(N) \subseteq Z(C)$.

Each submodule $V \leq N$ has three associated submodules which are defined next, and in this chapter will sometimes be denoted by V^c,V^{tc}, and V^*.

4.1.3. DEFINITION. For $V \leq N$, let $V^c \leq N$ denote a complement closure of V in N. By Zorn's Lemma there exists a submodule $P \leq N$ maximal with respect to $V \subseteq P$ and $V \| P$. Then $P \leq_t N$ is a type submodule, called the **type closure** of V in N. In general it is not unique, but any such one will be denoted by $P = V^{tc}$. Note that if $V \subseteq Y \subset V^{tc} \leq N$, then Y is not a type submodule of N. In general, V need not be essential in V^{tc}, but always $V \| V^{tc}$. Again by Zorn's Lemma, there exists a submodule $Q \leq N$ maximal with respect to $V \perp Q$. Any such a submodule Q is called a **type complement** of V, and will be denoted in this section by V^*. It is always true that $V^* \leq_t N$, $V^{tc} \perp V^* = 0$, and $V^{tc} \oplus V^* \leq_e N$.

4.1.4. PROPOSITION. Start with any $V \leq N$.

1. There exists a type submodule $P \leq_t N$ minimal with the property that $V \subseteq P$.

2. There exists a type submodule $Q \leq_t N$ maximal with respect to $V \cap Q = 0$.

3. $P \oplus Q \leq_e N$ and P and Q are complements of each other.

4. $P = V^{tc}$ and $Q = V^*$.

PROOF. Take $P = V^{tc}$ and $Q = V^*$ as in (4.1.3). If $V = 0$, then $V^{tc} = 0$ and $V^* = N$, so let $V \neq 0$.

(1) If $V \subseteq L \subset P$ with $L \leq_t N$, then L is a complement submodule of N. Thus $L \oplus C \leq P$ for some $0 \neq C \leq P$. Hence $L \perp C$ and also $V \perp C$. This contradicts that $V \| V^{ct}$.

(2) Suppose $Q \subset W \leq N$ with $W \leq_t N$ and $V \cap W = 0$. Then $V \perp W$ by (4.1.2)(1), contradicting the maximality of Q.

(3) follows from (4.1.3). □

The next lemma will allow us to start with a type submodule, and construct some new type submodules.

4.1.5. LEMMA. Let P, C be submodules of N. Then the following hold:

1. $P \leq_t N, P \subseteq C \Longrightarrow P \leq_t C$.

2. If $P \leq_t N$, then $P \cap C = 0$ if and only if $P \perp C$.

3. Let $X = \oplus_{i \in I} X_i$ be a direct sum of type submodules X_i of N. Then $X^c = X^{tc} \leq_t N$ is a type submodule of N.

4. $P \leq_t N \leq_t M \Longrightarrow P \leq_t M$.

PROOF. (1) follows from (4.1.2)(1).

(2) Clearly $P \perp C$ always implies $P \cap C = 0$. If $P \oplus C \leq N$, then $P \leq_t N$ implies $P \perp C$ by (4.1.2).

(3) Let Y be a complement of X in N. Then $X^c \oplus Y \leq_e N$. If Y, X^c are not orthogonal, then for some $0 \neq xR \leq X^c$, $xR \hookrightarrow Y$. By (2.3.3), for some $a \in R$, $0 \neq xaR$ embeds into X_i for some i. But by (2), $X_i \perp Y$, since $X_i \leq_t N$. This is a contradiction. Thus $X^c \perp Y$, and by (4.1.2)(3), $X^c \leq_t N$ is a type submodule.

(4) Let $P \oplus P' \leq_e N$ and $N \oplus N' \leq_e M$, where P' is a complement of P in N, and N' is a complement of N in M. Since $P \subseteq N$ and $N \perp N'$, $P \perp N'$. Also $P \perp P'$. Then $P \oplus P' \oplus N' \leq_e M$, with $P \perp (P' \oplus N')$. Since $P \leq M$ is a complement submodule, it now follows from (4.1.2)(3) that $P \leq_t M$. □

4.1.6. PROPOSITION. Let $A < P \leq N$ where A is a complement submodule of N.

1. If $P \leq_t N$, then $P/A \leq_t N/A$.
2. If $A <_t P$ and $P/A \leq_t N/A$, then $P \leq_t N$ or $P <_e P^{tc}$.

PROOF. (1) If not, P/A has a proper parallel extension H/A in N/A. Then A is properly contained in H. Let C be a complement of A in P, and X a complement of P in H. Then $A \oplus C \leq_e P$, $A \oplus C \oplus X \leq_e P \oplus X \leq_e H$ with $X \neq 0$ and $P \perp X$. Since A is a complement submodule of N, C is isomorphic to an essential submodule of P/A, while $C \oplus X$ is isomorphic to an essential submodule of H/A. Since P/A and H/A are parallel, C and $C \oplus X$ are parallel. This contradicts that $C \perp X$. So P/A is a type submodule of N/A.

(2) If not, there exist submodules $B \cong C$ of P^{tc} with $0 \neq B \leq P$ and $P \oplus C \leq P^{tc}$. If $A \cap B \neq 0$, then $A \cap B$ is isomorphic to a submodule of C, where $A \cap C = 0$, contradicting that $A <_t N$. But then the following shows that P/A is not a type submodule of N/A:

$$P/A \geq (A \oplus B)/A \cong (A \oplus C)/A \leq P/A \oplus (A \oplus C)/A \leq N/A.$$

□

We next discuss the type dimension of a module which is the analogue of the well-known uniform dimension or Goldie dimension. The type dimension can be used the same way that the uniform dimension can. Since the uniform dimension of a module might be infinite while the type dimension is finite, it is more widely applicable.

4.1.7. LEMMA. Suppose that $A_1, \cdots, A_n \leq M$ are pairwise orthogonal atomic submodules with $A_1 \oplus \cdots \oplus A_n \leq_e M$. If $B_1, \cdots, B_m \leq M$ are nonzero pairwise orthogonal, then $m \leq n$.

PROOF. By (2.3.3), each B_i contains an atomic submodule. Without loss of generality, let the B_i be atomic. By (2.3.3), after renumbering the A_i, we may assume that $B_1\|A_1$. Repetition of this process gives us that $B_2\|A_i$ for some i. Since $B_1 \perp B_2$, we have $i \neq 1$. Renumbering, take $i = 2$, if $2 \leq n$. Suppose that $n+1 \leq m$. In that case we are able to repeat this process n times obtaining $B_i\|A_i$ for $i = 1, \ldots, n$. But also for B_{n+1} we get that $B_{n+1}\|A_i$ for some $i \leq n$, so $B_{n+1}\|A_i$ and $A_i\|B_i$. It follows that $B_{n+1}\|B_i$. The contradiction shows that $m \leq n$. □

The previous lemma shows that the integer n there is unique for M, and does not depend on the choice of the A_i's. The next definition was first introduced in [140].

4.1.8. DEFINITION. A module M has finite **type dimension** n, denoted by $t.\dim M = n$, if M contains an essential direct sum of n pairwise orthogonal atomic submodules of M. If no such n exists, we say that the type dimension of M is infinite, and write $t.\dim M = \infty$. If $M = 0$, then $t.\dim M = 0$.

The finite uniform dimension of a module M will here be denoted by $u.\dim M$ or $u.\dim M = \infty$ if it is not finite. Note that like $u.\dim$, also $t.\dim M = t.\dim E(M)$. That is, the type dimension of any module is equal to that of any essential submodule.

4.1.9. EXAMPLES. As modules over $\mathbb{Z}$, the following hold:

1. If $M = \mathbb{Z}_2 \oplus \mathbb{Z}_3 \oplus \mathbb{Z}_4$, then $t.\dim M = 2$.

2. If $M = \oplus_{i>0}\mathbb{Z}_{p^i}$ where p is a prime number, then $t.\dim M = 1$.

3. If $M = \oplus_{i>0}\mathbb{Z}_{p_i}$ where p_i is the ith prime number, then $t.\dim M = \infty$.

□

Several useful properties of the type dimension, first observed in [140], are listed below. Below, the type dimension is allowed to be ∞, in which case $n + \infty = \infty$ for any n.

4.1.10. LEMMA. Any $N \leq M$ satisfies the following:

1. $t.\dim M \leq t.\dim N + t.\dim M/N$.

2. If $M = M_1 \oplus \cdots \oplus M_n$, then $t.\dim M \leq t.\dim M_1 + \cdots + t.\dim M_n$.

3. Let $M = M_1 \oplus \cdots \oplus M_n$. If $M_i \perp M_j$ for all $i \neq j$, then $t.\dim M = t.\dim M_1 + \cdots + t.\dim M_n$. The converse holds if $t.\dim M < \infty$.

4. $N\|M \Longrightarrow t.\dim M = t.\dim N$. In particular, $t.\dim N = t.\dim N^{tc}$.

5. $t.\dim M < \infty$, and $t.\dim M = t.\dim N \Longrightarrow M\|N$.

6. $N \leq_e M \Longrightarrow t.\dim N = t.\dim M$.

PROOF. The proofs of (1), (2), (4), and (6) are omitted. They can be proved using definition (4.1.8).

(3) Let $A_1^{(i)} \oplus \cdots \oplus A_{m(i)}^{(i)} \leq_e M_i$, $i = 1, \cdots, n$, be an essential orthogonal sum of atomic modules. Hence $t.\dim M_i = m(i)$. Put an equivalence relation on the $A_j^{(i)}$'s whereby any two of these, say $A_s^{(1)}$ and $A_t^{(2)}$, are equivalent if they have a common nonzero isomorphic submodule, equivalently, $A_s^{(1)} \| A_t^{(2)}$. Let there be k equivalence classes. Then $t.\dim M = k$. But $k = m(1) + \cdots + m(n)$ if and only if each equivalence class is a singleton if and only if $M_i \perp M_j$ whenever $i \neq j$.

(5) If N and M are not parallel, then there exists an $0 \neq X \subseteq M$ such that $N \perp X$. Thus, $t.\dim M \geq t.\dim(N \oplus X) = t.\dim N + t.\dim X > t.\dim N$. This is a contradiction. □

Recall that if N is any compressible module ($\forall\, 0 \neq V \leq N, N \hookrightarrow V$), then N is atomic. Consequently, any (noncommutative) domain has type dimension 1. For a direct product of rings $R = R_1 \times \cdots \times R_n$, as R-modules, $R_i \perp R_j$ for $i \neq j$. Consequently, $t.\dim R_R = t.\dim({R_1}_{R_1}) + \cdots + t.\dim({R_n}_{R_n})$, where $t.\dim({R_k}_{R_k}) = t.\dim(R_k)_R$. For an infinite product of rings $R = \prod_{i=1}^{\infty} R_i$, $\oplus_{i=1}^{\infty} R_i \leq_e R_R$ and as before, $R_i \perp R_j$ for $i \neq j$, $t.\dim R_R = \infty$.

The next two propositions explain how the type dimension $t.\dim M$ can be realized.

4.1.11. For a module M, $t.\dim M = \infty$ if and only if there exist an infinite number of pairwise orthogonal nonzero submodules of M.

PROOF. It follows directly from (4.1.8). □

The next result is due to [140].

4.1.12. PROPOSITION. The following hold for a module M:

1. For any type submodule $N \leq_t M$, $t.\dim M = t.\dim N + t.\dim(M/N)$.
2. The following are all equivalent:
   (a) $t.\dim M < \infty$.
   (b) M has ACC on type submodules.
   (c) M has DCC on type submodules.

PROOF. (1) Let $N \oplus P \leq_e M$, so $N \perp P$ by (4.1.5), and P embeds as an essential submodule in M/N. Consequently, $t.\dim M = t.\dim(N \oplus P) = t.\dim N + t.\dim P = t.\dim N + t.\dim(M/N)$.

(2)$(a) \Longrightarrow$ (2)(b). Suppose that $M_1 \subset M_2 \subset \cdots \subset M$ is a strictly ascending countable infinite chain of type submodules of M. Let X_i be a complement of M_i in M_{i+1}. Then $X_i \neq 0$ and $M_i \perp X_i$ by (4.1.2)(2) and (4.1.5)(2). It follows that $X_i \perp X_j$ for all $i \neq j$. Hence by (4.1.11), $t.\dim M = \infty$, a contradiction.

(2)(b) $\Longrightarrow$ (2)(c). Suppose there is a strictly descending countable infinite chain $M \supseteq M_1 \supset M_2 \supset \cdots$ with $M_i \leq_t M$ for all i. For each $i \geq 1$, let P_i be a complement of M_{i+1} in M_i. Then $P_i \neq 0$ and $P_i \perp M_{i+1}$ by (4.1.3)(2) and (4.1.5)(2). So $P_i \leq_t M_i$ by (4.1.2)(3); hence $P_i \leq_t M$ by (4.1.5)(4). Thus $\oplus_{i=1}^{\infty} P_i$ is a direct sum of pairwise orthogonal type submodules. Let $X_1 = P_1^c = P_1$ and $X_2 = (X_1 \oplus P_2)^c$. Generally, for each $i \geq 1$, let $X_{i+1} = (X_i \oplus P_{i+1})^c$ be a complement closure of $X_i \oplus P_{i+1}$ in M. Then $X_1 \subset X_2 \subset \cdots$ and each X_i is a type submodule of M by (4.1.5)(3).

(2)(c) $\Longrightarrow$ (2)(a). If (2)(a) does not hold, then by (4.1.11) there is a family $\{Y_i : i = 1, 2, \cdots\}$ of pairwise orthogonal nonzero submodules of M. Since $Y_i^{tc} \perp Y_j^{tc}$ whenever $i \neq j$, replacing Y_i by Y_i^{tc} when necessary, we may assume that all Y_i are type submodules of M. Let $X_1 = M$ and, for each $i > 1$, let X_i be a complement closure of $\oplus_{j \geq i} Y_j$ in X_{i-1}. Then we have a chain $X_1 \supset X_2 \supset \cdots$ such that each X_i is a type submodule of M by (4.1.5)(3). □

4.1.13. PROPOSITION. For a natural class $\mathcal{K}$, suppose that N_1, N_2 are type submodules of type $\mathcal{K}$ of M. Then the following hold:

1. $E(N_1) \cong E(N_2)$.
2. There exist $D_1 \leq_e N_1$, $D_2 \leq_e N_2$ such that $D_1 \cong D_2$.
3. $E(M/N_1) \cong E(M/N_2)$.

PROOF. (1) Let C be a complement of N_2 in M. Since N_2 is a type submodule of M of type $\mathcal{K}$, $C \in c(\mathcal{K})$, and C is a type submodule of M by (4.1.2). So $N_1 \perp C$. Let $N_1 \oplus C \oplus D \leq_e M$ for some $D \leq M$. Then $D \perp N_1$ and $D \in c(\mathcal{K})$. Since C is a type submodule of M and $C \cap D = 0$, $C \perp D$. But $N_2 \oplus C \leq_e M$, so $D \in d(N_2) \in \mathcal{K}$. Thus, $D \in \mathcal{K} \cap d(\mathcal{K}) = \{0\}$. Consequently, $E(M) = E(N_1) \oplus E(C)$. Also $E(M) = E(N_2) \oplus E(C)$. So it follows that $E(N_1) \cong E(N_2)$.

(2) Let $g : E(N_1) \longrightarrow E(N_2)$ be an isomorphism. Then $g(N_1) \leq_e E(N_2)$, and $g(N_1) \cap N_2 \leq_e E(N_2)$. Since the inverse image of an essential submodule is essential, $D_1 = N_1 \cap g^{-1}(gN_1 \cap N_2) \leq_e N_1$, and $D_1 \cong g(D_1) \subseteq g(N_1) \cap N_2 \subseteq N_2$. But clearly, $D_2 = g(D_1) \leq_e E(N_2)$.

(3) Let C_i be a complement of N_i in M ($i = 1, 2$). Then C_1, C_2 are type submodules of type $c(\mathcal{K})$ of M. By (1), $E(C_1) \cong E(C_2)$, and hence $E(M/N_1) \cong E(C_1) \cong E(C_2) \cong E(M/N_2)$. □

4.1.14. PROPOSITION. Let M be a nonsingular module and $\mathcal{K}$ a natural class. Then the following hold:

1. $N = \Sigma\{V : V \subseteq M, V \in \mathcal{K}\}$ is the unique type submodule of type $\mathcal{K}$.
2. $N \leq M$ is fully invariant.

PROOF. (1) Let $f : P = \oplus\{V : V \subseteq M, V \in \mathcal{K}\} \longrightarrow N$ be the surjective sum map, and K the kernel of f. Since $Z(N) = 0$, K is a complement submodule of P. Let $K \oplus W \leq_e P$, where $W \in \mathcal{K}$. Then $W \cong (K \oplus W)/K \leq_e P/K \cong f(P) = N$ shows that $N \in \mathcal{K}$.

(2) For a nonsingular module M, and any $N \leq M$ and $x \in M$, $x^{-1}N \leq_e R$ if and only if $N \leq_e xR+N$ (see [24, 1.1, p.52]). Hence if $N < M$ is a complement submodule and $x \notin N$, then $x^{-1}N$ is not essential in R. Suppose that for some homomorphism $\phi : M \longrightarrow M$, $\phi(N) \not\subseteq N$ and $x = \phi(n) \notin N$ for some $n \in N$ where N is given as in (1). Then $x^{-1}N \oplus C \leq R$ for some $0 \neq C \leq R$. Since $n^{\perp} \subseteq x^{\perp} \subseteq x^{-1}N$ and $x^{-1}N \cap C = 0$, $C \cong xC \cong nC \in \mathcal{K}$. If $xc_0 \in xC \cap N$ with $c_0 \in C$, then $c_0 \in x^{-1}N \cap C = 0$. Thus $N + xC = N \oplus xC \in \mathcal{K}$, contradicting the maximality of N in $\mathcal{K}$. □

4.1.15. COROLLARY. For a natural class $\mathcal{K}$ and a module M, assume that $Z(M) \in \mathcal{K}$.

1. There exists a unique type submodule $N \leq M$ of type $\mathcal{K}$.
2. $Z_2(M) \subseteq N$ for N as in (1).
3. $N \leq M$ is fully invariant.

PROOF. Now also $Z_2(M) \in \mathcal{K}$, and consequently there exists a type submodule $N \leq_t M$ with $Z_2(M) \subseteq N \in \mathcal{K}$.

(1) and (2) Let $N_2 \leq_t M$ be of type $\mathcal{K}$. Then

$$(ZM + N_2)/N_2 \subseteq Z(M/N_2) \subseteq M/N_2 \subseteq E(M/N_2)$$

shows that $(ZM + N_2)/N_2 \subseteq Z(E(M/N_2))$. But $E(M/N) \cong E(M/N_2)$ by (4.1.13); this shows that $Z(E(M/N_2)) = 0$ since $Z(M/N) = 0$. Hence $ZM \subseteq N_2$. So, being a quotient of $(Z_2M + N_2)/ZM$, $(Z_2M + N_2)/N_2$ is singular. Then we have

$$(Z_2M + N_2)/N_2 \subseteq Z(M/N_2) \subseteq M/N_2 \subseteq E(M/N_2).$$

As above, we have $Z_2M \subseteq N_2$. Since Z_2M is a complement submodule of M and $N \leq_t M$, $N/Z_2M \leq_t M/Z_2M$ by (4.1.6). Similarly, we have $N_2/Z_2M \leq_t M/Z_2M$. Since Z_2M is a complement submodule of M, both N/Z_2M and N_2/Z_2M are type submodules of the same type $\mathcal{K}$ of M/Z_2M. Thus, $N/Z_2M = N_2/Z_2M$, or $N = N_2$ by (4.1.14).

(3) By the last proposition and the first part of the proof, we see that $N/Z_2M \leq M/Z_2M$ is fully invariant. If $\phi : M \longrightarrow M$ is an endomorphism, then $\phi(Z_2M) \subseteq Z_2M$. Hence ϕ induces a map $\bar{\phi} : M/Z_2M \longrightarrow M/Z_2M$. Then $\bar{\phi}(N/Z_2M) \subseteq N/Z_2M$ which implies that $\phi(N) \subseteq N$. □

The concluding example shows that type submodules of a module are not necessarily invariant.

4.1.16. EXAMPLE. Let $R = \mathbb{Z}$ and $M = \mathbb{Z} \oplus X$ where $X_{\mathbb{Z}}$ is a nonzero torsion module. Fix $0 \neq x_0 \in X$ and let $f : M \longrightarrow M$ be defined by $f(n, x) = n(0, x_0)$ $(n \in \mathbb{Z}, x \in X)$. Then $f(\mathbb{Z} \oplus 0) \not\subseteq \mathbb{Z} \oplus 0$, but clearly $\mathbb{Z} \oplus 0$ is a type submodule of M. □

4.1.17. REFERENCES. Dauns [24,32]; Dauns and Zhou [42]; Zhou [140, 141].

4.2 Several Type Dimension Formulas

For a module M, there are several computational formulas relating the uniform dimension $u.\dim M$ of M to that of any submodule $A \leq M$, its complement closure A^c, and quotient modules like M/A. The analogues of these for the type dimension were developed in [42]. Usually, simple replacement of "A^c" by "A^{tc}" and "$u.\dim$" by "$t.\dim$" does not give correct formulas. This section begins with formulas which are applicable to all modules.

Type dimension formulas for several special classes of rings and modules, such as formal triangular rings, polynomial rings $R[x]$, and laurent polynomial rings $R[x, x^{-1}]$, were invented in [140]. The arguments there immediately carry over to formal modules $M[x]$ and $M[x, x^{-1}]$ over these rings.

For any submodule A of M, A^c/A is a complement submodule of M/A, and hence

$$\begin{aligned} u.\dim M/A &= u.\dim A^c/A + u.\dim M/A^c \\ &= u.\dim A^c/A + u.\dim M - u.\dim A. \end{aligned}$$

The formula

$$u.\dim M + u.\dim A^c/A = u.\dim A + u.\dim M/A$$

is due to Camillo and Zelmanowitz [17] as are several others. We begin by showing that the parallel formula

$$t.\dim A + t.\dim M/A = t.\dim M + t.\dim A^{tc}/A$$

is false in general, but does hold if $A \leq M$ is a complement submodule.

4.2.1. EXAMPLE. For $R = \mathbb{Z}$, set $M = \mathbb{Z} \oplus \mathbb{Z}_2$ and $A = 2\mathbb{Z} \oplus (0)$. Then $A^{tc} = \mathbb{Z} \oplus (0)$. Moreover, $M/A = \mathbb{Z}_2 \oplus \mathbb{Z}_2$, $A^{tc}/A = \mathbb{Z}_2 \oplus (0)$, and hence $t.\dim A + t.\dim M/A = 1 + 1 \neq 2 + 1 = t.\dim M + t.\dim A^{tc}/A$. □

In subsequent formula, if one side equals "∞", then the other side also necessarily does so.

4.2.2. LEMMA. For a complement submodule A of M, let A^{tc} be any type closure of A in M. Then

$$t.\dim M + t.\dim A^{tc}/A = t.\dim A + t.\dim M/A.$$

PROOF. Let $A^{tc} \oplus P \leq_e M$ where $P \leq_t M$ is a (type) complement of A^{tc} in M. Then $t.\dim M = t.\dim A^{tc} + t.\dim P$. Since P embeds as an essential submodule of M/A^{tc}, $t.\dim P = t.\dim M/A^{tc}$. As noted in (4.1.10), $t.\dim A^{tc} = t.\dim A$. The formula at the beginning of this argument now says that

$$t.\dim M = t.\dim A + t.\dim M/A^{tc}\ . \tag{1}$$

So far we have not used the fact that A is a complement submodule of M, but now it is needed to conclude from (4.1.6) that $A^{tc}/A \leq_t M/A$. The formula at the beginning of this argument now says that

$$t.\dim M/A = t.\dim A^{tc}/A + t.\dim \left[(M/A)/(A^{tc}/A)\right].$$

Thus

$$t.\dim M/A = t.\dim A^{tc}/A + t.\dim M/A^{tc}. \tag{2}$$

Add "$t.\dim A^{tc}/A$" to both sides of equation (1) and then use (2) to simplify and get

$$\begin{aligned} t.\dim M + t.\dim A^{tc}/A &= t.\dim A + t.\dim M/A^{tc} + t.\dim A^{tc}/A \\ &= t.\dim A + t.\dim M/A. \end{aligned}$$

□

4.2.3. EXAMPLE. Return to the previous Example (4.2.1). There A was not a complement submodule of M, and the previous lemma failed. The complement closure of $A^{tc}/A = \mathbb{Z}_2 \oplus (0)$ in $M/A = \mathbb{Z}_2 \oplus \mathbb{Z}_2$ is $(A^{tc}/A)^{tc} = \mathbb{Z}_2 \oplus \mathbb{Z}_2$. Adding the term

$$\begin{aligned} t.\dim \left[(A^{tc}/A)^{tc}/(A^{tc}/A)\right] &= t.\dim \left[(\mathbb{Z}_2 \oplus \mathbb{Z}_2)/(\mathbb{Z}_2 \oplus (0))\right] \\ &= 1 \end{aligned}$$

to the left side of the formula in the last lemma gives a correct answer. □

4.2.4. THEOREM. If A is a submodule of M, let A^{tc} be any type closure of A in M, and let $(A^{tc}/A)^{tc}$ be any type closure of A^{tc}/A in M/A. Then

$$\begin{aligned} t.\dim M + t.\dim(A^{tc}/A) = t.\dim A &+ t.\dim(M/A) \\ &+ t.\dim \left[(A^{tc}/A)^{tc}/(A^{tc}/A)\right]. \end{aligned}$$

PROOF. From the proof of the last lemma the formula (1) shows that $t.\dim M = t.\dim A + t.\dim M/A^{tc}$. Since A^{tc} is a complement submodule of M, A^{tc}/A is a complement submodule of M/A. By (4.2.2),

$$t.\dim M/A + t.\dim \left[(A^{tc}/A)^{tc}/(A^{tc}/A)\right] = t.\dim A^{tc}/A + t.\dim M/A^{tc}.$$

Add "$t.\dim A$" to both sides of the last equation, and use the above formula (1) to prove the equation in the theorem. □

For submodules A and B of some bigger module M, Camillo and Zelmanowitz [17] developed formulas for $u.\dim(A + B)$ in terms of uniform dimensions of A, B and some other modules. Later, Dauns and Zhou [42] found the analogues for $t.\dim(A + B)$. We now turn to type dimensions of specific classes of rings. The type dimension formulas for a triangular matrix ring were discovered in [140].

4.2.5. PROPOSITION. Let $R = \begin{pmatrix} S & B \\ 0 & T \end{pmatrix}$ be the formal triangular matrix ring where S, T are rings and B is an (S,T)-bimodule. Let $\mathbf{l}(B) = \{s \in S : sB = 0\}$ and $\mathbf{r}(B) = \{t \in T : Bt = 0\}$. Then

(1) $t.\dim R_R = t.\dim \mathbf{l}(B)_S + t.\dim(B \oplus T)_T$;

(2) $t.\dim {}_RR = t.\dim_T \mathbf{r}(B) + t.\dim {}_S(S \oplus B)$.

PROOF. As right R-submodules of R, it is asserted that

$$\begin{pmatrix} \mathbf{l}(B) & 0 \\ 0 & 0 \end{pmatrix} \perp \begin{pmatrix} 0 & B \\ 0 & T \end{pmatrix} \quad \text{and} \quad \begin{pmatrix} \mathbf{l}(B) & 0 \\ 0 & 0 \end{pmatrix} \oplus \begin{pmatrix} 0 & B \\ 0 & T \end{pmatrix} \leq_e R_R .$$

The orthogonality follows from the fact that any two nonzero elements from the two right ideals of R have different right annihilators in R; and the equality $\begin{pmatrix} s & 0 \\ 0 & 0 \end{pmatrix}\begin{pmatrix} 0 & B \\ 0 & 0 \end{pmatrix} = \begin{pmatrix} 0 & sB \\ 0 & 0 \end{pmatrix}$ shows that the sum of the above two right ideals is essential in R_R. There is a bijection f from the set of all right S-submodules of $\mathbf{l}(B)_S$ onto the set of all right R-submodules of $\begin{pmatrix} \mathbf{l}(B) & 0 \\ 0 & 0 \end{pmatrix}$ given by $X \longmapsto f(X) = \begin{pmatrix} X & 0 \\ 0 & 0 \end{pmatrix}$. Furthermore, $X_S \cong Y_S$ if and only if $f(X)_R \cong f(Y)_R$. Therefore, $t.\dim \mathbf{l}(B)_S = t.\dim \begin{pmatrix} \mathbf{l}(B) & 0 \\ 0 & 0 \end{pmatrix}_R$.

Similarly, there is a bijection g from the set of all right T-submodules of $(B \oplus T)_T$ onto the set of all right R-submodules of $\begin{pmatrix} 0 & B \\ 0 & T \end{pmatrix}$ given by

$$X \longmapsto g(X) = \left\{ \begin{pmatrix} 0 & b \\ 0 & t \end{pmatrix} : (b,t) \in X \right\}.$$

Again $X_T \cong Y_T$ if and only if $g(X)_R \cong g(Y)_R$. Thus $t.\dim(B \oplus T)_T = t.\dim \begin{pmatrix} 0 & B \\ 0 & T \end{pmatrix}_R$. By (4.1.10)(3),

$$\begin{aligned} t.\dim R_R &= t.\dim \begin{pmatrix} \mathbf{l}(B) & 0 \\ 0 & 0 \end{pmatrix}_R + t.\dim \begin{pmatrix} 0 & B \\ 0 & T \end{pmatrix}_R \\ &= t.\dim \mathbf{l}(B)_S + t.\dim(B \oplus T)_T. \end{aligned}$$

The proof of (2) is similar and is omitted. □

The last proposition is used to show the left-right asymmetry of the concept of type dimension in the next two examples from [140].

4.2.6. EXAMPLE. (1) Set $R = \begin{pmatrix} \mathbb{Z} & 2\mathbb{Z}_4 \\ 0 & \mathbb{Z}_4 \end{pmatrix}$. Here

$$\begin{aligned} \mathbf{l}(B) &= 2\mathbb{Z}, \\ t.\dim \mathbf{l}(B)_S &= t.\dim \mathbf{l}(B)_{\mathbb{Z}} = 1, \\ t.\dim(B \oplus T)_T &= t.\dim(2\mathbb{Z}_4 \oplus \mathbb{Z}_4)_{\mathbb{Z}_4} = 1, \\ t.\dim R_R &= 2. \end{aligned}$$

On the other side,

$$\begin{aligned} \mathbf{r}(B) &= 2\mathbb{Z}_4, \\ t.\dim {}_T\mathbf{r}(B) &= t.\dim {}_{\mathbb{Z}_4}2\mathbb{Z}_4 = 1, \\ t.\dim {}_S(S \oplus B) &= t.\dim {}_{\mathbb{Z}}(\mathbb{Z} \oplus 2\mathbb{Z}_4) = 2, \\ t.\dim {}_R R &= 3. \end{aligned}$$

(2) Let $\{F_i : i \in I\}$ be a set of fields, and $|I|$ the cardinality of I. Define S to be the subring of $\Pi_{i \in I} F_i$ generated by $B = \oplus_{i \in I} F_i$ and the identity element of the full product. Let $e_i \in S$ be the element whose ith component is 1 and all other components are 0. Note that $e_i S = e_i F_i \perp e_j F_j = e_j S$ for $i \neq j$ as S-modules. Hence $t.\dim {}_S S = |I|$.

Set $R = \begin{pmatrix} S & B \\ 0 & \mathbb{Z} \end{pmatrix}$. Now $\mathbf{l}(B)_S = 0$ and ${}_T\mathbf{r}(B) = 0$. Since ${}_S B \leq_e {}_S S$, $t.\dim {}_S(S \oplus B) = t.\dim {}_S S = |I|$. Therefore,

$$t.\dim {}_R R = t.\dim {}_T r(B) + t.\dim {}_S(S \oplus B) = 0 + |I|.$$

Next,

$$t.\dim(B \oplus T)_T = t.\dim[(\oplus_{i \in I} F_i) \oplus \mathbb{Z}]_T = k + 1,$$

where $0 \leq k \leq \infty$ is the number of distinct nonzero characteristics $\text{char}(F_i)$ of the fields. Therefore,

$$t.\dim R_R = t.\dim \mathbf{l}(B)_S + t.\dim(B \oplus T)_T = 0 + k + 1 .$$

□

If R^{op} is the **opposite ring** of R, then $t.\dim R_R = t.\dim_{R^{op}} R^{op}$. For any k (an integer or ∞) and any set I with $0 \leq k \leq |I|$, a family of $|I|$ fields can be found containing k distinct nonzero characteristics. Consequently, the above shows that for any $1 \leq n, m \leq \infty$, there exists a ring R with $t.\dim {}_R R = n$ and $t.\dim R_R = m$.

The type dimension formulas for formal polynomial and Laurent polynomial rings were discovered by [140].

4.2.7. DEFINITION. Let M be any right R-module, and let $M[x, x^{-1}]$ be the Abelian group of all expressions ξ of the form

$$\xi = \Sigma\{x^i m_i : -p \leq i \leq q, m_i \in M\}$$

where p, q are non-negative integers, and addition is componentwise. Identify M with the abelian subgroup $M = x^0M \hookrightarrow M[x, x^{-1}]$. Then $M[x, x^{-1}]$ becomes a right module over the Laurent polynomial ring $R[x, x^{-1}]$ as follows. For $m \in M$, $r \in R$, and $i, j \in \mathbb{Z}$ first define $x^i m x^j r = x^{i+j} mr$, and then simply extend this by distributivity and linearity. Restricting ourselves to non-negative powers of x we get the right module $M[x]$ over the polynomial ring $R[x]$.

The above construction can be iterated a finite or infinite number of times to yield the modules (1) and (3) below. In the case of an infinite set of variables in (2) and (4) below, we first well order the set of variables $\{x_1, x_2, \ldots\}$. The next theorem was first established in [140]. Although there $M = R_R$, the proofs for the two cases are identical.

4.2.8. THEOREM. Let M be a right R-module, and let N_S be any one of the following modules over the following rings S.

1. $M[x_1, \cdots, x_n]$ over $R[x_1, \cdots, x_n]$.
2. $M[x_1, x_2, \cdots]$ over $R[x_1, x_2, \cdots]$.
3. $M[x_1, \cdots, x_n, x_1^{-1}, \cdots, x_n^{-1}]$ over $R[x_1, \cdots, x_n, x_1^{-1}, \cdots, x_n^{-1}]$.
4. $M[\cdots, x_2, x_1, x_1^{-1}, x_2^{-1}, \cdots]$ over $R[\cdots, x_2, x_1 \cdots, x_1^{-1}, x_2^{-1}, \cdots]$.

Then $t.\dim N_S = t.\dim M_R$.

PROOF. To prove (1) and (3) it suffices to take $n = 1$, and then iterate. That is, if (1) holds for $n = 1$, replace M and R by $M[x_1]$ and $R[x_1]$ and consider $M[x_1][x_2] = M[x_1, x_2]$ and $R[x_1][x_2] = R[x_1, x_2]$.

(1) A result from Goodearl [66, Lemma 3.21, p.88] implies the following:

(i) For any $0 \neq f \in M[x]$, there exists a $u \in R$ such that $fu \neq 0$, and the right annihilators in R of the nonzero coefficients of fu are all equal.

(ii) Assume that the right annihilators in R of the nonzero coefficients of $0 \neq f \in M[x]$ are all equal. Then the right annihilator $Ann(f)$ of f in $R[x]$ is the same as the right annihilator in $R[x]$ of the leading coefficient of f.

(iii) Hence if in (ii), the leading term of f is $x^k m$, $m \in M$, then $Ann(f) = Ann(m) = m^{\perp}[x]$ where $m^{\perp} = \{r \in R : mr = 0\}$. Moreover, for any $0 \neq s \in R$ with $ms \neq 0$, $Ann(fs) = (ms)^{\perp}[x] = Ann(ms)$.

First it will be shown that for any atomic submodule A of M, $A[x]$ is an atomic submodule of $M[x]$. Let f and g be any two nonzero elements in $A[x]$ with leading coefficients a and b, respectively. We want to show that $fR[x]$ and $gR[x]$ contain isomorphic nonzero submodules. To this end, by (i) above, it may be assumed that $Ann(f) = Ann(a)$ and $Ann(g) = Ann(b)$. Since aR and bR contain isomorphic nonzero cyclic submodules, there exist $s, t \in R$ with $0 \neq asR \cong btR$ via $asr \longmapsto btr$, $r \in R$. From $(as)^{\perp} = (bt)^{\perp}$ it follows that $(as)^\perp[x] = (bt)^\perp[x]$. Since f satisfies (ii), fs satisfies (iii), and similarly for g and gt. Hence

$$\begin{aligned} Ann(fs) &= Ann(as) = (as)^\perp[x] = (bt)^\perp[x] \\ &= Ann(bt) = Ann(gt). \end{aligned}$$

Consequently, the $R[x]$-homomorphism

$$fsR[x] \longrightarrow gtR[x], \ fsh \longmapsto gth, \ h \in R[x]$$

is an isomorphism. Hence $A[x]$ is an atomic $R[x]$-module.

Suppose A, B are any two R-submodules of M with $A \perp B$, but $A[x], B[x]$ are not orthogonal as $R[x]$-submodules of $M[x]$. Then there is an $R[x]$-isomorphism

$$\psi : fR[x] \longrightarrow gR[x] \neq 0, \ \text{ via } \psi(fh) = gh \ (h \in R[x])$$

for some $f \in A[x]$ and $g \in B[x]$. Let a and b be the leading coefficients of f and g, respectively. By (i) above, for some $u \in R$ with $fu \neq 0$, $Ann(fu) = Ann(au) = (au)^\perp[x]$, and still $fuR[x] \cong guR[x]$. Next, again by (i), for some $v \in R$ with $guv \neq 0$, $Ann(guv) = Ann(buv) = (buv)^\perp[x]$. Furthermore, $Ann(fuv) = Ann(auv) = (auv)^\perp[x]$. But $\psi(fuv) = guv$, and hence $Ann(fuv) = Ann(guv)$, or $(auv)^\perp[x] = (buv)^\perp[x]$. Thus, $(auv)^\perp = (buv)^\perp$; so $0 \neq auvR \cong buvR$ gives a contradiction.

If $t.\dim M_R = \infty$, then $t.\dim M[x]_{R[x]} = \infty$. Now if $t.\dim M_R = n$, i.e., there exist pairwise orthogonal atomic submodules $A_1, \ldots, A_n$ of M such that $A_1 \oplus \cdots \oplus A_n \leq_e M_R$, then, from above, $A_1[x], \cdots, A_n[x]$ are pairwise orthogonal atomic submodules of $M[x]$. So it suffices to show that $A_1[x] \oplus \cdots \oplus A_n[x]$ is essential in $M[x]$. If not then $A_1[x] \oplus \cdots \oplus A_n[x] \oplus H \leq_e M[x]$ where H is a nonzero submodule of $M[x]$. Take a nonzero element $f = \Sigma_{k=1}^{l} x^{i(k)} m_k \in H$, where all $0 \neq m_k \in M$. Since $0 \neq m_1 \in M$ and $A = A_1 \oplus \cdots \oplus A_n \leq_e M$, there exists an $r_1 \in R$ such that $0 \neq m_1 r_1 \in A$. Hence $fr_1 \neq 0$, and $m_2 r_1 \neq 0$. Again there exists an $r_2 \in R$ such that $0 \neq m_2 r_1 r_2 \in A$. Consequently $fr_1 r_2 \neq 0$. Continuing this way we obtain an element

$$0 \neq fr_1 r_2 \cdots r_m \in (A_1[x] \oplus \cdots \oplus A_n[x]) \cap H = 0,$$

a contradiction.

(2) The proof of (1) can be adapted to show that if A, B are orthogonal atomic submodules of M then $A[x_1, x_2, \cdots], B[x_1, x_2, \cdots]$ are orthogonal atomic submodules of $N_S = M[x_1, x_2, \cdots]$, where $S = R[x_1, x_2, \cdots]$. The same argument as before now shows that if $A_1 \oplus \cdots \oplus A_k \leq_e M_R$, then $A_1[x_1, x_2, \cdots] \oplus \cdots \oplus A_k[x_1, x_2, \cdots]$ is essential in $M[x_1, x_2, \cdots]$.

(3) and (4). Take $n = 1$ for simplicity; the proof for any $n \leq \infty$ will be the same. Set $N = M[x, x^{-1}]$ and $S = R[x, x^{-1}]$. Note that the positive powers x^n of x in R have two remarkable properties. First, x^n does not annihilate

any nonzero elements of either N or S. Secondly, for any $0 \neq v \in N$ and $0 \neq s \in S$, for some sufficiently big n, $0 \neq vx^n \in M[x]$ while $0 \neq sx^n \in R[x]$. We will simply reduce the proof in the case of N_S to that of $M[x]_{R[x]}$ and (1), from which $t.\dim M[x]_{R[x]} = t.\dim M_R$.

To prove that $t.\dim N_S \leq t.\dim M[x]_{R[x]}$, it suffices to show that if $fS \perp gS$ for $f, g \in N$, then also $(fS \cap M[x]) \perp (gS \cap M[x])$ over $R[x]$. If not, there exist $s_1, s_2 \in S$, with $fs_1, gs_2 \in M[x]$ and an $R[x]$-isomorphism

$$\phi : fs_1R[x] \longrightarrow gs_2R[x], \; fs_1p \longmapsto gs_2p \; p \in R[x].$$

Then it follows that

$$fs_1S \longrightarrow gs_2S, \; fs_1s \longmapsto gs_2s, \; s \in S,$$

is a well defined S-isomorphism. For if $fs_1s = 0$ but $gs_2s \neq 0$, simply take an x^n such that $0 \neq s_1sx^n \in R[x]$, $0 \neq s_2sx^n \in R[x]$, and $gs_2sx^n \neq 0$. Then $0 = \phi(fs_1sx^n) = gs_2sx^n \neq 0$ is a contradiction.

To establish the converse that $t.\dim M[x]_{R[x]} \leq t.\dim N_S$, it has to be shown that for any $fR[x] \perp gR[x]$ over $R[x]$ for $f, g \in M[x]$, also $fS \perp gS$ over S. If not, there are $s_1, s_2 \in S$ and an S-isomorphism $0 \neq fs_1S \to gs_2S$, $fs_1s \longmapsto gs_2s$, $s \in S$. There is an x^n such that $s_1x^n, s_2x^n \in R[x]$. But then the above isomorphism induces an $R[x]$-isomorphism $fs_1x^nR[x] \cong gs_2x^nR[x]$, a contradiction. Hence $t.\dim M_R = t.\dim M[x]_{R[x]} = t.\dim N_S$. □

4.2.9. COROLLARY. For a ring R, let S be any one of the following rings:

1. $R[x_1, \cdots, x_n]$,
2. $R[x_1, x_2 \cdots]$,
3. $R[x_1, \cdots, x_n, x_1^{-1}, \cdots, x_n^{-1}]$,
4. $R[\cdots, x_2, x_1, x_1^{-1}, x_2^{-1}, \cdots]$.

Then $t.\dim S_S = t.\dim R_R$.

Recall that, for two rings $R \subseteq S$ with the same identity, S is a **ring of right quotients** of R if for every $0 \neq s_1 \in S$, and for every $s_2 \in S$, there exists an $r \in R$ such that $s_1r \neq 0$ and $s_2r \in R$.

The next result due to [140] is stated without proof, which is similar to that of (4.2.8)(3,4). Note that in (4.2.9), the ring $R[x_1, \cdots, x_n, x_1^{-1}, \cdots, x_n^{-1}]$ is a ring of right (and left) quotients of $R[x_1, \cdots, x_n]$, and similarly for (4.2.9)(4).

4.2.10. PROPOSITION. Let R be a ring and S a ring of right quotients of R. Then $t.\dim R_R = t.\dim S_S$.

4.2.11. REFERENCES. Camillo and Zelmanowitz [17]; Dauns [32]; Dauns and Zhou [42]; Goodearl [66]; Zhou [140,141].

4.3 Some Non-Classical Finiteness Conditions

In this section, we discuss several classes of rings which are defined in terms of type dimension, in particular, rings satisfying the type ascending and descending chain conditions on right ideals (see Definitions (4.3.3) and (4.3.22)). The classical analogs of these conditions have been very interesting subjects of research.

We first consider rings over which every cyclic module has finite type dimension. A module M_R is called **weakly injective** (respectively, **weakly R-injective**) if for any finitely generated (respectively, cyclic) submodule $Y \subseteq E(M)$, there exists a submodule $X \subseteq E(M)$ such that $Y \subseteq X \cong M$ (see Jain and López-Permouth [74]). Recall that a ring R is called **right QFD** if every cyclic R-module has finite uniform dimension. It is known that R is right QFD iff every direct sum of weakly injective (respectively, weakly R-injective) modules is weakly injective (respectively, weakly R-injective) (see Al-Huzali, Jain and López-Permouth [4]). The next theorem is a type analog of this result. The following lemma due to Kamal and Müller [77, Lemma 17] is needed.

4.3.1. LEMMA. If $M = M_1 \oplus M_2$ is a direct sum of submodules M_1 and M_2 with M_2 being M_1-injective, then for each submodule $N \leq M$ with $N \cap M_2 = 0$, there exists a submodule $N^{'} \leq M$ with $N \subseteq N^{'}$ and $M = N^{'} \oplus M_2$.

PROOF. Let $\pi_i : M \longrightarrow M_i$ be the projections ($i = 1, 2$). The map

$$\pi : \pi_1(N) \longrightarrow M_2, \ \pi_1 n \longmapsto \pi_2 n \ (n \in N)$$

extends to $f : M_1 \to M_2$. Set

$$N^{'} = \{x + f(x) : x \in M_1\} \leq M_1 \oplus M_2.$$

Then $N^{'} + M_2 = M$, and $N^{'} \cap M_2 = 0$. Thus $M = N^{'} \oplus M_2$. For $n \in N$, $n = \pi_1 n + \pi_2 n = \pi_1 n + f(\pi_1 n) \in N^{'}$. □

A direct sum of modules $\oplus_{i \in I} N_i$ is said to be a **type direct sum** if $N_i \perp N_j$ whenever $i \neq j$ in I.

4.3.2. THEOREM. The following are equivalent for a ring R:

1. Every cyclic module has finite type dimension.
2. Every finitely generated module has finite type dimension.
3. Every type direct sum of weakly injective modules is weakly injective.
4. Every type direct sum of weakly R-injective modules is weakly R-injective.

5. Every type direct sum of injective modules is weakly R-injective.

PROOF. It is obvious that either of (3) and (4) implies (5).

(1) $\Longrightarrow$ (2). By noting that the injective hull of a finitely generated module is a finite direct sum of injective hulls of cyclic modules and then applying (4.1.10).

(2) $\Longrightarrow$ (3). Let $N = \oplus_{i \in I} N_i$ be a type direct sum, where each N_i is weakly injective. Let Y be a finitely generated submodule of $E(N)$. Since Y has finite type dimension, there exists a finite set $\{Y_1, \cdots, Y_n\}$ of pairwise orthogonal atomic submodules of Y such that $Y_1 \oplus \cdots \oplus Y_n \leq_e Y$. Hence, $E(Y_1) \oplus \cdots \oplus E(Y_n) = E(Y)$. For each i $(1 \leq i \leq n)$, $E(Y_i) \cap N \neq 0$. Since $E(Y_i)$ is atomic, it follows that there exists a unique number t_i such that Y_i and $E(N_{t_i})$ have nonzero isomorphic submodules and $E(Y_i) \perp E(\oplus_{j \neq t_i} EN_j)$. Therefore, $E(Y) \perp E(\oplus_{j \notin J} EN_j)$, where $J = \{t_1 \cdots, t_n\}$, implying that $E(Y) \cap E(\oplus_{j \notin J} EN_j) = 0$. Note that

$$E(N) = E(N_{t_1}) \oplus \cdots \oplus E(N_{t_n}) \oplus E(\oplus_{j \notin J} EN_j).$$

By (4.3.1), there exists a submodule P of $E(N)$ such that $E(Y) \subseteq P$ and $E(N) = P \oplus E(\oplus_{j \notin J} EN_j)$. Thus, $P \cong E(N_{t_1}) \oplus \cdots \oplus E(N_{t_n})$. Let

$$X = P + (\oplus_{j \notin J} EN_j) = P \oplus (\oplus_{j \notin J} EN_j) \subseteq E(N).$$

Then $Y \subseteq E(Y) \subseteq P \subseteq X \cong \oplus_{i \in I} E(N_i)$. Write $X = \oplus_{i \in I} E(X_i)$, where $X_i \cong N_i$ for each i. Then there exists a finite subset K of I such that $Y \subseteq \oplus_{i \in K} E(X_i) = E(\oplus_{i \in K} X_i)$. Since all X_i are weakly-injective, it can easily be proved that $\oplus_{i \in K} X_i$ is weakly injective. Thus there exists a $Z \subseteq E(\oplus_{i \in K} X_i)$ such that $Y \subseteq Z \cong \oplus_{i \in K} X_i$. Let $Q = Z + (\oplus_{i \notin K} X_i)$. Since $Z \perp (\oplus_{i \notin K} X_i)$, $Q = Z \oplus (\oplus_{i \notin K} X_i) \subseteq E(N)$ and $Y \subseteq Q \cong N$. Therefore, N is weakly injective.

(2) $\Longrightarrow$ (4). Similar to the proof of "(2) $\Longrightarrow$ (3)".

(5) $\Longrightarrow$ (1). If a cyclic module X does not have finite type dimension, then X has an essential submodule $\oplus_{i=1}^{\infty} X_i$ with each $X_i \neq 0$ and $X_i \perp X_j$ whenever $i \neq j$. Since $X \subseteq E(X) = E(\oplus_i X_i) = E(\oplus_i EX_i)$ and, by (5), $\oplus_i E(X_i)$ is weakly R-injective, we have an embedding $X \hookrightarrow \oplus_{i=1}^{\infty} E(X_i)$, giving $X \hookrightarrow \oplus_{i=1}^{m} E(X_i)$ for some $m > 0$. Thus, X_{m+1} is embeddable in $\oplus_{i=1}^{m} E(X_i)$, implying X_{m+1} and X_i have nonzero isomorphic submodules for some $i \leq m$. This is a contradiction. □

A classical result says that a ring R is right Noetherian iff every direct sum of injective R-modules is injective. Next, we characterize the rings R for which every type direct sum of injective R-modules is injective. We need the following definition, due to [141].

4.3.3. DEFINITION. A module M is said to satisfy t-**ACC** if, for any ascending chain

$$X_1 \subseteq X_2 \subseteq \cdots \subseteq X_n \subseteq \cdots$$

of submodules of M, $t.\dim(\oplus_i M/X_i) < \infty$, and the ring R is said to satisfy (right) t-ACC if R_R satisfies t-ACC.

4.3.4. LEMMA. Let N be a submodule of M. Then M satisfies t-ACC if and only if N and M/N both satisfy t-ACC.

PROOF. One direction is obvious. Suppose that N and M/N satisfy t-ACC. Let $M_1 \subseteq M_2 \subseteq \cdots$ be a chain of submodules of M. Then

$$N \cap M_1 \subseteq N \cap M_2 \subseteq \cdots,$$
$$(N + M_1)/N \subseteq (N + M_2)/N \subseteq \cdots.$$

Note that $N/(N \cap M_i) \cong (N + M_i)/M_i$ and

$$(M/N)/[(N + M_i)/N] \cong M/(N + M_i).$$

We see that $t.\dim[\oplus_i(N + M_i)/M_i] < \infty$ and $t.\dim[\oplus_i M/(N + M_i)] < \infty$. Consider the exact sequence

$$0 \longrightarrow \oplus_i(N + M_i)/M_i \longrightarrow \oplus_i M/M_i \longrightarrow \oplus_i M/(N + M_i) \longrightarrow 0.$$

By (4.1.10)(1), $t.\dim(\oplus_i M/M_i) < \infty$. So M satisfies t-ACC. □

4.3.5. DEFINITION. Recall that a **local direct summand** of a module M is a set $\{X_\lambda : \lambda \in \Lambda\}$ of submodules of M such that the sum $\Sigma\{X_\lambda : \lambda \in \Lambda\} = \oplus_{\lambda\in\Lambda} X_\lambda$ is direct and every finite (direct) sum of the X_λ is a direct summand of M. If, in addition, all the X_λ are type submodules of M, then $\{X_\lambda : \lambda \in \Lambda\}$ is called a **local type direct summand**, or just a **local type summand**.

A ring R is said to be **indecomposable** if R cannot be decomposed as a direct sum of two nonzero ideals. A module M is said to be t-**indecomposable** if M is not a direct sum of two nonzero orthogonal submodules. Every atomic module is t-indecomposable; but the converse is clearly false.

4.3.6. PROPOSITION. If every local type summand of M is a summand, then M is a direct sum of pairwise orthogonal t-indecomposable modules.

PROOF. By Zorn's Lemma, there exists a maximal local type summand $\mathcal{F} = \{X_\lambda : \lambda \in \Lambda\}$ of M, where each X_λ is t-indecomposable. By our assumption, $M = X \oplus Y$, where $X = \sum_{\lambda\in\Lambda} X_\lambda$. Note that for each λ, $X_\lambda \perp Y$ and hence $X \perp Y$. By (4.1.2)(3), we have $Y \leq_t M$. If $Y \neq 0$, then we take $0 \neq y \in Y$. By Zorn's Lemma, there exists a maximal local type summand $\mathcal{G}$ of Y such that $y \notin A = \Sigma_{N\in\mathcal{G}} N$. By (4.1.5)(4), $\mathcal{G}$ is a local type summand of M and hence $Y = A \oplus B$ for some $B(\neq 0)$. Note that $A \perp B$ and hence $B \leq_t Y$. It follows that $B \leq_t M$ by (4.1.5)(4). By the maximality of $\mathcal{F}$, B is not t-indecomposable, and thus $B = B_1 \oplus B_2$ with $B_i \neq 0$ and $B_1 \perp B_2$. Noting $\mathcal{G} \cup \{B_i\}$ is a local type summand of Y, we have $y \in A \oplus B_i$ for $i = 1, 2$. It follows that $y \in (A \oplus B_1) \cap (A \oplus B_2) = A$, a contradiction. □

4.3.7. ORTHOGONAL PROJECTION ARGUMENT. Suppose that M_i, $i \in I$, is a family of pairwise orthogonal modules, and that V is a submodule of $E(\oplus_{i\in I} M_i)$. Then

1. $\oplus_{i\in I}(V \cap M_i) \leq_e V$.

2. For any $0 \neq x \in E(\oplus_{i\in I} M_i)$, there exists an $i \in I$, $r_0 \in R$ with $0 \neq xr_0 R \subseteq M_i$.

PROOF. (2) Choose $r_0 \in \mathcal{S} := \{r \in R : 0 \neq xr \in \oplus_{i\in I} M_i\}$ such that the length n of xr_0 is a minimum, where $0 \neq xr_0 = x_1 + \cdots + x_n$, $0 \neq x_k \in M_{i(k)}$ for $1 \leq k \leq n$, and $i(1), \ldots, i(n)$ are all distinct. Then $(xr_0)^\perp = \bigcap_{i=1}^n x_i^\perp \neq R$. Suppose that $2 \leq n$, and that for $i \neq j \leq n$, there exists $b \in x_i^\perp \backslash x_j^\perp$. Then $0 \neq xr_0 b$ has length $\leq n-1$, a contradiction. Consequently, $x_i^\perp = x_j^\perp = (xr_0)^\perp$ for all $i, j \leq n$. Thus, $x_1^\perp = x_2^\perp$ and

$$M_{i(1)} \supseteq x_1 R \cong R/x_1^\perp = R/x_2^\perp \cong x_2 R \subseteq M_{i(2)}$$

contradicts that $M_{i(1)} \perp M_{i(2)}$. Hence $n = 1$, and $0 \neq xr_0 = x_1 \in M_{i(1)}$, or $0 \neq xr_0 R \subseteq M_{i(1)}$ as required.

(1) If (1) is false, then $xR \oplus [\oplus_{I\in I}(V \cap M_i)] \leq V$ for some $0 \neq x \in V$. By (2), for some $r_0 \in R$ and $i \in I$, $0 \neq xr_0 \in M_i$. But then $0 \neq xr_0 \in M_i \cap V$ is a contradiction. Consequently, $\oplus_{i\in I}(V \cap M_i) \leq_e V$ is essential. □

4.3.8. REMARKS. (a) The above can be applied in the category $\sigma[M]$, when $M_i \in \sigma[M]$ and $V \leq E_M(\oplus_{i\in I} M_i) \leq_e E(\oplus_{i\in I} M_i)$.

(b) In the ordinary projection argument (2.3.3), the orthogonality hypothesis $M_i \perp M_j$ is dropped. But then two things happen. In (2), we get only the weaker conclusion that $0 \neq xr_0 R \cong yR \subseteq M_i$ for some $0 \neq y \in M_i$. And conclusion (1) is false. □

The next result is not only used later, but is of independent interest.

4.3.9. PROPOSITION. Let $\{X_\lambda : \lambda \in \Lambda\}$ be a local type summand of a module M and $X = \oplus\{X_\lambda : \lambda \in \Lambda\}$. If $X^{tc} \neq X$, then there exists a sequence $\{y_i \in X^{tc} : i = 0, 1, \cdots\}$ such that $y_0^\perp \subset y_1^\perp \subset y_2^\perp \subset \cdots$ and $t.\dim(\oplus_{i\geq n} R/y_i^\perp) = \infty$ for all $n \geq 0$. In particular, if R satisfies t-ACC, then every local type summand of M is a type submodule of M.

PROOF. By (4.1.5)(3), $X^{tc} = X^c$ and so $X \neq X^c$. There exists a $y_0 \in X^c \backslash X$. By (4.3.7), $\oplus_{\lambda\in\Lambda}(y_0 R \cap X_\lambda) \leq_e y_0 R$, and there is a $\lambda(1) \in \Lambda$ and an $r_1 \in R$ with $0 \neq y_0 r_1 \in y_0 R \cap X_{\lambda(1)}$. Since $X_{\lambda(1)} \subseteq^\oplus M$, write

$$\begin{aligned} &X^c = X_{\lambda(1)} \oplus Y_1 \text{ and} \\ &y_0 = x_1 + y_1, \quad \text{where} \\ &x_1 \in X_{\lambda(1)},\, y_1 \in Y_1,\, y_1 \notin X, \text{ and } y_0^\perp \subseteq y_1^\perp. \end{aligned}$$

Since $y_1r_1 = y_0r_1 - x_1r_1 \in X_{\lambda(1)} \cap Y_1 = 0$, it follows that $r_1 \in y_1^{\perp} \backslash y_0^{\perp}$, and the inclusion $y_0^{\perp} \subset y_1^{\perp}$ is proper. Since $y_1R \cap X \neq 0$, again by (4.3.7), $0 \neq y_1r_2 \in y_1R \cap X_{\lambda(2)}$ where $r_2 \in R$ and $\lambda(2) \in \Lambda$. Since $y_1R \perp X_{\lambda(1)}$, $\lambda(2) \neq \lambda(1)$. Write as before

$$\begin{aligned} &X^c = X_{\lambda(1)} \oplus X_{\lambda(2)} \oplus Y_2, \\ &y_1 = x_2 + y_2, \\ &x_2 \in X_{\lambda(1)} \oplus X_{\lambda(2)},\ y_2 \in Y_2, \text{ and } y_1^{\perp} \subseteq y_2^{\perp}. \end{aligned}$$

If $y_2 \in X$, then $y_1 \in X$, a contradiction. So $y_2 \notin X$. From $y_2r_2 = y_1r_2 - x_2r_2 \in (X_{\lambda(1)} \oplus X_{\lambda(2)}) \cap Y_2 = 0$, it follows that $r_2 \in y_2^{\perp} \backslash y_1^{\perp}$. Again, $y_1^{\perp} \subset y_2^{\perp}$ is proper.

Repetition of the same process will yield $0 \neq y_2r_3 \in y_2R \cap X_{\lambda(3)}$ where $r_3 \in R$ and $\lambda(3) \in \Lambda$ such that

$$\begin{aligned} &X^c = X_{\lambda(1)} \oplus X_{\lambda(2)} \oplus X_{\lambda(3)} \oplus Y_3, \\ &y_2 = x_3 + y_3, \\ &x_3 \in X_{\lambda(1)} \oplus X_{\lambda(2)} \oplus X_{\lambda(3)},\ y_3 \in Y_3. \end{aligned}$$

Note that $y_0 = x_1 + y_1 = x_1 + x_2 + y_2 = x_1 + x_2 + x_3 + y_3$. As before, $r_3 \in y_3^{\perp} \backslash y_2^{\perp}$, and $y_1^{\perp} \subset y_2^{\perp} \subset y_3^{\perp}$. Furthermore, $y_0R \cap X_{\lambda(1)}, y_1R \cap X_{\lambda(2)}$, and $y_2R \cap X_{\lambda(3)}$ are pairwise orthogonal.

Repeating this process we obtain a sequence $\{y_0, y_1, \cdots\}$ of elements in X^c and a sequence $\{\lambda(1), \lambda(2), \cdots\} \subseteq \Lambda$ of distinct elements, with

$$0 \neq y_{i-1}R \cap X_{\lambda(i)} \subseteq X_{\lambda(i)} \text{ and } y_i^{\perp} \backslash y_{i-1}^{\perp} \neq \emptyset \text{ for all } i \geq 1.$$

Since $\{y_{i-1}R \cap X_{\lambda(i)}\}_{i=1}^{\infty}$ are pairwise orthogonal,

$$t.\dim(\oplus_{i \geq n} R/y_i^{\perp}) = t.\dim(\oplus_{i \geq n} y_iR) = \infty$$

for all $n \geq 0$. If R satisfies the t-ACC, this is a contradiction, in which case $X = X^c = X^{tc}$. □

A decomposition $M = \oplus_{i \in I} M_i$ is said to **complement type summands** if, for every type summand N of M (i.e., a type submodule and a summand), there exists a subset $J \subseteq I$ such that $M = N \oplus M(J)$, where $M(J) = \oplus_{j \in J} M_j$.

4.3.10. LEMMA. The following are equivalent for an injective module M:

1. M is a direct sum of atomic modules.
2. M has a decomposition that complements type summands.
3. Every local type summand of M is a summand.

PROOF. (1) $\Longrightarrow$ (2). We may assume $M = \oplus_I M_i$, where each M_i is atomic and $M_i \perp M_j$ if $i \neq j$. Let N be a type summand of M and let

$$J = \{j \in I : M_j \cap N = 0\}.$$

Then $N \cap M(J) = 0$ by (4.1.5)(2). Thus, $M = N \oplus M(J) \oplus X$ for some X. If $X \neq 0$, then $0 \neq Y \hookrightarrow M_i$ for some $Y \subseteq X$ and some i. Since $M(J) \perp X$ ($M(J)$ is a type submodule), $i \in I \backslash J$. It follows that X has a nonzero submodule embeddable in N. Then $N \cap X \neq 0$ by (4.1.5)(2). This is a contradiction. So $X = 0$ and $M = N \oplus M(J)$.

(2) $\Longrightarrow$ (3). Let $M = \oplus_I M_i$ be a decomposition that complements type summands. We first show that each M_i is an atomic module. If M_k is not atomic, then there exist nonzero submodules A and B of M_k such that $A \perp B$. Let $\mathcal{K} = d(A)$. For each $i \in I$, let $X_i \oplus Y_i \leq_e M_i$ with $X_i \in \mathcal{K}$ and $Y_i \in c(\mathcal{K})$. Set $X = \oplus_I X_i$ and $Y = \oplus_I Y_i$. Then $X \perp Y$ and $X \oplus Y \leq_e M$. So the complement closure X^c of X in M is a type submodule of M by (4.1.2)(3), and hence a type summand of M. Therefore, by (2), $M = X^c \oplus M(J)$ for some $J \subseteq I$. Note that $k \notin J$ since $X^c \perp M(J)$. This implies that $B \hookrightarrow M(I \backslash J) \cong X^c$, contradicting the fact that $A \perp B$. So each M_i is atomic. We now can choose a partition of I: $\{I_s : s \in S\}$, such that $M = \oplus_S N_s$ with each $N_s = \oplus_{I_s} X_i$ atomic and $N_{s_1} \perp N_{s_2}$ whenever $s_1 \neq s_2$. From the proof of "(1) $\Longrightarrow$ (2)", the decomposition $M = \oplus_S N_s$ complements type summands.

Let $Z = \oplus_{\lambda \in \Lambda} Z_\lambda$ be any local type summand of M. Then any complement closure Z^c of Z in M is a type submodule by (4.1.5)(3), and $M = Z^c \oplus C$ for some $C \leq_t M$. It follows that $(\oplus_{\lambda \in \Lambda} Z_\lambda) \oplus C \leq_e M$ is also a local type summand of M. If $Z \oplus C$ is a summand of M, then so is Z. Therefore, without loss of generality, assume that $Z \leq_e M$, and we need to show that $Z = M$.

Assume $Z \neq M$. We then construct a sequence $m_i \in N_{s(i)}$ for distinct $s(i)$ where $m_i \notin Z$ with $m_1^\perp \subset m_2^\perp \subset \ldots \subset m_i^\perp \subset \ldots \subseteq R$ a strictly ascending chain. Since $M \backslash Z = (\oplus_{s \in S} N_s) \backslash Z \neq \emptyset$, there exists a $0 \neq m_1 \in N_{s(1)} \backslash Z$. Let $0 \neq m_1 r_1 \in Z_F$ for $r_1 \in R$, where $Z_F = \oplus\{Z_\lambda : \lambda \in F\}$ for some finite subset $F \subseteq \Lambda$, and $M = Z_F \oplus N(J)$ for some $J \subseteq S$ where $N(J) = \oplus\{N_s : s \in J\}$. Write $m_1 = z + \Sigma n_s$ for $z \in Z_F$ where $\Sigma n_s = \Sigma_{s \in J} n_s \in N(J)$ with only a finite number of nonzero $n_s \in N_s$ if $s \in J$. Then $m_1^\perp \subseteq n_s^\perp$ for all $s \in J$. Since $m_1 \notin Z$, there exists an $s(2) \in J$ such that $n_{s(2)} \notin Z$. Also $(\Sigma_J n_s)^\perp = \bigcap_{s \in J} n_s^\perp \subseteq n_s^\perp$ for all $s \in J$. Furthermore,

$$(\Sigma n_s) r_1 = m_1 r_1 - z r_1 \in Z_F \cap N(J) = 0,$$

and therefore,

$$r_1 \in (\Sigma n_s)^\perp \backslash m_1^\perp \subseteq n_{s(2)}^\perp \backslash m_1^\perp .$$

Set $m_2 = n_{s(2)} \in N_{s(2)} \backslash Z$. We claim $s(1) \neq s(2)$. If $s(1) = s(2) \in J$, $m_1 \in N(J)$. Then

$$0 \neq m_1 r_1 = z r_1 + (\Sigma_{s \in J} n_s) r_1 \in Z_F \oplus N(J)$$

implies $z r_1 = 0$, and $0 \neq m_1 r_1 \in Z_F \cap N(J)$, a contradiction. Thus so far we have $m_i \in N_{s(i)} \backslash Z$ $(i = 1, 2)$ for $s(1) \neq s(2)$ with $m_1^\perp \subset m_2^\perp$ properly. The reader should be warned from repeating the above argument verbatim with m_1 replaced by m_2 to obtain an $m_3 \in N_{s(3)}$ with $s(2) \neq s(3)$ and $m_1^{\perp} \subset m_2^{\perp} \subset m_3^{\perp}$. The problem is that perhaps $s(3) = s(1)$.

However, $s(1) \notin J$ as shown above, and $N(J) = \oplus_{s \in J} N_s$ complements type summands of $N(J)$ (as proved in the proof of (1) $\Longrightarrow$ (2)). Notice $Z = Z_F \oplus Z_{\Lambda \backslash F}$ where $Z_{\Lambda \backslash F} = \oplus\{Z_\lambda : \lambda \in \Lambda \backslash F\}$. We have

$$Z_{\Lambda \backslash F} \cap N(J) = (Z_{\Lambda \backslash F} \cap Z_F) \oplus (Z_{\Lambda \backslash F} \cap N(J))$$

is essential in $Z_{\Lambda \backslash F}$ by (4.3.7). So

$$Z_{\Lambda \backslash F} \subseteq E(Z_{\Lambda \backslash F}) = E(Z_{\Lambda \backslash F} \cap N(J)) \subseteq N(J),$$

since $N(J)$ is injective. Thus $Z_{\Lambda \backslash F} = \oplus\{Z_\lambda : \lambda \in \Lambda \backslash F\}$ is a local type summand of $N(J)$. Now $s(2) \in J$, $m_2 \in N_{s(2)} \subseteq N(J)$, and proceeding as before we get $m_3 \in N_{s(3)}$ with $m_2^{\perp} \subset m_3^{\perp}$ and $s(2) \neq s(3) \in J$. Since $s(1) \notin J$ and $s(2), s(3) \in J$, they are all distinct. Now an induction shows that there exists a sequence $m_i \in N_{s(i)}$ for distinct $s(i)$ where $m_i \notin Z$ with $m_1^{\perp} \subset m_2^{\perp} \subset \cdots \subset m_i^{\perp} \subset \cdots \subseteq R$ a strictly ascending chain.

Now let $x = (m_i) \in \Pi_{i=1}^{\infty} N_{s(i)}$ and $I = \cup_{i=1}^{\infty} m_i^{\perp}$ and consider the homomorphism $f : m_1 I \longrightarrow \oplus_{i=1}^{\infty} N_{s(i)}$ defined by $f(m_1 r) = xr$ for $r \in I$. Then f is well-defined. Since $\oplus_{i=1}^{\infty} N_{s(i)}$ is injective (a summand of M), f can be extended to a map $g : m_1 R \longrightarrow \oplus_{i=1}^{\infty} N_{s(i)}$. Thus, $xI = f(m_1 I) \subseteq g(m_1 R) \subseteq \oplus_{i=1}^{k} N_{s(i)}$ for some $k > 0$. It follows that $m_i I = 0$ for all $i > k$. So $m_{k+1}^{\perp} = m_{k+2}^{\perp} = \cdots$. The contradiction shows that $Z = M$.

(3) $\Longrightarrow$ (1). By (4.3.6). □

4.3.11. LEMMA. If a module M does not satisfy t-ACC, then there exist two sequences $\{X_i : i \in \mathbb{N}\}$ and $\{Y_i : i \in \mathbb{N}\}$ of submodules of M, such that $X_1 \subset X_2 \subset \cdots$ is a strictly ascending chain and, for each i, $X_i \subset Y_i \subseteq X_{i+1}$ and $(Y_i/X_i) \perp (Y_j/X_j)$ whenever $i \neq j$.

PROOF. We proceed by considering two cases.

Case 1: $t.\dim(M/N) = \infty$ for some $N \subseteq M$. Then there exists a sequence $\{N_i : i \in \mathbb{N}\}$ of submodules of M such that $(N_i/N) \perp (N_j/N)$ whenever $i \neq j$ and all $N_i/N \neq 0$. If we let $X_1 = N_1$ and, for each $n > 1$, choose $X_n = N_1 + \cdots + N_n$ and $Y_{n-1} = X_n$, then those X_n's and Y_n's are as required.

Case 2: $t.\dim(M/N) < \infty$ for all $N \subseteq M$. Since M does not satisfy t-ACC, there exists a sequence $\{N_i : i \in \mathbb{N}\}$ of submodules of M such that $N_1 \subseteq N_2 \subseteq \cdots$ and $t.\dim(\oplus_{i=1}^{\infty} M/N_i) = \infty$. Since $t.\dim(M/N_i) < \infty$ for each i, we may assume without loss of generality that we have a strictly ascending chain $N_1 \subset N_2 \subset \cdots$ and, for each i, there exists a $0 \neq P_i/N_i \subseteq M/N_i$ such that $(P_i/N_i) \perp (P_j/N_j)$ if $i \neq j$. Let $L = \cup N_i$ and $V = \{i \in \mathbb{N} : P_i \cap L = N_i\}$. If V is an infinite set, then $P_i/N_i \cong (P_i + L)/L \hookrightarrow M/L$ for all $i \in V$, and hence $t.\dim(M/L) = \infty$. This is a contradiction. Therefore, there exists a number n such that $N_i \subset P_i \cap L$ for all $i \geq n$. Then, if we choose $A_i = N_{i+n}$

and $B_i = P_{i+n} \cap L$ for each $i > 0$, we have $A_1 \subset A_2 \subset \cdots$ such that $A_i \subset B_i \subseteq \cup A_j$ for each i and $(B_i/A_i) \perp (B_j/A_j)$ when $i \neq j$. Choose a sequence $\{1 = n_1 < n_2 < \cdots\}$ of positive integers, such that there exists a $b_i \in B_{n_i} \cap A_{n_{i+1}}$ but $b_i \notin A_{n_i}$ for each i. Then $X_i = A_{n_i}$ and $Y_i = b_i R + A_{n_i}$ $(i = 1, 2, \cdots)$ are required submodules. □

The next theorem, which appeared in [141], gives type analogs of some well-known characterizations of right Noetherian rings. A **TS-module** is any module whose type submodules are summands.

4.3.12. THEOREM. The following are equivalent for a ring R:

1. R satisfies t-ACC.
2. For every (countable) family $\{M_i : i \in I\}$ of pairwise orthogonal modules, $\oplus_{i \in I} E(M_i)$ is injective.
3. Every injective module is a direct sum of atomic modules.
4. Every injective module has a decomposition that complements type summands.
5. Every TS-module is a direct sum of atomic modules.
6. Every module contains a maximal injective type submodule.

PROOF. (1) $\Longrightarrow$ (5). By (4.3.6) and (4.3.9).

(5) $\Longrightarrow$ (3). It is clear.

(3) $\Longleftrightarrow$ (4). By (4.3.10).

(4) $\Longrightarrow$ (6). Let M be a module. Note that 0 is an injective type submodule of M. By Zorn's Lemma, there exists a maximal independent set $\mathcal{F}$ of injective type submodules of M. Write $A = \oplus_{\mathcal{F}} X$. Let A^c be a complement closure of A in M. Note that each X is a type summand of $E(A^c)$ and hence A is a local type summand of $E(A^c)$. By (4) and (4.3.10), A is a summand of $E(A^c)$, implying $A = A^c = E(A^c)$. Therefore, A is an injective type submodule of M (by 4.1.5(3)). By the choice of $\mathcal{F}$, A is a maximal injective type submodule of M.

(6) $\Longrightarrow$ (2). Let $\{M_i : i \in I\}$ be a family of pairwise orthogonal modules and $M = \oplus_{i \in I} E(M_i)$. Let N be a maximal injective type submodule of M. For each i, write $E(M_i) = E(EM_i \cap N) \oplus X_i$. Then $N \cap X_i = 0$ and hence $N \perp X_i$ since $N \leq_t M$. Therefore, $E(EM_i \cap N) \perp X_i$. By (4.1.2)(3), X_i is an injective type submodule of $E(M_i)$. Since $E(M_i) \leq_t M$, X_i is an injective type submodule of M by (4.1.5)(4). Write $M = (N \oplus X_i) \oplus Y$. Since $N \leq_t M$ and $X_i \leq_t M$, $Y \perp N$ and $Y \perp X_i$, and thus $Y \perp (N \oplus X_i)$. By (4.1.2)(3), $N \oplus X_i \leq_t M$. From the choice of N, we have $X_i = 0$ and hence $E(M_i) \cap N \leq_e E(M_i)$ for each i. This implies that $N \leq_e M$. Therefore, $M = N$ is injective.

(2) $\Longrightarrow$ (1). If R does not satisfy t-ACC, then, by (4.3.11), there exist two sequences $\{I_i : i \in \mathbb{N}\}$ and $\{J_i : i \in \mathbb{N}\}$ of right ideals of R such that

$I_1 \subset I_2 \subset \cdots$, and for each i, $I_i \subset J_i \subseteq I_{i+1}$ and $(J_i/I_i) \perp (J_j/I_j)$ if $i \neq j$. Then $E = \oplus_{i=1}^{\infty} E(J_i/I_i)$ is injective by (2). Let l_i be the inclusion of J_i/I_i to $E(J_i/I_i)$. Then there exists a homomorphism $f_i : R/I_i \longrightarrow E(J_i/I_i)$ that extends l_i. Let $I = \cup I_i$ and define $f : I \longrightarrow E$ by $(\pi_i f)(a) = f_i(a + I_i)$, where π_i is the projection of E onto $E(J_i/I_i)$. Since E is injective, there exists a $g : R \longrightarrow E$ that extends f. It follows that $f(I) \subseteq g(R) \subseteq \oplus_{i=1}^{m} E(J_i/I_i)$ for some m. Then, for any $a \in J_{m+1}$, we have

$$0 = \pi_{m+1} f(a) = f_{m+1}(a + I_{m+1}) = l_{m+1}(a + I_{m+1}) = a + I_{m+1},$$

implying $J_{m+1} = I_{m+1}$. This is a contradiction. □

It is easy to give a right QFD-ring that is not right Noetherian; for example, the trivial extension of $\mathbb{Z}$ and the $\mathbb{Z}$-module $\mathbb{Z}_{2^\infty}$. Below, we construct a class of rings whose cyclics have finite type dimension but they do not satisfy t-ACC.

4.3.13. EXAMPLE. Let F be a field. Let R be the formal power series ring of all $r = \Sigma\{c_q x^q : q \in \mathbb{Q}^+\}$, where $c_q \in F$, $\mathbb{Q}$ is the set of the rationals and $\mathbb{Q}^+ = \{q \in \mathbb{Q} : 0 \leq q\}$, such that the support of $r = supp(r) = \{q : c_q \neq 0\}$ satisfies the descending chain condition. (Need the DCC in order for power series multiplication to make sense.) Define $v(r) =$ minimum of $supp(r)$ for all $0 \neq r \in R$. Then $v : R\backslash\{0\} \longrightarrow \mathbb{Q}^+$ satisfies the conditions:

1. $v(kr) = v(r)$ for $k \in F$ with $kr \neq 0$.
2. $v(r_1 + r_2) \geq min\{v(r_1), v(r_2)\}$ provided that $r_1 + r_2 \neq 0$.
3. $v(r_1 r_2) = v(r_1) + v(r_2)$.

Thus, R is a **valuation domain**. (Note that v can be uniquely extended to a **valuation** w of the quotient field of R defined by the formula $w(r_1/r_2) = w(r_1)/w(r_2)$.) Since the ideals of R are totally ordered by inclusion, every cyclic R-module has uniform dimension 1 (hence has finite type dimension). Define $v(0) = \infty$. For any $\eta \in \mathbb{R}^+$ where $\mathbb{R}$ is the set of the real numbers and $\mathbb{R}^+ = \{r \in \mathbb{R} : 0 \leq r\}$, let

$$P_\eta = \{r \in R : v(r) > \eta\}.$$

Then P_η is an ideal of R. We prove that $R/P_{\sqrt{2}} \perp R/P_{\sqrt{3}}$. In fact, if not then there exist $a \in R\backslash P_{\sqrt{2}}$ and $b \in R\backslash P_{\sqrt{3}}$ such that

$$(aR + P_{\sqrt{2}})/P_{\sqrt{2}} \overset{\phi}{\cong} (bR + P_{\sqrt{3}})/P_{\sqrt{3}} \text{ and } \phi(a + P_{\sqrt{2}}) = b + P_{\sqrt{3}}.$$

Then $q_1 = v(a) \leq \sqrt{2}$ and $q_2 = v(b) \leq \sqrt{3}$. Note that,

$$\begin{aligned} (a + P_{\sqrt{2}})^\perp &= \{r \in R : ar \in P_{\sqrt{2}}\} \\ &= \{r \in R : v(a) + v(r) > \sqrt{2}\} \\ &= \{r \in R : v(r) > \sqrt{2} - q_1\} \end{aligned}$$

and, similarly,

$$(b+P_{\sqrt{3}})^{\perp}=\{r\in R: v(r)>\sqrt{3}-q_2\}.$$

It follows that

$$\{r\in R: v(r)>\sqrt{2}-q_1\}=\{r\in R: v(r)>\sqrt{3}-q_2\}.$$

Note that $v(r)$ attains every non-negative rational and $\mathbb{Q}$ is dense in $\mathbb{R}$. It follows $\sqrt{2}-q_1=\sqrt{3}-q_2$, and so $\sqrt{2}-\sqrt{3}=q_1-q_2\in\mathbb{Q}$, a contradiction. Hence $R/P_{\sqrt{2}}\perp R/P_{\sqrt{3}}$. Let $\{\eta_n : n=1,2,\cdots\}\subseteq\mathbb{R}^+\backslash\mathbb{Q}^+$ be pairwise rationally independent, i.e., $\eta_i-\eta_j\notin\mathbb{Q}$ whenever $i\neq j$. We can assume that $\eta_i-\eta_j>0$ whenever $i<j$. Set $P_{\eta_i}=\{r\in R: v(r)>\eta_i\}$. Then we have a strictly ascending chain of ideals:

$$P_{\eta_1}\subset P_{\eta_2}\subset\cdots\subset P_{\eta_n}\subset\cdots$$

with, as above, $R/P_{\eta_i}\perp R/P_{\eta_j}$ if $i\neq j$. Thus, R does not satisfy t-ACC. □

Note that (4.3.13) also holds for any commutative valuation domain whose **value group** Γ is a subgroup of a totally ordered abelian group G such that Γ is "dense" in G, i.e., for any $g_1<g_2$ in G there exists a $g\in\Gamma$ with $g_1<g<g_2$ and there exists an infinite chain $g_1<g_2<\cdots$ from G with $g_j-g_i\notin\Gamma$ for all $j>i$. In above example, these $P_{\sqrt{2}}, P_{\sqrt{3}}, P_{\eta_n}$ are not prime.

Suppose that there is a strictly ascending chain $P_1\subset P_2\subset\cdots$ of prime ideals of a commutative ring R. Note that, for any prime ideal P and $a\in R\backslash P$, $(a+P)^{\perp}=P$. Thus, if $i\neq j$, $R/P_i\perp R/P_j$. So R does not satisfy t-ACC. The next example is a valuation domain containing a strictly ascending chain of prime ideals.

4.3.14. EXAMPLE. Let Λ be a **totally ordered abelian group** ($a<b\Longrightarrow a+c<b+c$, $a,b,c\in\Lambda$). Let $\Lambda^+=\{g\in\Lambda: 0\leq g\}$, a **semigroup**. Let F be a field and R be the formal power series ring of all $r=\Sigma_{s\in\Lambda^+}r(s)x^s$, where $r(s)\in F$, such that the support of $r=supp(r)=\{s\in\Lambda^+: r(s)\neq 0\}$ satisfies the DCC. Define $v(r)=min\{supp(r)\}$ for all $0\neq r\in R$. Then $v: R\backslash\{0\}\longrightarrow\Lambda^+$ is a valuation satisfying (1), (2), and (3) in (4.3.13). So R is a valuation domain and hence every cyclic R-module has finite type dimension. Define $v(0)=\infty$. For a **convex subsemigroup** H of Λ^+ ($s\leq h$ with $s\in\Lambda^+, h\in H\Longrightarrow s\in H$),

$$P=\{r\in R: v(r)\in(\Lambda^+\backslash H)\cup\{\infty\}\}$$

is a prime ideal. (Also every prime ideal is of this kind.) To see this, let $r,s\in R$ and $a,b\in P$. Since H is convex, the fact that ar and $a+b$ are in P follows from that $v(ar)=v(a)+v(r)\geq v(a)$ (since $v(r)\geq 0$) $\notin H$ and that $v(a+b)\geq min\{v(a),v(b)\}=v(a)$ (say $v(a)\leq v(b)$)$\notin H$. If $rs\in P$, $r\notin P$, and $s\notin P$, then $v(r)$ and $v(s)$ are in H and $v(rs)=v(r)+v(s)\notin H$. The fact that H is a **subsemigroup** gives $v(r) + v(s) \in H$, a contradiction. So P is prime. Below we construct a totally ordered abelian group Λ with an infinite descending chain of **convex subgroups**:

$$\cdots \subset \Lambda_n \subset \cdots \subset \Lambda_1 \subset \Lambda_0 = \Lambda.$$

Thus,

$$\Lambda^+ \backslash \Lambda_1^+ \subset \Lambda^+ \backslash \Lambda_2^+ \subset \Lambda^+ \backslash \Lambda_3^+ \subset \cdots$$

is a chain of convex subsemigroups. Hence

$$P_n = \{r \in R : v(r) \in (\Lambda^+ \backslash \Lambda_n^+) \cup \{\infty\}\} \ (n = 1, 2, \cdots)$$

are the required prime ideals.

Our totally ordered abelian group Λ is given by $\Lambda = \Pi_1^\infty A_i$, where each $A_i = \mathbb{Z}$, the abelian group of the integers, lexicographically ordered. Write an element of Λ as a mapping from $\mathbb{N}$ to $\mathbb{Z}$. Then, for $f, g \in \Lambda$, $f > g$ means there exists some $i \in \mathbb{N}$ such that $f(j) = g(j)$ for all $j < i$ and $f(i) > g(i)$. Let $\Lambda_n = \{f \in \Lambda : f(i) = 0 \text{ for all } i \leq n\}$. Then $\Lambda_n \subset \cdots \subset \Lambda_2 \subset \Lambda_1$ gives a chain of **convex subgroups** of Λ. □

Teply [115] characterized the rings R for which every direct sum of nonsingular injective R-modules is injective and proved they are just those rings R for which $R/Z_2(R_R)$ is of finite uniform dimension. The type analog of this can be stated as follows.

4.3.15. THEOREM. The following are equivalent for a ring R:

1. $R/Z_2(R_R)$ is of finite type dimension.
2. Every cyclic nonsingular R-module is of finite type dimension.
3. Every finitely generated nonsingular R-module is of finite type dimension.
4. Every type direct sum of nonsingular weakly injective modules is weakly injective.
5. Every type direct sum of nonsingular weakly R-injective modules is weakly R-injective.
6. Every type direct sum of nonsingular injective modules is weakly R-injective.
7. For any chain $I_1 \subseteq I_2 \subseteq \cdots$ of right ideals of R such that each R/I_i is nonsingular, we have $t.\dim(\oplus_i R/I_i) < \infty$.
8. Every type direct sum of nonsingular injective R-modules is injective.
9. Every nonsingular injective R-module is a direct sum of atomic modules.

10. Every nonsingular injective module has a decomposition that complements type summands.

11. Every nonsingular TS-module is a direct sum of atomic modules.

12. Every nonsingular R-module contains a maximal injective type submodule.

PROOF. (1) $\Longrightarrow$ (2). If $xR \cong R/x^{\perp}$ is a cyclic nonsingular module, then $x^{\perp}$ must be a complement right ideal of R and $Z_2(R_R) \subseteq x^{\perp}$. Let K be a right ideal of R such that $x^{\perp} \oplus K \leq_e R_R$. Then K is isomorphic to an essential submodule of $R/x^{\perp}$ and $K \hookrightarrow R/Z_2(R_R)$. Hence

$$t.\dim(xR) = t.\dim(R/x^{\perp}) = t.\dim(K) \leq t.\dim(R/Z_2(R_R)) < \infty.$$

(2) $\Longrightarrow$ (1) and (8) $\Longrightarrow$ (6). They are clear.

(2) $\Longrightarrow$ (3) $\Longrightarrow$ (4) $\Longrightarrow$ (6) $\Longrightarrow$ (2) and (3) $\Longrightarrow$ (5) $\Longrightarrow$ (6). Similar to the proofs of (4.3.2).

(7) $\Longrightarrow$ (11) $\Longrightarrow$ (9) $\Longrightarrow$ (10) $\Longrightarrow$ (12) $\Longrightarrow$ (8). Similar to the proofs of (4.3.12).

(2) $\Longrightarrow$ (7). If (7) does not hold, then there exists a chain $I_1 \subseteq I_2 \subseteq \cdots$ of right ideals of R such that each R/I_i is nonsingular and $t.\dim(\oplus_i R/I_i) = \infty$. By (2) each R/I_i is of finite type dimension. Then, by the proof of (4.3.11), there exists a sequence $\{1 = n_1 < n_2 < \cdots\}$ of natural numbers and, for each i, there exists a right ideal X_i of R such that $I_{n_i} \subset X_i \subseteq I_{n_{i+1}}$ and $(X_i/I_{n_i}) \perp (X_j/I_{n_j})$ if $i \neq j$. For $i > 1$, there is an R-homomorphism from R/I_{n_1} to R/I_{n_i} with kernel $I_{n_i}/I_{n_1} \neq 0$. Because all R/I_{n_i} are nonsingular, we have that I_{n_i}/I_{n_1} have no proper essential extensions in R/I_{n_1} for all $i > 1$ (i.e., I_{n_i}/I_{n_1} is a complement submodule of R/I_{n_1}). Thus, for each i there exists a $0 \neq Y_i/I_{n_1} \subseteq X_i/I_{n_1}$ such that $(Y_i/I_{n_1}) \cap (I_{n_i}/I_{n_1}) = 0$. We see that

$$0 \neq Y_i/I_{n_1} = Y_i/(Y_i \cap I_{n_i}) \cong (Y_i + I_{n_i})/I_{n_i} \subseteq X_i/I_{n_i}.$$

So $(Y_i/I_{n_1}) \perp (Y_j/I_{n_1})$ for all $i \neq j$. This shows that R/I_{n_1} does not have finite type dimension. □

It is well-known that a nonsingular ring has finite uniform dimension iff its maximal quotient ring is a semisimple ring (see Sandomierski [106]). There is a similar result for a nonsingular ring with finite type dimension.

4.3.16. LEMMA. Let M be an R-module.

1. M is an atomic module iff $E(M)$ is t-indecomposable.

2. If $E = E(M)$ is t-indecomposable, then $\mathrm{End}(E_R)$ is indecomposable. The converse holds if M is nonsingular.

PROOF. Part (1) is obvious. For (2), suppose $S = \text{End}(E)$ is not indecomposable. Then there exists a central idempotent $e \in S$ such that $e \neq 0$ and $e \neq 1$. Then $E = e(E) \oplus (1-e)(E)$. We show that $e(E) \perp (1-e)(E)$. If not, then we have an R-module isomorphism $\pi : e(X) \longrightarrow (1-e)(Y) \neq 0$ for some $X \subseteq E$ and some $Y \subseteq E$. There exists an $f \in S$ such that f extends π. Thus, we have $(1-e)(Y) = f(e(X)) = f(e^2(X)) = e(f(e(X))) = e((1-e)(Y)) = 0$, a contradiction. So $e(E) \perp (1-e)(E)$ and hence E is not t-indecomposable.

For the converse, let $E = E_1 \oplus E_2$ with each $E_i \neq 0$ and $E_1 \perp E_2$. Since E is nonsingular, we have $\text{End}(E) \cong \text{End}(E_1) \times \text{End}(E_2)$. Therefore, $\text{End}(E)$ is not indecomposable. □

4.3.17. PROPOSITION. A nonsingular ring R has $t.\dim(R) = n$ iff the maximal right quotient ring of R is a finite direct sum of n indecomposable right self-injective (von Neumann) regular rings.

PROOF. "$\Longrightarrow$". It is well-known that the maximal right quotient ring of R is $Q = \text{End}(ER)$ ($\cong ER$)). Suppose that $I_1 \oplus \cdots \oplus I_n \leq_e R_R$, where each I_i is a nonzero atomic right ideal of R and $I_i \perp I_j$ (if $i \neq j$). Then $E(R) = E(I_1) \oplus \cdots \oplus E(I_n)$ and $Q \cong \text{End}(EI_1) \times \cdots \times \text{End}(EI_n)$. By [66, 2.22] and (4.3.16), each $\text{End}(EI_i)$ is a regular right self-injective indecomposable ring.

"$\Longleftarrow$". Suppose $Q = E(R) = E_1 \oplus \cdots \oplus E_n$ is a direct sum of regular right self-injective indecomposable rings. Then E_i is an atomic E_i-module by (4.3.16). Therefore, as Q-modules, E_i is atomic and $E_i \perp E_j$ (if $i \neq j$). By [66, 2.7], as R-modules, each $R \cap E_i$ is atomic and $(R \cap E_i) \perp (R \cap E_j)$ (if $i \neq j$). Since $(R \cap E_1) \oplus \cdots \oplus (R \cap E_n) \leq_e R_R$, we have $t.\dim(R) = n$. □

If R has finite type dimension, then R is a finite direct product of indecomposable rings. But, a nonsingular indecomposable ring may not have finite type dimension as the following example shows.

4.3.18. EXAMPLE. Let $Q = \Pi_{i=1}^{\infty} R_i$ be a direct product of rings R_i with $R_i = \mathbb{Z}$ for each i, and R be the subring generated by $\oplus_{i=1}^{\infty} 2R_i$ and 1_Q. Clearly, if $i \neq j$, then $2R_i \perp 2R_j$ as R-modules. So $t.\dim(R) = \infty$. It is easy to see that R is nonsingular and R has no nontrivial idempotents. Therefore, R is indecomposable. □

Next, we will show that a nonsingular ring has the property that every nonzero right ideal contains an atomic right ideal iff its maximal right quotient ring is a direct product of indecomposable right self-injective regular rings. It is interesting to compare this to a well-known result which states that, for a nonsingular ring R, every nonzero right ideal of R contains a uniform right ideal iff the maximal right quotient ring of R is a direct product of **right full linear rings** (see [66, p.92]).

4.3.19. LEMMA. Let M be a module and N a submodule of M.

1. Let $N \leq_t M$ and $N \subseteq X \subseteq M$. Then $X/N \leq_t M/N$ iff $X \leq_t M$; in particular, M/N is atomic iff N is a maximal type submodule of M.

2. M has an atomic submodule iff M has a maximal type submodule; more precisely, N is a maximal type submodule of M iff $N <_t M$ and any (or one) of its complements in M is atomic.

3. If every nonzero submodule of M contains an atomic submodule, then the intersection of all maximal type submodules of M is equal to 0.

4. If $M = X \oplus Y$ is nonsingular and $X \perp Y$, then $\text{Hom}_R(X, Y) = 0$.

5. Let M be nonsingular and N_i $(i = 1, 2)$ be maximal type submodules of M. If $N_1 \neq N_2$, then $(M/N_1) \perp (M/N_2)$.

PROOF. (1) Suppose X is not a type submodule of M. Then there exists $Y \subseteq M$ such that $X \subset Y$ and $X \| Y$. Thus, $X/N \subset Y/N$. For any $0 \neq A/N \subseteq Y/N$, we have $N \perp B$ for some $0 \neq B \subseteq A$ since N is a type submodule of M. It follows from $X \| Y$ that $0 \neq C \cong D$ for some $C \subseteq X$ and $D \subseteq B$. Then $N \perp C$ and hence $0 \neq (C + N)/N \cong (D + N)/N$. Therefore, $(X/N) \| (Y/N)$ and so X/N is not a type submodule of M/N. Conversely, if X/N is not a type submodule of M/N, then there exists $Y/N \subseteq M/N$ such that $X/N \subset Y/N$ and $(X/N) \| (Y/N)$. We prove that X is not a type submodule of M by showing $X \| Y$. For any $0 \neq A \subseteq Y$, we show that A is not orthogonal to X. We may assume $A \cap N = 0$. Then $A \hookrightarrow Y/N$. Because $(X/N) \| (Y/N)$, we have $0 \neq B/N \hookrightarrow A$ for some $B \subseteq X$. Since N is a complement submodule of M, there exists $0 \neq C \subseteq B$ such that $C \cap N = 0$. Thus, $C \hookrightarrow A$. So A is not orthogonal to X, and hence $X \| Y$.

(2) It is routine.

(3) For any $0 \neq x \in M$, choose an atomic submodule A in xR. By Zorn's Lemma, there exists a submodule B of M maximal with respect to the property that $A \subseteq B$ and $A \| B$. It follows that B is a type submodule of M and is atomic. Let N be a complement of B in M. Then N is a maximal type submodule of M by (2). Note that $N \perp B$ and hence $A \perp N$, implying $x \notin N$.

(4) It is clear.

(5) For each i, let A_i be a complement of N_i in M. Then A_i is atomic by (2) and $A_i \| (M/N_i)$. So we only need to show $A_1 \perp A_2$. If not, then $A_1 \| A_2$. Note that each A_i is a type submodule of M (by 4.1.2(3)). It follows that $A_1 \cap A_2 \leq_e A_i$ for each i. Therefore, A_1 and A_2 are complement closures of $A_1 \cap A_2$ in M. But since M is nonsingular, it must be $A_1 = A_2$ by (1.1.7). This implies that $N_1 \| N_2$. Since N_1 and N_2 are type submodules, we have $N_1 \cap N_2 \leq_e N_i$ for each i. So, N_1 and N_2 are complement closures of $N_1 \cap N_2$ in M. Thus, $N_1 = N_2$. □

4.3.20. THEOREM. The following are equivalent for a nonsingular ring R:

1. Every nonzero right ideal of R contains an atomic right ideal.

2. The maximal right quotient ring of R is isomorphic to a direct product of indecomposable right self-injective regular rings.

PROOF. (1) $\Longrightarrow$ (2). Let $\{E_t : t \in I\}$ be the set of all maximal type submodules of $E = E(R)$ and $E_* = \Pi_t(E/E_t)$. We first prove $\text{End}(E) \cong \text{End}(E_*)$. For any $f \in \text{End}(E)$, we have $f(E_t) \subseteq E_t$ for all $t \in I$ by (4.3.19)(4). Define $\phi(f)$ as follows: If $(x_t + E_t) \in E_*$ with $x_t \in E$ for each t, then we let $\phi(f)((x_t + E_t)) = (f(x_t) + E_t)$. We have $\phi(f) \in \text{End}(E_*)$ and ϕ gives a ring homomorphism. By (4.3.19)(3), $\cap_t E_t = 0$. This implies that ϕ is one to one. In order to show ϕ is onto, we write $E = E_t \oplus E_t^{'}$ for each t and use π_t to indicate the canonical isomorphism from $E_t^{'}$ onto E/E_t (i.e., $\pi_t(x_t) = x_t + E_t$ with $x_t \in E_t^{'}$) and let $\pi = \oplus\pi_t : \oplus E_t^{'} \longrightarrow \oplus E/E_t$. For any $\theta \in \text{End}(E_*)$, we have $\theta(E/E_t) \subseteq E/E_t$ by (4.3.19)(4,5). Hence $\theta(\oplus E/E_t) \subseteq \oplus E/E_t$. There exists a homomorphism $h \in \text{End}(E)$ such that h extends the map $\pi^{-1}\theta\pi : \oplus E_t^{'} \longrightarrow \oplus E_t^{'}$. Note that for all $t \in I$, $h(E_t) \subseteq E_t$ and $h(E_t^{'}) \subseteq E_t^{'}$ by (4.3.19)(4). It can easily be checked that $\phi(h) = \theta$. Therefore, $\text{End}(E) \cong \text{End}(E_*)$. Next, we note that $\text{End}(E)$ is the maximal right quotient ring of R and, because of (4.3.19)(4),

$$\text{End}(E_*) \cong \Pi_t\text{End}(E/E_t) \cong \Pi_t\text{End}(E_t^{'}).$$

By (4.3.16) and [66, 2.22], each $\text{End}(E_t^{'})$ is an indecomposable right self-injective regular ring.

(2) $\Longrightarrow$ (1). It is similar to the proof of "$\Longleftarrow$" of (4.3.17). □

4.3.21. EXAMPLE. Let $F_1, F_2, \cdots$ be fields, $R = (\Pi F_n)/(\oplus F_n)$. Then $Z(R_R) = 0$ by [66, Ex.6, p.94]. It is easy to show that every nonzero principal right ideal of R contains two nonzero right ideals which are orthogonal, and therefore R has no atomic right ideals.

The dual of a right Noetherian ring is the notion of a right Artinian ring. In the remaining part of this section, we consider the dual of the t-ACC.

4.3.22. DEFINITION. A module M is said to satisfy t-**DCC** if for any descending chain $X_1 \supseteq X_2 \supseteq \cdots \supseteq X_n \supseteq \cdots$ of submodules, we have $t.\dim(\oplus_i M/X_i) < \infty$. A ring R is said to satisfy (right) t-DCC if R_R satisfies t-DCC.

4.3.23. LEMMA. Let N be a submodule of M. Then M satisfies t-DCC iff N and M/N both satisfy t-DCC.

PROOF. One direction is obvious. Suppose that N and M/N satisfy t-DCC. Let $M_1 \supseteq M_2 \supseteq \cdots$ be a chain of submodules of M. Then

$$N \cap M_1 \supseteq N \cap M_2 \supseteq \cdots,$$
$$(N + M_1)/N \supseteq (N + M_2)/N \supseteq \cdots.$$

Note that $N/(N \cap M_i) \cong (N + M_i)/M_i$ and

$$(M/N)/[(N + M_i)/N] \cong M/(N + M_i).$$

We see that $t.\dim[\oplus_i(N+M_i)/M_i] < \infty$ and $t.\dim[\oplus_i M/(N+M_i)] < \infty$. Consider the exact sequence

$$0 \longrightarrow \oplus_i(N+M_i)/M_i \longrightarrow \oplus_i M/M_i \longrightarrow \oplus_i M/(N+M_i) \longrightarrow 0.$$

By (4.1.10), $t.\dim(\oplus_i M/M_i) < \infty$. So M satisfies t-DCC. □

4.3.24. LEMMA. Let M be a module satisfying t-ACC. If $N \in \sigma[M]$ is a countably infinite direct sum of mutually orthogonal nonzero modules, then there exist $y_i \in N, i = 1, 2, \cdots$, such that $y_1^\perp \supset y_2^\perp \supset \cdots$ and $t.\dim(\oplus_{i=1}^\infty y_i R) = \infty$.

PROOF. Because of (4.3.4), every cyclic submodule module in $\sigma[M]$ satisfies t-ACC. In particular, every nonzero cyclic module in $\sigma[M]$ has finite type dimension. Write $N = \oplus_{i=1}^\infty N_i$ where each $N_i \neq 0$ and $N_i \perp N_j$ when $i \neq j$. For each $i \geq 1$, take $0 \neq x_i \in N_i$. For $i \neq j$, since $x_i R$ and $x_j R$ are orthogonal, $x_i^\perp \neq x_j^\perp$. There does not exist an infinite chain

$$x_{i_1}^\perp \subset x_{i_2}^\perp \subset \cdots \subset x_{i_n}^\perp \subset \cdots,$$

for, otherwise, $x_{i_1} R \cong R/x_{i_1}^\perp$ does not satisfy the t-ACC. Thus, there exists an x_{i_1} such that

$$X_1 = \{x_j : x_{i_1}^\perp \not\subseteq x_j^\perp\}$$

is an infinite set. Note that for $x_j, x_k \in X_1$,

$$x_{i_1}^\perp \cap x_j^\perp = x_{i_1}^\perp \cap x_k^\perp \Longleftrightarrow x_j = x_k \Longleftrightarrow j = k.$$

Suppose that there exists an infinite chain

$$x_{i_1}^\perp \cap x_{j_1}^\perp \subset x_{i_1}^\perp \cap x_{j_2}^\perp \subset \cdots$$

where all $x_{j_k} \in X_1$. Then

$$0 \neq x_{i_1}^\perp/(x_{i_1}^\perp \cap x_{j_n}^\perp) \cong (x_{i_1}^\perp + x_{j_n}^\perp)/x_{j_n}^\perp \subseteq R/x_{j_n}^\perp \cong x_{j_n} R.$$

This implies that

$$t.\dim[\oplus_n R/(x_{i_1}^\perp \cap x_{j_n}^\perp)] = \infty.$$

This shows that $R/(x_{i_1}^\perp \cap x_{j_1}^\perp) \in \sigma[M]$ does not satisfy the t-ACC, a contradiction. So, there exists $x_{i_2} \in X_1$ such that

$$X_2 = \{x_j \in X_1 : x_{i_1}^\perp \cap x_{i_2}^\perp \not\subseteq x_j^\perp\}$$

is an infinite set. Continuing this manner, we obtain a sequence

$$\{x_{i_1}, x_{i_2}, \cdots, x_{i_n}, \cdots\}$$

such that $x_{i_1}^\perp \not\subseteq x_{i_j}^\perp$ for all $j > 1$ and, for each $n \geq 2$, $x_{i_1}^\perp \cap \cdots \cap x_{i_n}^\perp \not\subseteq x_{i_j}^\perp$ for all $j > n$. Now let $y_n = x_{i_1} + \cdots + x_{i_n}$. Then

$$y_1^\perp \supseteq y_2^\perp \supseteq \cdots, \ \oplus_i R/y_i^\perp \cong \oplus_i y_i R.$$

Take $r_n \in x_{i_1}^{\perp} \cap \cdots \cap x_{i_{n-1}}^{\perp}$ but $r_n \notin x_{i_n}^{\perp}$. Then $y_1 R = x_{i_1} R$, $y_2 r_2 R = x_{i_2} r_2 R \neq 0, \cdots$, and $y_n r_n R = x_{i_n} r_n R \neq 0$ for all n. Thus, $t.\dim(\oplus_{n=1}^{\infty} y_n R) = \infty$. □

4.3.25. THEOREM. The following are equivalent for a ring R:

1. R satisfies both right t-DCC and right t-ACC.

2. There does not exist an infinite set of pairwise orthogonal nonzero modules.

3. Every module has finite type dimension.

4. Every injective module is a finite direct sum of atomic modules.

5. Every module satisfies both t-ACC and t-DCC.

6. There exists an $n > 0$ such that, for any descending chain $I_1 \supseteq I_2 \supseteq \cdots$ of right ideals of R, we have $t.\dim(\oplus_i R/I_i) < n$.

7. There exists an $n > 0$ such that, for any ascending chain $I_1 \subseteq I_2 \subseteq \cdots$ of right ideals of R, we have $t.\dim(\oplus_i R/I_i) < n$.

PROOF. (1) $\Longrightarrow$ (2). Suppose that (2) does not hold. Then some module N is a countably infinite direct sum of mutually orthogonal nonzero modules. By (4.3.24), there exists a chain $y_1^{\perp} \supset y_2^{\perp} \supset \cdots$ where all $y_i \in N$, such that $t.\dim(\oplus_{i=1}^{\infty} R/y_i^{\perp}) = t.\dim(\oplus_{i=1}^{\infty} y_i R) = \infty$. This shows that $(R/\cap_{i=1}^{\infty} y_i^{\perp})_R$ does not satisfy the t-DCC. It follows from (4.3.23) that R does not satisfy t-DCC. This is a contradiction.

(2) $\Longrightarrow$ (3) $\Longrightarrow$ (5) $\Longrightarrow$ (1). They are clear.

(3) $\Longleftrightarrow$ (4). Just note that a module and its injective hull have the same type dimension.

(3) $\Longrightarrow$ (6) and (3) $\Longrightarrow$ (7). They are clear.

(6) $\Longrightarrow$ (1). Let n be the fixed number in (6). It is clear that (6) implies that R satisfies t-DCC, and hence every cyclic module has finite type dimension. Suppose R does not satisfy the t-ACC. Then there exists a chain $I_1 \subseteq I_2 \subseteq \cdots$ of right ideals of R such that $t.\dim(\oplus_i R/I_i) = \infty$. Since $t.\dim(R/I_i) < \infty$ for all i, we may assume, without loss of generality, that there exist right ideals K_i of R such that $K_i/I_i \neq 0$ and $(K_i/I_i) \perp (K_j/I_j)$ whenever $i \neq j$. It follows that $t.\dim(\oplus_{i=1}^{n} R/I_i) \geq n$. This is contradiction to (6).

(7) $\Longrightarrow$ (1). Similar to the proof of "(6) $\Longrightarrow$ (1)". □

4.3.26. EXAMPLE.

1. Every right Artinian ring satisfies t-DCC.

2. Every right Noetherian ring satisfies t-ACC. The ring $\mathbb{Z}$ is a Noetherian ring but it does not satisfy t-DCC.

3. Let R be the trivial extension of the ring $\mathbb{Z}_2$ and the $\mathbb{Z}_2$-module $\oplus_{i=1}^{\infty} M_i$ with all $M_i = \mathbb{Z}_2$. Then R satisfies t-ACC and t-DCC, but R is not Noetherian (hence not Artinian).

4. Let R be the trivial extension of the ring $\mathbb{Z}$ and the $\mathbb{Z}$-module $\mathbb{Z}_{p^\infty}$, where p is a prime number. Then R satisfies t-ACC but not t-DCC.

5. If R satisfies t-ACC or t-DCC, then every cyclic R-module has finite type dimension. The ring R in (4.3.13) satisfies neither t-ACC nor t-DCC, but every cyclic R-module has finite type dimension.

The well-known Hopkins–Levitzki Theorem states that every right Artinian ring is right Noetherian. Our concluding result shows that any ring with t-DCC which is Morita equivalent to a finite direct product of domains satisfies t-ACC, but we have been unable to answer the question whether every ring with t-DCC always satisfies t-ACC. The **Morita equivalent property** of rings is used only in the next proposition. For **Morita equivalences** as well as their properties, we refer to the book of Anderson and Fuller [8].

4.3.27. PROPOSITION. If R is Morita equivalent to a finite direct product of (not necessarily commutative) domains, then t-DCC of R implies t-ACC.

PROOF. We first assume that R is a domain. Note that, for any $0 \neq x \in R$ and any right ideal I, $R/I \cong xR/xI \subseteq R/xI$. If t-ACC fails, then there exists a chain of right ideals

$$I_0 \subseteq I_1 \subseteq \cdots \subseteq I_n \subseteq \cdots$$

such that $t.\dim(\oplus_{n=1}^{\infty} R/I_n) = \infty$. Select $0 \neq x_n \in I_n$. Thus,

$$I_0 \supseteq x_0 I_1 \supseteq x_0 x_1 I_2 \supseteq \cdots \supseteq x_0 x_1 \cdots x_{n-1} I_n \supseteq x_0 x_1 \cdots x_n I_{n+1} \supseteq \cdots,$$
$$R/I_n \hookrightarrow R/x_0 \cdots x_{n-1} I_n \ \forall n,$$

and hence

$$\infty = t.\dim(\oplus_{n=1}^{\infty} R/I_n) \leq t.\dim(\oplus_{n=1}^{\infty} R/x_0 \cdots x_{n-1} I_n).$$

So R does not satisfy t-DCC.

Next assume that $R = R_1 \times \cdots \times R_n$ where $n \geq 1$ and all R_i are domains. Since R satisfies t-DCC, so does $(R_i)_R$ by (4.3.23). Thus, $(R_i)_{R_i}$ satisfies t-DCC. Since R_i is a domain, $(R_i)_{R_i}$ satisfies t-ACC from above. It follows that $(R_i)_R$ satisfies t-ACC (for $i = 1, \cdots, n$). By (4.3.4), R_R satisfies t-ACC.

Now suppose that R is Morita equivalent to S, a finite direct product of domains. We can assume $S = \mathrm{End}(P_R)$ where P_R is a **progenerator** of Mod-R. Then

$$\mathrm{Hom}_R(P, -) : N_R \longmapsto \mathrm{Hom}_R({}_S P_R, N_R)$$

gives a Morita equivalence between Mod-R and Mod-S with **inverse equivalence**

$$- \otimes_S P : M_S \longmapsto M \otimes P.$$

Since R satisfies t-DCC, so does P_R by (4.3.23). Thus, S_S satisfies t-DCC by the Morita equivalences. Since S is a finite direct product of domains, S satisfies t-ACC from above. By (4.3.12), t-ACC is a Morita invariant property of rings. So R satisfies t-ACC. □

4.3.28. REFERENCES. Al-Huzali, Jain and López-Permouth [4]; Dauns and Zhou [39,42]; Goodearl [66]; Teply [115]; Zhou [140,141].

Chapter 5

Type Theory of Modules: Decompositions

Section 5.1 explains in greater detail how type submodules are used to obtain direct sum decompositions of modules. It shows that the set of natural classes forms a complete Boolean lattice. Also, certain universal natural classes which are available for every ring are derived, such as type I, II, and III modules, and others.

Type I, II, and III submodules were defined only for injective nonsingular modules in Goodearl and Boyle [68], and they had to give three separate proofs that an injective nonsingular module M is a direct sum $M_I \oplus M_{II} \oplus M_{III}$ of type I, II, and III submodules. Later in Dauns [25, 3.3, p.107, and 3.16, p.112], their definitions were extended to all nonsingular modules and it was shown that they form pairwise orthogonal natural classes with join equal to $\mathbf{1} \in \mathcal{N}(R)$. Hence the direct sum decomposition $M = M_I \oplus M_{II} \oplus M_{III}$ follows at once from our Theorem 5.1.7, and one proof suffices. In Dauns [32, 4.5, p.528], types I, II and III were extended to all modules, even singular ones.

While section 5.1 developed methods of decomposing injective modules, section 5.2 does the same for other classes of modules far less restrictive than the injectives.

Section 5.3 discusses a class of modules which includes the nonsingular modules properly, namely the unique type closure modules. These are the modules all of whose submodules have unique type closures.

A module M is called a **CS-module** (or extending) if every complement submodule of M is a direct summand of M. The module M is called a TS-module if every type submodule of M is a direct summand of M. Zhou demonstrated in [141] that much of the rich structure theory of CS-modules could be developed for TS-modules. Although TS-modules inevitably came up previously in section 5.2, section 5.4 covers the TS-modules.

5.1 Type Direct Sum Decompositions

Natural classes and the corresponding type submodules are extraordinarily well suited for decomposing injective modules into finite direct sums. In the infinite case, for any module M, it will be shown here how to find essential direct sums $\oplus\{M_\gamma \mid \gamma \in \Gamma\} \leq_e M$, where the $M_\gamma \in \gamma \in \Gamma \subseteq \mathcal{N}(R)$ are type submodules of type γ. Then the injective hull of M is $E(M) = E\left(\oplus\{E(M_\gamma) \mid \gamma \in \Gamma\}\right)$. For fixed γ, the type direct summands $E(M_\gamma)$ of $E(M)$ will be unique up to superspectivity (see [87, Def.1.30]).

One of the main problems here is to logically construct many useful sets $\Gamma \subseteq \mathcal{N}(R)$ suited for various diverse kinds of decompositions. Here some basic facts about the lattice $\mathcal{N}(R)$ are proved because they are useful not only in constructing the above sets Γ, but also in understanding various direct sum decompositions. Also, lattice operations can lead to the construction of certain modules that are of interest. Of particular interest and usefulness are the so-called "universal natural classes" which are defined for every ring, e.g., see (5.1.8) and "A,B,C,D." Others depend entirely on the ring.

The following facts on natural classes can easily be verified.

5.1.1. LEMMA. Let $\{\mathcal{K}_i : i \in I\}$ be a set of natural classes. Then the following hold:

1. $\cap_I \mathcal{K}_i$ is a natural class.
2. $d(\cup_I \mathcal{K}_i)$ is the smallest natural class containing all $\mathcal{K}_i$.
3. If $\mathcal{K}_i = d(X_i)$, then $d(\oplus_I X_i)$ is the smallest natural class containing all $\mathcal{K}_i$. □

From (2.3.9), the collection $\mathcal{N}(R)$ of all natural classes is a set. It will be shown next that $\mathcal{N}(R)$ is indeed a complete Boolean lattice. The next proposition follows from (5.1.1).

5.1.2. PROPOSITION. For a ring R, $\mathcal{N}(R)$ is a complete lattice with smallest element $\mathbf{0} = \{0\}$ and greatest element $\mathbf{1} = \text{Mod-}R$ under the partial ordering and lattice operations as below:

1. For $\mathcal{K}, \mathcal{L} \in \mathcal{N}(R)$, $\mathcal{K} \leq \mathcal{L} \Longleftrightarrow \mathcal{K} \subseteq \mathcal{L}$.
2. For any set of natural classes $\{\mathcal{K}_i : i \in I\}$, $\wedge \mathcal{K}_i = \cap_i \mathcal{K}_i$; and $\vee \mathcal{K}_i = d(\cup_I \mathcal{K}_i)$. □

5.1.3. LEMMA. For any class $\mathcal{F}$ of modules, we have $d(\mathcal{F}) \wedge c(\mathcal{F}) = \mathbf{0}$ and $d(\mathcal{F}) \vee c(\mathcal{F}) = \mathbf{1}$. In particular, if $\mathcal{K} \in \mathcal{N}(R)$, then $\mathcal{K} \wedge c(\mathcal{K}) = \mathbf{0}$ and $\mathcal{K} \vee c(\mathcal{K}) = \mathbf{1}$.

PROOF. By (2.3.4), $d(\mathcal{F}) \wedge c(\mathcal{F}) = d(\mathcal{F}) \cap c(\mathcal{F}) = \mathbf{0}$. For any $M \in \mathbf{1}$, let $M_1 \leq_t M$ be a type submodule of type $d(\mathcal{F})$ and let M_2 be a complement of M_1 in M. Then M_2 is a type submodule of M by (4.1.2)(3) and $M_2 \in c(\mathcal{F})$. Since $M_1 \oplus M_2 \leq_e M$ and $M_1 \oplus M_2 \in d(\mathcal{F}) \vee c(\mathcal{F}) \in \mathcal{N}(R)$, we have $M \in d(\mathcal{F}) \vee c(\mathcal{F})$. So $d(\mathcal{F}) \vee c(\mathcal{F}) = \mathbf{1}$. □

The next lemma appeared in [32, Lemma 2.14].

5.1.4. LEMMA. For any $\mathcal{K}, \mathcal{L}, \mathcal{H} \in \mathcal{N}(R)$, $\mathcal{K} \wedge (\mathcal{L} \vee \mathcal{H}) = (\mathcal{K} \wedge \mathcal{L}) \vee (\mathcal{K} \wedge \mathcal{H})$.

PROOF. Always $(\mathcal{K} \wedge \mathcal{L}) \vee (\mathcal{K} \wedge \mathcal{H}) \subseteq \mathcal{K} \wedge (\mathcal{L} \vee \mathcal{H})$. For any $M \in \mathcal{K} \wedge (\mathcal{L} \vee \mathcal{H})$, let $L \leq_t M$ be a type submodule of type $\mathcal{L}$ and let H be a complement of L in M. Then, by (4.1.2), $L \leq_t M$ and hence $L \perp H$. It follows that no nonzero submodules of H belong to $\mathcal{L}$. But, from $H \in \mathcal{L} \vee \mathcal{H}$, every nonzero submodule of H contains a nonzero submodule in $\mathcal{L} \cup \mathcal{H}$. It follows that every nonzero submodule of H contains a nonzero submodule in $\mathcal{H}$. This shows that $H \in \mathcal{H}$. Thus, $L \oplus H \in (\mathcal{K} \wedge \mathcal{L}) \vee (\mathcal{K} \wedge \mathcal{H}) \in \mathcal{N}(R)$. Hence $M \in (\mathcal{K} \wedge \mathcal{L}) \vee (\mathcal{K} \wedge \mathcal{H})$ because $L \oplus H \leq_e M$. □

The next result, due to [32], follows from (5.1.2), (5.1.3), and (5.1.4).

5.1.5. THEOREM. For a ring R, $\mathcal{N}(R)$ is a complete Boolean lattice. □

One reason for introducing the lattice $\mathcal{N}(R)$ is that lattice direct sum decompositions of $\mathcal{N}(R)$ will always give corresponding direct sum decompositions of injective modules.

5.1.6. DEFINITION. A subset $\{\mathcal{K}_i : i \in I\}$ of $\mathcal{N}(R)$ is said to be a **maximal set of pairwise orthogonal types** if $\mathcal{K}_i \wedge \mathcal{K}_j = \mathbf{0}$ for $i \neq j$ in I and $\vee_{i \in I} \mathcal{K}_i = \mathbf{1}$.

For a module M and $\gamma \in \mathcal{N}(R)$, we let $M_{(\gamma)} \leq_t M$ be any type submodule of type γ. The next result is contained in [31].

5.1.7. THEOREM. Let Γ be a maximal set of pairwise orthogonal types. Then the following hold:

1. $\sum_{\gamma \in \Gamma} M_{(\gamma)} = \oplus_{\gamma \in \Gamma} M_{(\gamma)} \leq_e M$, $E(M) = E\left(\oplus_{\gamma \in \Gamma} M_{(\gamma)}\right)$.
2. Suppose that $N_{(\gamma)} \leq M$ is any other choice of type submodules of type γ, $\gamma \in \Gamma$. Then $E(M_\gamma)$ and $E(N_\gamma)$ are superspective ([87, Def.1.30]).

PROOF. (1) If $\gamma_1 \neq \gamma_2$ in Γ, then $M_{(\gamma_1)} \perp M_{(\gamma_2)}$ since $\gamma_1 \wedge \gamma_2 = \mathbf{0}$. Suppose by induction that the sum of any n distinct $M_{(\gamma)}$'s is direct but that $M_{(\gamma_0)} \cap \left(M_{(\gamma_1)} \oplus \ldots \oplus M_{(\gamma_n)}\right) \neq 0$. By (2.3.3), a nonzero submodule of $M_{(\gamma_0)}$ is isomorphic to a submodule of $M_{(\gamma_i)}$ for some $i > 0$. But then $M_{(\gamma_0)}$ and $M_{(\gamma_i)}$ are not orthogonal, a contradiction. Next, let $\left(\oplus_{\gamma \in \Gamma} M_{(\gamma)}\right) \oplus P \leq_e M$. If $P \neq 0$, then P has a nonzero submodule contained in some γ of Γ since $P \in \bigvee_\Gamma \gamma$. Thus P and $M_{(\gamma)}$ are not orthogonal. So $P \cap M_{(\gamma)} \neq 0$ since $M_{(\gamma)} \leq_t M$. This is a contradiction.

(2) For any $\gamma \in \Gamma$, suppose that $E(M) = E(M_{(\gamma)}) \oplus X$ for some $X \leq E(M)$. It has to be shown that also $E(M) = E(N_{(\gamma)}) \oplus X$. By (4.1.2)(3), $X \in c(\gamma)$ is a type submodule of $E(M)$, and thus $E(N_{(\gamma)}) \cap X = 0$. Let $E(N_{(\gamma)}) \oplus X \oplus C = E(M)$. If $C \neq 0$, by (2.3.3), for some $0 \neq V \leq C$, V is either isomorphic to a submodule of $E(M_{(\gamma)})$ or isomorphic to a submodule of X. Both contradict the fact that $E(M_{(\gamma)})$ and X are type submodules of $E(M)$. Thus $E(M) = E(N_{(\gamma)}) \oplus X$ and $E(M_{(\gamma)})$ is superspective to $E(N_{(\gamma)})$ in $E(M)$. □

Our next objective is to give several natural classes which are available over every ring.

5.1.8. Definition. Recall that a nonzero module V is **atomic** if $d(V)$ is an atom of $\mathcal{N}(R)$. Let A, B, C, D be modules. Then D is **discrete** if it contains as an essential submodule a direct sum of uniform modules; C is **continuous** if C contains no uniform modules. The module A is **molecular** if every nonzero submodule of A contains an atomic submodule. At the opposite end of the module spectrum, B is **bottomless** if B contains no atomic submodules. The above four classes of modules are examples of natural classes. Moreover, c("discrete") = "continuous", and c("molecular") = "bottomless." Lastly, a nonzero module N is **compressible** if for any $0 \neq V \leq N$, there exists an embedding $N \hookrightarrow V$. Note that the uniform, atomic, or compressible modules do not form a natural class. From now on, A, B, C, D stand for the classes of all molecular, bottomless, continuous, and discrete modules, respectively.

5.1.9. EXAMPLE. (1) Uniform, homogeneous semisimple modules and compressible modules are atomic. If R is a (noncommutative) domain and $0 \neq L \leq R_R$, then L is atomic. Moreover, for a uniform module M and any family $\{M_i : i \in I\}$ of nonzero submodules of M, $\oplus M_i$ is atomic. It is easy to see that submodules and injective hulls of atomic modules are atomic.

(2) Let $R = (\Pi_{i=1}^{\infty}\mathbb{Z}_2)/(\oplus_{i=1}^{\infty}\mathbb{Z}_2)$. Then R is bottomless. To see this, take any $0 \neq x + \oplus_{i=1}^{\infty}\mathbb{Z}_2 \in R$. Since the support $supp(x)$ is infinite, write $x = y + z$ where both $supp(y)$ and $supp(z)$ are infinite. Then $yR = (xy)R$ and $zR = (xz)R$ are submodules of xR and $yR \perp zR$. Thus $V = xR$ is not an atomic module. Hence R is bottomless.

(3) For any infinite set X, let $\mathcal{P}(X)$ be the Boolean ring where $A \cdot B = A \cap B$ and $A + B = (A \cup B) \backslash (A \cap B)$ for $A, B \subseteq X$. Let

$$\mathcal{F}(X) = \{A : A \subseteq X, |A| < \aleph_0\} \triangleleft \mathcal{P}(X)$$

be the ideal of all finite subsets. Then $R = \mathcal{P}(X)/\mathcal{F}(X)$ is a nonsingular ring with R_R bottomless. For $X = \{1, 2, \cdots\}$, the later ring is isomorphic to the one in (2). □

The definitions of "square", "square free", and "square full" come from [87, Def.2.34 and Def.1.32]. Their definition of "P is purely infinite" is exactly the same as saying that "P is an idempotent square." The reader should be aware that in the literature for a long time a different concept of "purely infinite" has been used (e.g., see Kaplansky [78], and Goodearl and Boyle [68, pp.40-41]). Here we follow Mohamed and Müller [87] for this concept.

5.1.10. DEFINITION. A module N is a **square** if $N \cong P \oplus P$ for some $P \hookrightarrow N$; N is **purely infinite** if $N \cong N \oplus N$. A module M_1 is **square free** if $X \oplus X \not\hookrightarrow M_1$ for any nonzero module X. A module M_2 is **square full** if for any $0 \neq N \leq M_2$, there exists a $0 \neq X \leq N$ such that $X \oplus X \hookrightarrow M_2$.

5.1.11. DEFINITION. First, M is any injective module and N stands for any arbitrary nonzero direct summand of M. The module M is **Abelian** if no nonzero direct summand N of M is a square, i.e., $M = P_1 \oplus P_2 \oplus C$ with $P_1 \cong P_2$ is possible only if $P_1 = P_2 = 0$. The injective module M is **idempotent square free** if $N \ncong N \oplus N$ for all N.

Next, M is **type I** if every N contains a nonzero injective Abelian submodule. It is **type III** if every N contains a nonzero injective direct summand $0 \neq P \subseteq {}^{\oplus}N$ such that $P \cong P \oplus P$. Lastly, M is of **type II**, provided that for every $0 \neq N \subseteq {}^{\oplus}M$, N is not Abelian, and there exists a nonzero direct summand $0 \neq X \subseteq {}^{\oplus}N$ such that $P \ncong P \oplus P$ for any nonzero summand $0 \neq P \subseteq {}^{\oplus}X$. In other words, X is idempotent square free.

An arbitrary module M will be said to have any of the above listed properties (Abelian, idempotent square free, and types I, II, III) if and only if its injective hull $E(M)$ does. A module M is **locally idempotent square free** if every nonzero submodule of M contains a nonzero idempotent square free submodule.

The next definition of directly finite module has been used by [68, p.16 and Thm.3.1] and by [87, Def.1.24 and Prop.1.25]. For injective modules these coincide with [32, Def.4.5].

5.1.12. DEFINITION. A module D is **directly finite** if D is not isomorphic to any proper direct summand of D.

5.1.13. REMARKS. (1) It would not have been enough to define the previous concepts in (5.1.11) for injective modules only as had been done previously. Otherwise these classes would not be closed under submodules.

(2) By [87, Prop.1.27], a module M is locally idempotent square free if and only if for any $0 \neq V \leq M$, there exists a $0 \neq W \leq V$ such that $E(W)$ is directly finite.

(3) The square free modules and the Abelian modules are one and the same. In the literature, the traditional term "Abelian" has been used longer and more frequently, while the newer "square free" might be more descriptive, at least in a module context.

5.1.14. Let $\mathcal{A}$ denote the class of Abelian modules, and $\mathcal{P}(\infty)$ the class of purely infinite modules. For any ring the following classes of modules are natural classes:

1. square full modules

2. locally idempotent square free modules $= c(\mathcal{P}(\infty))$
 $= \{V \in \text{Mod-}R : \exists\ V \hookrightarrow E(\oplus_{i\in I} N_i),\ N_i$ are idempotent square free$\}$

3. type I $= d(\mathcal{A})$

4. type III $= d(\mathcal{P}(\infty))$

5. type II $= c(\text{type I}) \cap c(\text{type III}) = c(\mathcal{A}) \cap c(\mathcal{P}(\infty))$

PROOF. (1) and (2). From the way they are defined, these two classes are closed under submodules and injective hulls. Closure under direct sums follows from the projection argument. (3) Type I $= d(\mathcal{A}) \in \mathcal{N}(R)$ as in (5.1.11). (4) Similarly, type III $= d(\mathcal{P}(\infty)) \in \mathcal{N}(R)$. (5) For type II, that every submodule is not Abelian means that type II $\subseteq c(\mathcal{A}) = c(\text{type I})$. Also by (5.1.11), type II $\subseteq d(\mathcal{P}(\infty))$, and type II $= c(\mathcal{A}) \cap c(\mathcal{P}(\infty))$. □

At the expense of a more demanding proof, below (5.1.15)(3) can be improved to hold for certain non-injective modules.

5.1.15. For any ring R, abbreviate type I, II, and III modules as $\mathcal{I}_1, \mathcal{I}_2, \mathcal{I}_3 \in \mathcal{N}(R)$. Let M be any R-module. Then the following hold:

1. $\mathcal{I}_1 \wedge \mathcal{I}_2 = \mathbf{0}$, $\mathcal{I}_1 \wedge \mathcal{I}_3 = \mathbf{0}$, $\mathcal{I}_2 \wedge \mathcal{I}_3 = \mathbf{0}$; $\mathcal{I}_1 \vee \mathcal{I}_2 \vee \mathcal{I}_3 = \mathbf{1} \in \mathcal{N}(R)$.

2. $\exists\ M_{(\mathcal{I}_1)} \oplus M_{(\mathcal{I}_2)} \oplus M_{(\mathcal{I}_3)} \leq_e M$.

3. $E(M) = E(M_{(\mathcal{I}_1)}) \oplus E(M_{(\mathcal{I}_2)}) \oplus E(M_{(\mathcal{I}_3)})$ is unique up to superspectivity.

4. If $Z(M) = 0$, then the sum in (3) is unique, and $\mathcal{I}_1, \mathcal{I}_2, \mathcal{I}_3$ as in (5.1.11) are the usual types I, II, and III.

PROOF. (1) First note that $d(\mathcal{P}(\infty)) \subseteq c(\mathcal{A})$. There is a pairwise orthogonal decomposition

$$\mathbf{1} = d(\mathcal{A}) \vee c(\mathcal{A}) = \mathcal{I}_1 \vee \Big\{[c(\mathcal{A}) \wedge d(\mathcal{P}(\infty))] \vee [c(\mathcal{A}) \wedge c(\mathcal{P}(\infty))]\Big\}$$

in the distributive lattice $\mathcal{N}(R)$. But $d(\mathcal{P}(\infty)) \subseteq c(\mathcal{A})$ implies that

$$c(\mathcal{A}) \wedge d(\mathcal{P}(\infty)) = \mathcal{I}_3, \quad \text{and } \mathbf{1} = \mathcal{I}_1 \vee \mathcal{I}_2 \vee \mathcal{I}_3.$$

□

Next, some criteria are given in order for a module to belong to some natural classes.

5.1.16. (1) A module A is atomic if and only if for any $0 \neq x \in A$, there exists an index set I such that $A \hookrightarrow E\left((xR)^{(I)}\right)$. Equivalently, for any $y, 0 \neq x \in A$, there exists a set I with $yR \subseteq E\left((xR)^{(I)}\right)$.

(2) A module V is Abelian if and only if for any $P_1 \oplus P_2 \leq V$, $P_1 \cong P_2$ implies $P_1 = P_2 = 0$.

PROOF. $\Longrightarrow$. This follows from (5.1.11). $\Longleftarrow$. Suppose that $E(V) = V_1 \oplus V_2 \oplus C$ and $0 \neq f : V_1 \longrightarrow V_2$ is an isomorphism. Set $0 \neq P_2 = f(V_1 \cap V) \cap V \leq_e V_2$, and $P_2 \cong P_1 = f^{-1}(P_2) \leq_e V_1$.

(3) For a module M, $M \in \mathcal{I}_3$ if and only if for any $0 \neq N \leq M$, there exist submodules P_1, P_2 of N such that $0 \neq P_1 \oplus P_2 \leq N$ and $E(P_1) \cong E(P_2) \cong E(P_1) \oplus E(P_2)$.

PROOF. $\Longleftarrow$. This follows from (5.1.11). $\Longrightarrow$. Given $N \leq M$, since $E(M) \in \mathcal{I}_3$, there exist modules Q_1, Q_2, and Q such that $0 \neq Q_1 \oplus Q_2 \cong Q \subseteq^{\oplus} E(N)$ and $Q_1 \cong Q_2 \cong Q$. As in the proof of (2) above, there exist $P_1 \leq_e Q_1$, $P_2 \leq_e Q_2$. Hence $P_1 \oplus P_2 \leq_e Q$ with $P_1 \cong P_2$. The rest is clear. □

Once early in the development of this subject it was thought that there might only be a finite number of certain kinds of natural classes defined for every ring like the classes A, B, C, D in (5.1.8), and $\mathcal{I}_1, \mathcal{I}_2, \mathcal{I}_3$ above. Next, it will be shown in (5.1.18) that they form a proper class that is not even a set.

5.1.17. DEFINITION. The **Goldie dimension** $Gd(M)$ of any module M is the finite or infinite cardinal number

$$Gd(M) = \sup\Big\{|I| : \exists \ \oplus \{M_i : i \in I\} \leq M, \ 0 \neq M_i \leq M, i \in I\Big\}.$$

If $Gd(M)$ is not inaccessible (= regular limit cardinal), then $Gd(M) = |I|$ is attained for some $\oplus\{M_i : i \in I\} \leq M$ (Dauns and Fuchs [38, p.297]).

Now let $\aleph$ be a cardinal, where either $\aleph = 1$, or $\aleph_0 \leq \aleph$. Define M to be of **local Goldie dimension** $\aleph$ if for any $0 \neq V \leq M$, there exists a $0 \neq W \leq V$ with $Gd(W) = \aleph$. Define $\Delta_\aleph(R)$ to be the class of all right R-modules of local Goldie dimension $\aleph$. Note that $\Delta_1(R) = D(R)$ are the discrete modules.

More details and proofs for some of the subsequent examples (5.1.18), (5.1.20), and (5.1.22) can be found in [32, sections 6 and 7].

5.1.18. For any ring R and cardinal number $\aleph = 1, \aleph_0, \aleph_1, \ldots$, the following hold:

1. There exists a unique smallest cardinal $\tau(R)$ such that $\Delta_\aleph(R) = 0$ for all $\aleph \geq \tau(R)$; $\tau(R) \leq 2^{|R|}$.

2. $\Delta_\aleph(R) \in \mathcal{N}(R)$.

3. $\Delta_\aleph(R) \cap \Delta_\kappa(R) = 0$ for $\kappa \neq \aleph$; $\bigvee_\aleph \Delta_\aleph(R) = \bigvee_{\aleph < \tau(R)} \Delta_\aleph(R) = \mathbf{1} \in \mathcal{N}(R)$. □

The $\Delta_\aleph$ as $\aleph$ ranges over all cardinals form a proper class, because for each $\aleph$, there is a ring R such that $\Delta_\aleph(R) \neq 0$.

5.1.19. EXAMPLE. For any division ring F and set X with $2 \le |X| = \aleph$, the free algebra $R = F\{X\}$ on X has $Gd(R) = \max\{\aleph_0, |X|\} = \aleph_0 \cdot \aleph$, and $R_R \in \Delta_{\aleph_0 \cdot \aleph}(R)$. In order to show that $R_R \in \mathcal{I}_3$, let $E(R) = N \oplus N$ as in (5.1.11). It suffices to show that $N = P \oplus P$ for some $0 \ne P \cong P \oplus P$. Recall that for any infinite set I, $|I \times I| = I$, and for any $z \in R$, $zR \cong R$. As a consequence of the last two facts, there exists an infinite direct sum of nonzero cyclics

$$(\oplus_{i \in I} z_i R) \oplus (\oplus_{i \in I} z_i R) \cong \oplus_{i \in I} z_i R \le_e N \cap R \le_e N.$$

Consequently, $N \cong N \oplus N$ and we may take $P = N$. Since R is a domain, R_R is nonsingular, continuous, and molecular. □

5.1.20. EXAMPLE. Let $\aleph_0 \le \aleph$, and let Y be an infinite set. An **almost disjoint family** is a subset $\mathcal{F} \subseteq \mathcal{P}(Y)$ such that for any $A \in \mathcal{F}$, $|A| = |Y|$, but for any $A, B \in \mathcal{F}$ with $A \ne B$, $|A \cap B| < |Y|$. The generalized continuum hypothesis (GCH) is the statement that for any infinite cardinal $\aleph$, $2^{\aleph} = \aleph^+$, where $\aleph^+$ is the successor cardinal of $\aleph$. We need the following fact: Either let $|Y| = \aleph_0$, or assume that $|Y|$ is regular as well as the GCH. Then there exists an almost disjoint family $\mathcal{F} \subseteq \mathcal{P}(Y)$ with $|\mathcal{F}| = 2^{|X|}$. (See Kunen [82, Thm.1.2, p.48] and/or Levy [83, 5.29(ii), p.194].)

Let X be an infinite set. In the Boolean ring $\mathcal{P}(X)$, where $A \cdot B = A \cap B$ and $A + B = (A \cup B) \backslash (A \cap B)$ for $A, B \subseteq X$, let $J \lhd \mathcal{P}(X)$ be the ideal $J = \{C \subset X : |C| < |X|\}$. Let $R = \mathcal{P}(X)/J$. Since $|R| \le |\mathcal{P}(X)| = 2^{|X|}$, $Gd(R) \le |R| \le 2^{|X|}$.

In order to see that R is of local Goldie dimension $2^{|X|}$, it suffices to show that for any cyclic $0 \ne yR \le R$, $Gd(yR) \ge 2^{|X|}$. Then $y = Y + J$, $Y \subseteq X$, $Y \notin J$, or $|Y| = |X|$. Assume either that $|X| = \aleph_0$, or that $|X|$ is regular and the GCH in order to obtain an almost disjoint family $\mathcal{F} \subseteq \mathcal{P}(Y)$ as above. Then the sum

$$\Sigma\{(A + J) \cdot R : A \in \mathcal{F}\} = \oplus\{(A + J) \cdot R : A \in \mathcal{F}\} \le (Y + J) \cdot R = yR$$

is direct. Hence $Gd(yR) \ge |\mathcal{F}| = 2^{|X|}$. Thus $R \in \Delta_{\aleph}(R)$ is of local Goldie dimension $\aleph = 2^{|X|}$.

Any direct sum of submodules inside R_R is an orthogonal direct sum. Thus R_R is nonsingular type I, and bottomless.

Let $|X| = \aleph$, and $F_x = \mathbb{Z}_2$ for $x \in X$. The support $supp\, t$ of $t = (t_x)_{x \in X} \in \prod_{x \in X} F_x$ is $supp\, t = \{x \in X \mid t_x \ne 0\}$. The $\aleph$-**product** is

$$\prod_{x \in X}^{<\aleph} F_x = \left\{ t \in \prod_{x \in X} F_x : |supp\, t| < \aleph \right\}.$$

Then $\prod_{x \in X} F_x \cong \mathcal{P}(X)$, $\prod_{x \in X}^{<\aleph} F_x \cong J$, and thus $\prod_{x \in X} F_x \Big/ \prod_{x \in X}^{<\aleph} F_x \cong \mathcal{P}(X)/J = R$. In particular for $\aleph = \aleph_0$, $\prod_1^{\infty} \mathbb{Z}_2$ is a ring of local Goldie dimension $2^{\aleph_0}$. □

The next examples show that Goldie torsion type I and type III modules exist.

5.1.21. EXAMPLE. For any ring R, suppose that the module M is a direct sum $M = \oplus_\alpha U_\alpha$ of uniform modules U_α. Since every nonzero submodule of M contains a uniform module and a uniform module is Abelian, M is always a type I module.

Now in addition, assume that R is a Goldie torsion ring, i.e., $0 \neq Z(R) \leq_e R = Z_2(R)$. Then every module is Goldie torsion, and in particular M is Goldie torsion, discrete, and of type I. For example, let $R = \mathbb{Z}_{p^2}$ for a prime p. A non-commutative Goldie torsion ring can be found in [24, 6.16, p.79]. □

In some of the following examples, some of the proofs are omitted.

5.1.22. EXAMPLE. For a prime p, let $R = \mathbb{Z}_{p^2}\{x, y\}$ be the **free algebra** on two non-commuting indeterminates. From $(pR)^\perp = pR \leq_e R$, we get that $pR \subseteq Z(R)$, and $R = Z_2(R)$.

It can be shown that actually $Z(R) = pR$, and that for any $E(R) = N \oplus N'$, there exists an essential direct sum $(\oplus_{i \in I} v_i R) \oplus (\oplus_{j \in J} w_j R) \leq_e N \cap R$, where $pv_i \neq 0$ and $pw_j \neq 0$ for all i and j. Furthermore, we may assume that $|J| = \aleph_0$, and either $I = \emptyset$ or $|I| = \aleph_0$. Now an argument similar to that of Example (5.1.19) shows that R_R is Goldie torsion, continuous, molecular, and of type III. □

5.1.23. CONSTRUCTION. For a family R_j, $j \in J$, of rings, set $P = \prod_{j \in J} R_j$, $S = \oplus_{j \in J} R_j$, and $R = P/S$. Then (i) $Z(R) = 0$, R_R is nonsingular, and (ii) R_R is bottomless.

So far, the R_j could have been Goldie torsion $Z_2(R_j) = R_j$. If in addition all R_j, $j \in J$, are countable domains, then

(iii) $|\{j \in J : R_j \notin \mathcal{I}_1\}| < \infty \iff R_R \in \mathcal{I}_1$;

(iv) $|\{j \in J : R_j \notin \mathcal{I}_3\}| < \infty \iff R_R \in \mathcal{I}_3$.

Now let the R_j be any Goldie torsion free rings, $Z(R_j) = 0$, $j \in J$. Then

(v) $\forall j, R_j \in \mathcal{I}_1 \implies P_P \in \mathcal{I}_1$;

(vi) $\forall j, R_j \in \mathcal{I}_3 \implies P_P \in \mathcal{I}_3$. □

Some examples of Goldie torsion rings R with a unique maximal ideal $P \triangleleft R$ such that $Z(R) = P \leq_e R$ are given.

5.1.24. EXAMPLES. (i) For any commutative principal domain D and prime element $p \in D$, set $R = D/(p^n)$ for some $n = 2, 3, \ldots$. Then $Z(R) = (p)/(p^n) \leq_e R$.

(ii) For a prime p, let R be the previous free algebra $\mathbb{Z}_{p^2}\{x, y\}$ but now subject to the relations $x^2 = xy = yx = y^2 = 0$. Then R is commutative with a unique maximal ideal $P = Z(R) = p\mathbb{Z}_{p^2} + xR + yR \leq_e R$, $Z(R)^3 = 0$. Note that R is not uniform, since $xR \cap yR = 0$. In the above examples, the elements of $R \backslash P$ are invertible. □

The next construction describes how to obtain bottomless Goldie torsion rings.

5.1.25. CONSTRUCTION. Let T_i, $i \in I$, be an infinite family of rings with a unique maximal ideal $P_i \triangleleft T_i$ where P_i is the right singular submodule of T_i and P_i is essential in T_i as a right ideal. Set $T = \prod_{i\in I} T_i$, $S = \oplus_{i\in I} T_i$, and $R = T/S$. Then it can be shown that

$$Z(R) = \left(\prod_{i\in I} P_i + S\right) /S \leq_e R \, .$$

This ring R is bottomless Goldie torsion. If all $T_i \in \mathcal{I}_1$, then it can be shown that $R_R \in \mathcal{I}_1$. □

5.1.26. REFERENCES. Dauns [25,31,32]; Goodearl-Boyle [68]; Mohamed-Müller [87].

5.2 Decomposability of Modules

The previous section gave many examples of type submodules, and how they can be constructed. It also showed that given a finite set of pairwise orthogonal types in the lattice $\mathcal{N}(R)$ whose join is $\mathbf{1} \in \mathcal{N}(R)$, then any injective module M can be written as a direct sum of submodules belonging to those types. In the last part of the last century, direct sum decompositions were developed for various classes of modules generalizing the injectives: extending, fully invariant extending, quasi-continuous, etc. For the most part, hypotheses had to be placed on all complement submodules of a module M. In Zhou [141] it was shown that it was enough to impose hypotheses only on the type submodules of M, which are automatically complement submodules. By that time it became apparent that what was going on in [141] was more than a generalization. The emphasis here is to find conditions on the submodule structure of a module M in order for M to have certain direct sum decompositions.

5.2.1. DEFINITION. For a positive integer n, a module M is said to be n-**decomposable** if for any maximal set $\{\mathcal{K}_i : i = 1, \ldots, n\}$ of pairwise orthogonal types, there is a decomposition $M = \oplus_{i=1}^{n} M_i$ where $M_i \in \mathcal{K}_i$. A module which is n-decomposable for all integers $n \geq 1$ is called **finitely decomposable**. If for every maximal set (respectively, every countable maximal set) $\{\mathcal{K}_i : i \in I\}$ of pairwise orthogonal natural classes, M has a decomposition $M = \oplus\{M_i : i \in I\}$ with $M_i \in \mathcal{K}_i$, then M will be said to be **fully decomposable** (respectively, **countably decomposable**).

The basic questions here are: (i) characterize n- (respectively, finitely, countably, fully) decomposable modules; (ii) characterize the rings R for which certain modules are n- (respectively, finitely, countably, fully) decomposable.

5.2.2. THEOREM. For a module M, the following are all equivalent:

1. M is 2-decomposable.
2. For any $N \leq M$, N has a type complement Q in M such that Q is a direct summand of M.
3. Every type submodule $N \leq_t M$ of M has a complement Q in M with Q a direct summand of M.

PROOF. (1) $\Longrightarrow$ (2). For $N \leq M$, set $\mathcal{K} = d(N)$. By (1), $M = M_1 \oplus M_2$, where $M_1 \in \mathcal{K}$ and $M_2 \in c(\mathcal{K})$. So $N \perp M_2$. Suppose that $N \perp P$ for some $M_2 \subset P \subseteq M$. Then, since $M_2 \leq_t M$, we have $M_2 \perp X$ for some $0 \neq X \subseteq P$. But then we have $Y \hookrightarrow M_1$ for some $0 \neq Y \subseteq X$ by (2.3.3). This contradicts the assumption that $N \perp P$. So M_2 is a type complement of N in M. Since (3) is a special case of (2), (2) $\Longrightarrow$ (3).

(3) $\Longrightarrow$ (1). Let $\mathcal{K}_1, \mathcal{K}_2 \in \mathcal{N}(R)$ be given with $\mathcal{K}_1 \vee \mathcal{K}_2 = \mathbf{1}$ and $\mathcal{K}_1 \wedge \mathcal{K}_2 = \mathbf{0}$. Then $\mathcal{K}_2 = c(\mathcal{K}_1)$ by (5.1.3) and (5.1.5).

Let $N \leq_t M$ be a type submodule of type $\mathcal{K}_1$. By the hypothesis (3), N has a complement Q such that $M = P \oplus Q$ for some $P \leq M$. Since N is a type submodule of type $\mathcal{K}_1$, $N \perp Q$ and $Q \in c(\mathcal{K}_1)$. Note that $N \oplus Q \leq_e M$, so N is isomorphic to an essential submodule of P. Hence $P \in \mathcal{K}_1$. □

5.2.3. THEOREM. Any direct sum of n-decomposable modules is n-decomposable. Note that n-decomposable can be replaced by any of the following: finitely decomposable, countably decomposable, or fully decomposable.

PROOF. We prove this only for n. Let $M = \oplus_{i \in I} M_i$ where M_i is n-decomposable. Suppose that $\mathcal{K}_1, \ldots, \mathcal{K}_n$ is a maximal set of pairwise orthogonal types. For each i, $M_i = M_{i1} \oplus \cdots \oplus M_{in}$ with $M_{ij} \in \mathcal{K}_j$. Set $M_j = \oplus_{i \in I} M_{ij}$ for $j = 1, \ldots, n$. Thus $M = M_1 \oplus \cdots \oplus M_n$ where each $M_j \in \mathcal{K}_j$. □

5.2.4. COROLLARY. Any direct sum of atomic modules is fully decomposable. In particular, direct sums of atomic modules are 2-decomposable.

PROOF. This follows from (5.2.3) because every atomic module is clearly fully decomposable. □

In Smith and Tercan [110], a module M is said to satisfy condition (C_{11}) if every submodule of M has a complement that is a summand of M. Therefore, by (5.2.2), every module with (C_{11}) is 2-decomposable, and so is every TS-module. Later we will see that there exist 2-decomposable modules which satisfy neither TS nor (C_{11}). In [141], M is defined to satisfy (T_3) if, for any type direct summands M_1 and M_2 with $M_1 \oplus M_2 \leq_e M$, $M = M_1 \oplus M_2$ holds.

We will use the concept of **superspectivity** ([87, Def.1.30]). The connection between (T_3) and superspectivity given below was discovered in [141, Lemma 6].

5.2.5. COROLLARY. The following are equivalent for an R-module M and a fixed $\mathcal{K} \in \mathcal{N}(R)$:

1. M has a decomposition $M = M_1 \oplus M_2$, unique up to superspectivity, with $M_1 \in \mathcal{K}$, $M_2 \in c(\mathcal{K})$.

2. M has a type summand of type $\mathcal{K}$, and for any type direct summands M_1, M_2 of M with $M_1 \oplus M_2 \leq_e M$ and with $M_1 \in \mathcal{K}$, $M_2 \in c(\mathcal{K})$, $M = M_1 \oplus M_2$ holds.

In particular, M satisfies (1) for all $\mathcal{K} \in \mathcal{N}(R)$ if and only if M is 2-decomposable and satisfies (T_3).

PROOF. Abbreviate "M_1 is superspective to A" by "$M_1 \sim A$" in this proof only.

(1) $\Longrightarrow$ (2). Let $M = M_1 \oplus Y$ and $M = X \oplus M_2$ be as in (2) with

$$M_1 \oplus M_2 \leq_e M, \; M_1 \in \mathcal{K}, \; M_2 \in c(\mathcal{K}).$$

Hence $Y \in c(\mathcal{K})$ and $X \in \mathcal{K}$. But then $M = M_1 \oplus Y$ and $M = X \oplus M_2$ are two decompositions which by hypothesis (1) are unique up to superspectivity. Thus $M_1 \sim X$ and $M = X \oplus M_2$ implies that $M = M_1 \oplus M_2$.

(2) $\Longrightarrow$ (1). By hypothesis (2), there exists a type summand M_1 of type $\mathcal{K}$ of M, with $M = M_1 \oplus M_2$, and hence $M_2 \in c(\mathcal{K})$. Suppose that $M = N_1 \oplus N_2$ with $N_1 \in \mathcal{K}$ and $N_2 \in c(\mathcal{K})$. First, our hypotheses are symmetric with respect to M_1 and M_2, and secondly, with respect to M_1 and N_1. Hence it suffices to prove that if $M = M_1 \oplus Y$ for some $Y \leq M$, then likewise $M = N_1 \oplus Y$. But since $N_1 \oplus Y \leq_e M$ (by 5.1.7), by (2), $M = N_1 \oplus Y$. □

The next result is pertinent to almost every type direct sum decomposition developed in this section.

5.2.6. UNIQUENESS. Let M be any 2-decomposable module satisfying (T_3), and $\{\mathcal{K}_i : i \in I\}$ any finite or infinite set of pairwise orthogonal types. Then any decomposition $M = \oplus_{i \in I} M_i$ of M with $M_i \in \mathcal{K}_i$ is unique up to superspectivity.

PROOF. Without loss of generality, we can assume that $\{\mathcal{K}_i : i \in I\}$ is a maximal set of pairwise orthogonal types. Suppose that $M = \oplus_{i \in I} N_i$, where $N_i \in \mathcal{K}_i$ for all $i \in I$. Then let for each $j \in I$,

$$M = M_j \oplus (\oplus_{i \neq j} M_i) = N_j \oplus (\oplus_{i \neq j} N_i)$$

with $M_j, N_j \in \mathcal{K}_j$ and $\oplus_{i \neq j} M_i, \oplus_{i \neq j} N_i \in c(\mathcal{K}_j)$. Then it follows from (5.2.5) that M_j and N_j are superspective. □

A submodule $N \leq M$ is **fully invariant** if $f(N) \subseteq N$ for all $f \in \text{End}(M_R)$. The next theorem will show that for any 2-decomposable module M, every fully invariant type submodule of M, such as $Z_2(M)$, is a summand. Consequently, the study of 2-decomposable modules can be conducted separately for nonsingular modules and Goldie torsion modules. Since for a nonsingular module N, type closures of submodules of N are unique, to say that N is 2-decomposable is the same as saying that N is TS.

5.2.7. THEOREM. The following are equivalent for a module M:

1. M is 2-decomposable.

2. $M = Z_2(M) \oplus K$ where $Z_2(M)$ is 2-decomposable and K is nonsingular TS.

3. There exists a fully invariant type submodule $F \leq_t M$ with $M = F \oplus K$, where both F and K are 2-decomposable.

4. For every fully invariant type submodule $F \leq_t M$, $M = F \oplus K$ with both F and K being 2-decomposable.

PROOF. (1) $\Longrightarrow$ (4). When $F \leq_t M$ is a fully invariant type submodule of M, then $M = X \oplus Y$ where $X \in d(F)$ and $Y \in c(F)$ by (1). Any fully invariant submodule F of a module $M = X \oplus Y$ always satisfies $F = (F \cap X) \oplus (F \cap Y)$. Then $Y \perp F$, hence $F \cap Y = 0$, and thus $F \subseteq X$. But then $F = X$ since F is a type submodule of type $d(F)$ of M. Thus $M = F \oplus K$, where $K = Y \in c(F)$ is orthogonal to F.

To see that F and K are 2-decomposable, let $\mathcal{K} \in \mathcal{N}(R)$ be any given natural class. Let $A \leq_t F$ be any type submodule of F of type $\mathcal{K}$. Since $A \in \mathcal{K}$, $d(A) \subseteq \mathcal{K}$. (Note that $c(A) = c(\mathcal{K}) \vee (c(A) \wedge \mathcal{K})$.) Use (1) to write $M = M_1 \oplus M_2$ with $M_1 \in d(A) \subseteq \mathcal{K}$ and $M_2 \in c(A)$. The full invariance of $F \leq M$ guarantees that $F = (F \cap M_1) \oplus (F \cap M_2)$, where $F \cap M_1 \in d(A) \subseteq \mathcal{K}$. From $M_2 \in c(A)$, it follows that $M_2 \perp A$, and $(F \cap M_2) \perp A$. Thus $A + (F \cap M_2) = A \oplus (F \cap M_2) \leq F$ is a direct sum. Since A is a type submodule of type $\mathcal{K}$ of F, necessarily $F \cap M_2 \in c(\mathcal{K})$. Hence F is 2-decomposable.

Next, let $B \leq_t K$ be a type submodule of type $\mathcal{K}$ of K. Again, $M = N_1 \oplus N_2$ where $N_1 \in d(B) \subseteq \mathcal{K}$ and $N_2 \in c(B)$, and $F = (F \cap N_1) \oplus (F \cap N_2)$. From $M = F \oplus K$ with $K \perp F$, and $F \cap N_1 \subseteq N_1 \in d(B) \subseteq \mathcal{K}$, it is asserted that $F \cap N_1 = 0$. If not, there exists a $0 \neq C_2 \leq B \leq K$ with $C_2 \cong C_1 \leq F \cap N_1$ for some $0 \neq C_1 \leq F \cap N_1$. This contradicts that $K \perp F$. Hence $F = F \cap N_2$. But then $N_2 = N_2 \cap (F \oplus K) = F \oplus (N_2 \cap K)$, and $M = N_1 \oplus F \oplus (N_2 \cap K)$. Let π be the projection of M onto K along F. Then $N_1 \oplus F = \pi(N_1) \oplus F$, and thus $M = \pi(N_1) \oplus F \oplus (N_2 \cap K) = F \oplus K$. Since $\pi(N_1) \oplus (N_2 \cap K) \subseteq K$, it follows that $K = \pi(N_1) \oplus (N_2 \cap K)$. This implies that

$$\pi(N_1) \cong K/(N_2 \cap K) \cong (K + N_2)/N_2 \leq (N_1 \oplus N_2)/N_2 \cong N_1 \in \mathcal{K}.$$

Next, if $N_2 \cap K \notin c(\mathcal{K})$, then there exists a $0 \neq V \leq N_2 \cap K$ with $V \in \mathcal{K}$. Since $N_2 \in c(B)$, we have $V \in c(B)$ and $V \perp B$. Then $B \subset B \oplus V \subseteq K$ and $B \oplus V \in \mathcal{K}$, contradicting that $B \leq_t K$ is a type submodule of type $\mathcal{K}$. Thus $N_2 \cap K \in c(\mathcal{K})$, and K is 2-decomposable.

(4) $\Longrightarrow$ (2). Since $Z_2(M) \leq M$ is fully invariant, by (4), we have $M = Z_2(M) \oplus K$ where both $Z_2(M)$ and K are 2-decomposable. Any nonsingular 2-decomposable module such as K is TS. (2) $\Longrightarrow$ (3) is trivial; and (3) $\Longrightarrow$ (1) is by (5.2.3). $\square$

5.2.8. COROLLARY. Suppose that $M = F \oplus K = G \oplus L$, $F, G \in \mathcal{K}$, $K, L \in c(\mathcal{K})$ with F fully invariant in M. Then $F = G$ and K and L are superspective.

PROOF. Clearly, $F = (F \cap G) \oplus (F \cap L)$, $F \cap L = 0$, and hence $F = F \cap G \subseteq G$ implies $F = G$.

Since our hypotheses on K and L have become symmetric, it suffices to show that if $M = X \oplus K$, then also $M = X \oplus L$. But $X \in \mathcal{K}$, and by the first sentence of this proof, necessarily $X = F = G$. But then $M = G \oplus L = X \oplus L$. □

The next lemma describes some facts about type and complement submodules of nonsingular modules which will be used to prove the next theorem.

5.2.9. LEMMA. Let N be a nonsingular module. Then the following hold:

1. If $K \leq N$ is fully invariant in N, then so is also its unique complement closure K^c.
2. If $N = X \oplus Y$ with $X \perp Y$, then $\mathrm{Hom}_R(X, Y) = 0$.
3. If N is 2-decomposable, then every type submodule of N is fully invariant in N.

PROOF. (1) In any nonsingular module such as N, for any submodule $K \leq N$, $K^c = \{x \in N : x^{-1}K \leq_e R\}$. Suppose that $f : N \longrightarrow N$ and $x \in K^c$. Then $f(x)(x^{-1}K) \subseteq f(K) \subseteq K$ shows that $(f(x))^{-1}K \leq_e R$ and $f(x) \in K^c$. Thus $f(K^c) \subseteq K^c$.

(2) This is (4.3.19)(4). (3) follows immediately from (2). □

5.2.10. THEOREM. Let $\mathcal{K}_0, \mathcal{K}_1, \cdots, \mathcal{K}_n \in \mathcal{N}(R)$ be a maximal set of pairwise orthogonal types, where $\mathcal{K}_0$ is the class of the Goldie torsion modules. Then the following hold for any module M:

1. M is 2-decomposable if and only if there exists a decomposition $M = A_0 \oplus A_1 \oplus \cdots \oplus A_n$ where $A_i \in \mathcal{K}_i$ is 2-decomposable for all $i = 0, 1, \cdots, n$. (Hence $A_1, \cdots, A_n$ are TS.)
2. The above decomposition is unique up to superspectivity.

PROOF. (1) To prove (1), by (5.2.3), it suffices to prove only the implication "$\Longrightarrow$". By (5.2.7), $M = A_0 \oplus A'$ where $A_0 = Z_2(M)$, and A' is 2-decomposable nonsingular. Consequently, $A' = A_1 \oplus A''$ where $A_1 \in \mathcal{K}_1$ and $A'' \in c(\mathcal{K}_1)$ are fully invariant type submodules of A' by (5.2.9)(3). By (5.2.7), A_1 and A'' are each 2-decomposable. Hence $A'' = A_2 \oplus A'''$, $A_2 \in \mathcal{K}_2$, $A''' \in c(\mathcal{K}_2)$ where both A_2, A''' are fully invariant 2-decomposable type submodules of A''. Repetition of this process proves (1).

(2) Suppose that $M = A'_0 \oplus A'_1 \oplus \cdots \oplus A'_n$ is a second decomposition satisfying (1). First of all, $A'_0 = Z_2(M) = A_0$. Without loss of generality, it suffices to show that if $M = A_1 \oplus X$, then also $M = A'_1 \oplus X$. Since $A_1 \parallel A'_1$, $(A'_2 \oplus \cdots \oplus A'_n) \perp A_1$. Let $\pi : M = A_1 \oplus X \longrightarrow A_1$ be the projection along X, and f the composite map

$$N = A_1 \oplus A'_2 \oplus \cdots \oplus A'_n \xrightarrow{\pi} A_1 \hookrightarrow N.$$

By (5.2.9)(2), $f(A'_2 \oplus \cdots \oplus A'_n) = 0$, and hence $A'_2 \oplus \cdots \oplus A'_n \subseteq Kerf = N \cap X \subseteq X$. Note that $Z_2(M) = Z_2(X) \subseteq X$. Thus $A'_0, A'_2 \oplus \cdots \oplus A'_n \subseteq X$, and hence $M = A_1' + X$. Since $X \perp A_1$ while $A_1 \parallel A_1'$, also $X \perp A_1'$. Hence $M = A_1' \oplus X$ as required. □

5.2.11. COROLLARY. The following hold for a module M:

1. M is 2-decomposable if and only if there exists a decomposition

$$M = A \oplus B \oplus C \oplus D \oplus E$$

satisfying the following properties:

   (a) A is Goldie torsion and $B \oplus C \oplus D \oplus E$ is nonsingular.

   (b) B has an essential socle.

   (c) C is socle free with an essential direct sum of uniform modules.

   (d) D contains no uniform submodules but has an essential direct sum of atomic modules.

   (e) E contains no atomic submodules.

   (f) A, B, C, D, E are each 2-decomposable (and hence B, C, D, E are TS).

2. The above decomposition of M is unique up to superspectivity. □

5.2.12. PROPOSITION. A module M is TS if and only if there exists a submodule $K \leq M$ such that $M = Z_2(M) \oplus K$, with both $Z_2(M)$ and K being TS, and furthermore, $Z_2(M)$ is K-injective.

PROOF. $\Longrightarrow$. Since $Z_2(M) \leq_t M$, and since M is TS, $M = Z_2(M) \oplus K$ for some $K \leq M$. But $K \perp Z_2(M)$, $K \leq_t M$. In general, since "$\leq_t$" is transitive, type submodules of any TS-module are TS. Hence $Z_2(M)$ and K are TS.

Next, let $X \leq K$ and $f : X \longrightarrow Z_2(M)$ be a module map. Define

$$N = \{x - f(x) : x \in X\} \leq K \oplus Z_2(M) = M.$$

Then $N \cap Z_2(M) = 0$. Let N^* be a complement of $Z_2(M)$ in M with $N \subseteq N^*$, i.e., $N^* \oplus Z_2(M) \leq_e M$. Since $N^* \leq_t M$ by (4.1.2)(3), $M = N^* \oplus V$ for some $V \leq_t M$. Then $Z_2(M) = Z_2(V) \leq_t V$. But V is TS, and thus $V = V_1 \oplus Z_2(M)$ for some $V_1 \leq V$. Let π be the projection $\pi : M = N^* \oplus V_1 \oplus Z_2(M) \longrightarrow Z_2(M)$ onto $Z_2(M)$ along $N^* \oplus V_1$. For $x \in X$,

$$x = (x - f(x)) + f(x) \quad \text{where} \quad x - f(x) \in N \subseteq N^* \subseteq N^* \oplus V_1,$$

and $f(x) \in Z_2(M)$. Consequently, $\pi(x) = f(x)$, and $\pi|_X = f$. But $X \subseteq K$ and $\pi|_K : K \longrightarrow Z_2(M)$ is an extension of f to K.

$\Longleftarrow$. In order to prove that $M = Z_2(M) \oplus K$ is TS, take any type submodule X of M. Then $Z_2(X) \leq_t X \leq_t M$, $Z_2(X) \leq_t M$, and $Z_2(X) \leq_t Z_2(M)$. Since $Z_2(M)$ is TS, $Z_2(M) = Z_2(X) \oplus P$ for some $P \leq Z_2(M)$. Hence

$M = Z_2(X) \oplus P \oplus K$ and $X = Z_2(X) \oplus Y$ where $Y = X \cap (P \oplus K)$ is nonsingular. Thus, $Y \cap Z_2(M) = 0$.

It is at this point that the K-injectivity of $Z_2(M)$ is used in (4.3.1) to obtain a summand Y^* of M with $Y \subseteq Y^*$ and $M = Y^* \oplus Z_2(M)$. Since $Y \leq_t X \leq_t M$, Y is a type submodule of M. But then $Y \leq_t Y^* \cong K$. Since K is TS, Y^* is TS and Y is a summand of Y^*, i.e., $Y^* = Y \oplus W$ for some $W \leq Y^*$. Therefore $M = (Y \oplus W) \oplus [Z_2(X) \oplus P]$ and $X = Z_2(X) \oplus Y$ is a summand of M. □

5.2.13. EXAMPLE. The module $A \oplus B$ will be 2-decomposable but not TS if A is atomic Goldie torsion, B atomic nonsingular such that A is not B-injective. For $R = \mathbb{Z}$, $\mathbb{Z}_n \oplus \mathbb{Z}$, where $1 < n \in \mathbb{Z}$, is such a module. For any field F, with $R = F[x]$ the polynomial ring, $(R/(x) \oplus R)_R$ is 2-decomposable but not TS, since $(R/(x))_R$ is not injective. □

Next we construct a 2-decomposable module M that does not satisfy either (C_{11}) or TS.

5.2.14. EXAMPLE. Let

$$R = \left\{ \begin{pmatrix} n & x \\ 0 & n \end{pmatrix} : n \in \mathbb{Z}, x \in \mathbb{Z}_2 \oplus \mathbb{Z}_2 \right\} \subset \begin{pmatrix} \mathbb{Z} & \mathbb{Z}_2 \oplus \mathbb{Z}_2 \\ 0 & \mathbb{Z} \end{pmatrix},$$

$$I_0 = \left\{ \begin{pmatrix} 2n & 0 \\ 0 & 2n \end{pmatrix} : n \in \mathbb{Z} \right\}, \quad I = \left\{ \begin{pmatrix} 4n & 0 \\ 0 & 4n \end{pmatrix} : n \in \mathbb{Z} \right\}, \text{ and}$$

$$J = \left\{ \begin{pmatrix} 0 & x \\ 0 & 0 \end{pmatrix} : x \in \mathbb{Z}_2 \oplus \mathbb{Z}_2 \right\}.$$

Then $J = Soc(R)$. For $M_1 = R/I$ and $M_2 = R/J$, set $M = M_1 \oplus M_2$. Let $U \leq_e R$. Then $J \subseteq U$ and $0 \neq I_0 \cap U \leq R$. So we see that $U \leq R$ is essential if and only if $U \supseteq V \oplus J$ for some $0 \neq V \subseteq I_0$.

If $r = \begin{pmatrix} n & x \\ 0 & n \end{pmatrix} \in R$ where $0 \neq n \in \mathbb{Z}$, then $r^\perp = 0$ if n is odd while $r^\perp = J$ if n is even. In either case, $r \notin Z(R)$. On the other hand, if $r = \begin{pmatrix} 0 & x \\ 0 & 0 \end{pmatrix} \in R$ where $x \neq 0$, then $r^\perp = I_0 \oplus J \leq_e R$. Thus $Z(R) = J$, and also $Z_2(R) = J$. Then $M_2 = R/J = R/Z_2(R)$ is nonsingular uniform.

Note that $M_1 = R/I$ contains an essential submodule

$$(I_0 + J)/I \cong (I_0/I) \oplus J,$$

and that I_0/I is embeddable in J. This shows that $(I_0 + J)/I$ is singular and atomic. It follows that M_1 is Goldie torsion and atomic. By (5.2.4), M is 2-decomposable. To show that M is not TS, it suffices to show that M_1 is not M_2-injective by (5.2.12). Let f be the R-homomorphism

$$f : (I + J)/J \longrightarrow R/I, \begin{pmatrix} 4n & 0 \\ 0 & 4n \end{pmatrix} + J \longmapsto \begin{pmatrix} 2n & 0 \\ 0 & 2n \end{pmatrix} + I.$$

If f extends to $g : R/J \longrightarrow R/I$. Then $g(1_R + J) = \begin{pmatrix} m & x \\ 0 & m \end{pmatrix} + I$, and

$$\begin{aligned}\begin{pmatrix} 2 & 0 \\ 0 & 2 \end{pmatrix} + I = f(4 \cdot 1_R + J) &= g(4 \cdot 1_R + J) \\ &= g(1_R + J)4 \cdot 1_R \\ &= \begin{pmatrix} m & x \\ 0 & m \end{pmatrix} 4 \cdot 1_R + I \\ &= \begin{pmatrix} 4m & 0 \\ 0 & 4m \end{pmatrix} + I = 0,\end{aligned}$$

a contradiction. So M_1 is not M_2-injective. Thus M is not TS.

To see M fails to satisfy (C_{11}), let $J_1 = \left\{ \begin{pmatrix} 0 & x \\ 0 & 0 \end{pmatrix} : x \in \mathbb{Z}_2 \oplus (0) \right\}$ and $J_2 = \left\{ \begin{pmatrix} 0 & x \\ 0 & 0 \end{pmatrix} : x \in (0) \oplus \mathbb{Z}_2 \right\}$, and form $N = [(J_2 + I)/I] \oplus (R/J)$. If M satisfies (C_{11}), then there exists a complement P of N in M such that P is a direct summand of M. Since $N \cap (J_1 + I)/I = 0$, $P \neq 0$. Since $P \cap M_2 \subseteq P \cap N = 0$, there exists an embedding $P \hookrightarrow M_1$. Thus P is Goldie torsion. Therefore, $P \subseteq Z_2(M) = M_1$, and thus P is a direct summand of M_1. But, since M_1 is indecomposable, it must be $P = M_1$. It follows that $(J_2 + I)/I \subseteq N \cap P = 0$, a contradiction. So M does not satisfy (C_{11}). □

The module decompositions given by the next two propositions were originally proved in [141]. Although stated for a TS-module M, the proofs there used only that M is 2-decomposable. The reader now should recall the concepts square free, square full, and directly finite from Definitions (5.1.10)-(5.1.12).

5.2.15. PROPOSITION. Every 2-decomposable module M has a decomposition $M = D \oplus P$ with D and $E(D)$ directly finite, $E(P)$ purely infinite, and $D \perp P$. If in addition M satisfies (T_3), then the decomposition is unique up to superspectivity.

PROOF. Define a non-empty class

$$\mathcal{F} = \{X \in \text{Mod-}R : X^{(\aleph_0)} \hookrightarrow EM\}.$$

Since $\mathcal{F}$ is closed under submodules, $c(c(\mathcal{F})) = d(\mathcal{F})$. Since M is 2-decomposable, $M = D \oplus P$ with $D \in c(\mathcal{F})$ and $P \in d(\mathcal{F})$. For a submodule N of EM, a result of [87, Lemma 1.26] says that if N is not directly finite, then for some nonzero module X, there exists an embedding $X^{(\aleph_0)} \hookrightarrow N$. This implies that $0 \neq X \in \mathcal{F} \subseteq d(\mathcal{F})$ and hence $N \notin c(\mathcal{F})$ because X and N are not orthogonal. We, therefore, conclude that both D and $E(D)$ are directly finite.

By [87, Thm.1.35], $E(P) = E_1 \oplus E_2$, where E_1 is directly finite, E_2 is purely infinite, and $E_1 \perp E_2$. If $E_1 \neq 0$, then $E_1 \subset E(P) \in d(\mathcal{F})$, or $E_1 \in d(\mathcal{F})$, and hence there exists a $0 \neq X \leq E_1 \subseteq E(M)$ with $X \in \mathcal{F}$. But $M \cap X \neq 0$, and it may be assumed that $0 \neq X \leq M$. Consequently, $X^{(\aleph_0)} \cong N \leq EM$ for some $N \in d(\mathcal{F})$, $N \perp D$. From $N \leq EM = ED \oplus EP$, $ED \perp EP$, it follows that $N \hookrightarrow EP$. Since $X \leq E_1$ and $E_1 \perp E_2$, also $N \perp E_2$. From $N \subseteq E(M) = E_1 \oplus E_2$ with $N \cap E_2 = 0$, it follows that $X^{(\aleph_0)} \cong N \hookrightarrow E_1$. By [87, Prop. 1.27], E_1 is not directly finite, which is a contradiction unless $E_1 = 0$. Therefore, $E(P) = E_2$ is purely infinite.

Suppose that $M = D_1 \oplus P_1$ is another decomposition as in the statement of the proposition. Thus $E(M) = E(D) \oplus E(P) = E(D_1) \oplus E(P_1)$. By the uniqueness in [87, Thm.1.35], $E(D_1) \cong E(D) \in c(\mathcal{F})$ and $E(P_1) \cong E(P) \in d(\mathcal{F})$. Since natural classes are closed under submodules, $D_1 \in c(\mathcal{F})$ and $P_1 \in d(\mathcal{F})$. Finally, uniqueness up to superspectivity follows now from (5.2.5) if M satisfies (T_3). □

Our next objective is to consider decompositions of a module as a direct sum of a square free module, and a square full module. The latter are characterized in the next proposition.

5.2.16. PROPOSITION. A module M is square free if and only if every complement submodule of M is a type submodule. In particular, a square free module is 2-decomposable if and only if it satisfies (C_{11}).

PROOF. $\Longrightarrow$. If not, then there exists a complement submodule $N \leq M$ which is not a type submodule. This means that N is properly contained in its type closure N^{tc} in M, and $N \| N^{tc}$. Since N is not essential in N^{tc}, there exists a $0 \neq W \leq N^{tc}$ with $N \cap W = 0$. Then there exist $0 \neq V_1 \leq N$ and $0 \neq V_2 \leq W$ with $V_1 \cong V_2$, contradicting that M is square free.

$\Longleftarrow$. If not, then there are submodules $A, B \leq M$, $A \cap B = 0$, and $0 \neq A \cong B$. Let A^c be any complement closure of A in M. Then $A^c \| (A^c \oplus B)$ contradicts that A^c is a type submodule of M. □

5.2.17. PROPOSITION. Any 2-decomposable module M has a decomposition $M = M_1 \oplus M_2$, where M_1 is square free, M_2 is square full, and $M_1 \perp M_2$. If M satisfies (T_3), this decomposition is unique up to superspectivity.

PROOF. Set $\mathcal{F} = \{X \in \text{Mod-}R : X \oplus X \hookrightarrow M\}$. Since M is 2-decomposable, $M = M_1 \oplus M_2$, where $M_1 \in c(\mathcal{F})$ is square free, $M_2 \in c(c(\mathcal{F})) = d(\mathcal{F})$, and $M_1 \perp M_2$. Assume $M_2 \neq 0$, and let $0 \neq N \leq M_2$. Since $0 \neq N \in d(\mathcal{F})$, there exists a $0 \neq Y \leq N$ such that $Y \oplus Y \cong Z \leq M$ for some Z. Since $Y \subseteq M_2$ and $M_2 \perp M_1$, also $Z \perp M_1$, and $Z \cap M_1 = 0$. Consequently $Y \oplus Y \cong Z \hookrightarrow M_2$. Therefore M_2 is square full.

Next it will be shown that if $M = A \oplus B$ is any other decomposition as in the statement of this proposition, then $A \in c(\mathcal{F})$ and $B \in d(\mathcal{F})$. Since B is square full, $B \in d(\mathcal{F})$. If $A \notin c(\mathcal{F})$, then for some $0 \neq X \leq A$, $X \in \mathcal{F}$, where $X \oplus X \cong N \leq M$. Since $0 \neq X \subseteq A$, $A \perp B$, also $X \perp B$ and $N \perp B$. But then $N \cap B = 0$, and $N \hookrightarrow A$, contradicting that A is square free. Lastly, if we now also assume (T_3), then (5.2.5) shows that these decompositions are unique up to superspectivity. □

5.2.18. EXAMPLE. A 2-decomposable module satisfying (T_3) which is not TS: For $R = \mathbb{Z}$, p a prime number, and $\mathbb{Q}$ the rationals, the module $M = \mathbb{Z}_p \oplus \mathbb{Q}$ is clearly 2-decomposable by (5.2.4). Since $Z_2(M) = \mathbb{Z}_p$ is not $\mathbb{Q}$-injective, M is not TS by (5.2.12). However, M satisfies (T_3), since M has exactly two nontrivial direct summands. □

Next, finitely and countably decomposable modules are discussed. Note that the hypothesis (i) below implies that every local type summand is a type submodule.

5.2.19. PROPOSITION. Assume that (i) every local type summand of a module M is a direct summand, and that (ii) every type direct summand of M is 2-decomposable. Then M is a direct sum of atomic modules.

PROOF. By (4.3.6), $M = \oplus_{\lambda\in\Lambda} X_\lambda$ where all X_λ are t-indecomposable and $X_{\lambda_1} \perp X_{\lambda_2}$ whenever $\lambda_1 \neq \lambda_2$. Suppose that for some $\lambda_0 \in \Lambda$, X_{λ_0} is not atomic. Then there exists a $0 \neq A < X_{\lambda_0}$ such that $X_{\lambda_0} \notin d(A)$. Since X_{λ_0} is 2-decomposable, $X_{\lambda_0} = B \oplus C$, where $B \in d(A)$ and $C \in c(A)$. Since $0 \neq A \in d(A)$, it follows that $B \neq 0$. Since $X_{\lambda_0} \notin d(A)$, $C \neq 0$. Then $X_{\lambda_0} = B \oplus C$ contradicts that X_{λ_0} is t-indecomposable. Hence all X_λ are atomic. □

Since a TS-module always satisfies (ii) of the last proposition, the next corollary follows from (5.2.19) and (4.3.9).

5.2.20. COROLLARY. Suppose that M is a TS-module and that for any chain $m_1^\perp \subseteq m_2^\perp \subseteq \cdots$ of right ideals of R, where $m_i \in M$, $t.\dim(\oplus_{i=1}^\infty R/m_i^\perp) < \infty$. Then M is a direct sum of atomic modules and hence fully decomposable (by 5.2.4). □

In all cases below, as a minimum, the module M satisfies the condition that every type direct summand of M is 2-decomposable. If R_R satisfies the t-ACC, then the condition in (2) holds, which guarantees that every local type summand of M is in fact a type submodule.

5.2.21. THEOREM. Suppose that every type direct summand of the module M is 2-decomposable. Then the following hold:

1. M is finitely decomposable. In particular, every TS-module is finitely decomposable.

2. Assume that for $y_i \in M$, $i = 0, 1, 2, \ldots$, for any countable chain of right annihilator ideals $y_0^\perp \subseteq y_1^\perp \subseteq y_2^\perp \subseteq \ldots$, $t.\dim\left(\oplus_{i=1}^\infty R/y_i^\perp\right) < \infty$. Then M is countably decomposable.

PROOF. For any finite or countable maximal set of pairwise orthogonal types $\left\{\mathcal{K}_i : i \in I\right\}$, let $I = \left\{1 < 2 < \ldots < n < \ldots\right\} \subseteq \left\{i : 1 \leq i < \omega\right\}$. By hypothesis, $M = M_1 \oplus N_1$, where $M_1 \in \mathcal{K}_1$, $N_1 \in c(\mathcal{K}_1)$, and $M_1 \perp N_1$. Assume by induction that for some $1 \leq j$, $M = (\oplus_{i=1}^j M_i) \oplus N_j$ is a type direct sum with $M_i \in \mathcal{K}_i$ for $i = 1, \cdots, j$, and $N_j \perp (\oplus_{i=1}^{j} M_i)$. Since N_j is a type direct summand of M, $N_j = M_{j+1} \oplus N_{j+1}$ with $M_{j+1} \in \mathcal{K}_{j+1}$ and $N_{j+1} \in c(\mathcal{K}_{j+1})$. Then again $M = (\oplus_{i=1}^{j+1} M_i) \oplus N_{j+1}$ is a type direct sum of the required kind with each $M_i \in \mathcal{K}_i$, and $N_{j+1} \perp (\oplus_{i=1}^{j+1} M_i)$.

(1) If the index set $I = \{1, \ldots, n\}$ is finite, since $\bigvee_{i \leq n} \mathcal{K}_i = \mathbf{1}$, $N_{n+1} = 0$, and so $M = \oplus_{i=1}^{n} M_i$

(2) Note that in view of (4.3.9), every local type summand of M is a type submodule of M. Now let $I = \{i : 1 \leq i < \omega\}$. By the above induction, $(\oplus_{i<\omega} M_i) \oplus C \leq_e M$ for some $C \leq M$, where M_i is a type submodule of type $\mathcal{K}_i$ of M, and $(\oplus_{i<\omega} M_i) \perp C$. Thus $\mathcal{K}_i \wedge d(C) = \mathbf{0}$. Since any complete Boolean lattice satisfies a limited infinite distributive law,

$$d(C) = d(C) \wedge \left(\bigvee_{i<\omega} \mathcal{K}_i \right) = \bigvee_{i<\omega} (d(C) \wedge \mathcal{K}_i) = \mathbf{0}.$$

This implies $C = 0$, and that the local type summand $\oplus_{i<\omega} M_i \leq_e M$ is a type submodule of M. Hence $M = \oplus_{i<\omega} M_i$. □

Note that for any module M over a commutative domain, the usual torsion submodule $\{m \in M : m^{\perp} \neq 0\}$ is $Z_2(M)$. The next theorem also holds for **Dedekind domains**.

5.2.22. THEOREM. For $R = \mathbb{Z}$, the following are equivalent for an Abelian group M:

1. M is 2-decomposable.

2. M is fully decomposable.

3. M is a direct sum of a torsion Abelian group, and a torsion free Abelian group.

4. Every direct summand of M is 2-decomposable.

PROOF. (3) $\Longrightarrow$ (2). Let $M = Z_2(M) \oplus K$. For a prime p, the p-torsion component of $Z_2(M)$ is atomic. Since also any torsion free Abelian group such as K is atomic, M is a direct sum of atomic modules. By (5.2.4), M is fully decomposable.

(3) $\Longrightarrow$ (4). By hypothesis, $M = Z_2(M) \oplus K$, and let N be a summand of M, where $M = N \oplus P$. Since $Z_2(M) \leq M$ is fully invariant,

$$Z_2(M) = [Z_2(M) \cap N] \oplus [Z_2(M) \cap P] = Z_2(N) \oplus Z_2(P).$$

Then $M = Z_2(N) \oplus Z_2(P) \oplus K$, and hence $N = Z_2(N) \oplus N'$ where $N' = N \cap [Z_2(P) \oplus K]$ is torsion free. Therefore, N satisfies (3), and by the above proof that (3) $\Longrightarrow$ (2), it follows that N is fully decomposable.

The implications (2) $\Longrightarrow$ (1) $\Longrightarrow$ (3), and (4) $\Longrightarrow$ (1) are immediate. Thus (1)-(4) are equivalent. □

The following proposition is a type analogue of a result of Müller and Rizvi [90].

5.2.23. PROPOSITION. Assume that every type direct summand of M is 2-decomposable. Then the following hold:

1. M has a decomposition $M = M_1 \oplus M_2$, where M_1 is molecular, M_2 is bottomless, and M_1 is essential over a type direct sum $\oplus_{i\in I} N_i$ of atomic type summands of M.

2. If in addition M satisfies (T_3), then the above decomposition is unique up to isomorphism. That is, if M has another decomposition $M = M_1^{'} \oplus M_2^{'}$, where $M_1^{'}$ is molecular, $M_2^{'}$ is bottomless, and $M_1^{'}$ is essential over a type direct sum $\oplus_{j\in J} N_j^{'}$ of atomic type summands of M, then $M_1 \cong M_1^{'}$, $M_2 \cong M_2^{'}$, and there is a bijection $\theta : I \longrightarrow J$ such that $N_i \cong N_{\theta(i)}^{'}$ for all $i \in I$.

PROOF. The existence: Let $\mathcal{K}$ be the class of molecular modules. Then $\mathcal{K}$ is a natural class and $c(\mathcal{K})$ is the class of bottomless modules by (5.1.8). Thus, there exist $M_1 \in \mathcal{K}$ and $M_2 \in c(\mathcal{K})$ such that $M = M_1 \oplus M_2$. It follows that M_1 contains an essential submodule $X = \oplus_{i\in I} X_i$ where each X_i is atomic. Without loss of generality, we may assume that $X_i \perp X_j$ for all $i \neq j$ in I. By the hypothesis, M_1 is 2-decomposable. For each $k \in I$, let $Z_k = \oplus\{X_i : i \in I, i \neq k\}$. Then, by (5.2.2), there exists a type complement N_k of Z_k in M_1 such that N_k is a summand of M_1 (and hence of M). Since $X_k \oplus Z_k \leq_e M_1$, it can easily be proved that $N_k \| X_k$. So N_k is an atomic type summand of M and $N_i \perp N_j$ whenever $i \neq j$ in I. To see that $\oplus_{i\in I} N_i$ is essential in M_1, let $(\oplus_{i\in I} N_i) \cap Y = 0$ where Y is a submodule of M_1. Since each $N_i \leq M$ is a type submodule, $Y \perp N_i$ for each $i \in I$. Thus, $Y \perp X_i$ for all $i \in I$. It follows that $Y \perp (\oplus_{i\in I} X_i)$ and so $Y \cap (\oplus_{i\in I} X_i) = 0$. Since $\oplus_{i\in I} X_i$ is essential in M_1, we have $Y = 0$.

The uniqueness: Suppose $M_1^{'}, M_2^{'}$, and $\oplus_{j\in J} N_j^{'}$ are as assumed in (2). Since M is a 2-decomposable module satisfying (T_3), by (5.2.6), $M_1 \cong M_1^{'}$ and $M_2 \cong M_2^{'}$. Since both $\oplus_{i\in I} N_i$ and $\oplus_{j\in J} N_j^{'}$ are essential in M_1 and all $N_i, N_j^{'}$ are atomic, for each $i \in I$, there exists a least one $j(i) \in J$ such that $N_i \| N_{j(i)}^{'}$. However, since $\oplus_{j\in J} N_j^{'}$ is a type direct sum, $j(i)$ is unique. And similarly, for each $j \in J$, there exists a unique $i(j) \in I$ such that $N_j^{'} \| N_{i(j)}$. This means that $\theta : I \longrightarrow J$, $\theta(i) = j(i)$ is a bijection. Since N_i and $N_{\theta(i)}^{'}$ both are type summands of M_1, write $M_1 = N_i \oplus A = N_{\theta(i)}^{'} \oplus B$ with $N_i \perp A$ and $N_{\theta(i)}^{'} \perp B$. Then N_i and $N_{\theta(i)}^{'}$ are in $d(N_i)$ and A, B are in $c(N_i)$. By (5.2.6), $N_i \cong N_{\theta(i)}^{'}$.
□

The rings R satisfying the right t-ACC were discussed and characterized in (4.3.12). The present section has given us a new concept - countable decomposability - in terms of which we are now able to give a new characterization of these rings.

5.2.24. THEOREM. The following are all equivalent for a ring R:

1. R_R satisfies t-ACC.
2. Every injective right R-module is countably (or fully) decomposable.
3. Every TS-module is countably (or fully) decomposable.
4. If M is any module whose type direct summands are 2-decomposable, then M is countably decomposable.

PROOF. (1) $\Longrightarrow$ (3) is by (5.2.20), (1) $\Longrightarrow$ (4) follows by (5.2.21), and (3) $\Longrightarrow$ (2) is obvious.

(2) $\Longrightarrow$ (1) and (4) $\Longrightarrow$ (1). Either of (2) and (4) guarantees that every injective module is countably decomposable. By (4.3.12), R_R satisfies t-ACC if and only if for any countable family $\{F_i : i \in I\}$ of pairwise orthogonal injective modules, with I countable, $\oplus_{i\in I} F_i$ is injective. Set $F = E(\oplus_{i\in I} F_i)$. Then $\{c(F)\} \cup \{d(F_i) : i \in I\}$ is a countable maximal set of pairwise orthogonal types. Since F is countably decomposable, $F = A \oplus (\oplus_{i\in I} A_i)$ for some $A \in c(F)$ and $A_i \in d(F_i)$. Trivially, $A = 0$, and for any $j \in I$, $F = A_j \oplus (\oplus_{j\neq i\in I} A_i)$ with $A_j \in d(F_j)$ and $\oplus_{j\neq i\in I} A_i \in c(F_j)$. Since F is injective, by (5.1.7), F_j is superspective to A_j, and consequently $F_j \cong A_j$ for all $j \in I$. But then $F = E(\oplus_{i\in I} F_i) = \oplus_{i\in I} A_i \cong \oplus_{i\in I} F_i$ shows that $\oplus_{i\in I} F_i$ is injective as required in order to show that R satisfies (1). □

The next lemma and proposition give a general and flexible method of constructing fully decomposable TS-modules satisfying (T_3).

5.2.25. LEMMA. Let $N \leq_t M = \oplus_{i\in I} M_i$ be a type submodule of M, where the M_i, $i \in I$, are atomic. Then for any $i \in I$, either $N \cap M_i = 0$, or $N \cap M_i \leq_e M_i$.

PROOF. If $(N \cap M_i) \oplus D \leq M_i$ with both $N \cap M_i$ and D nonzero, then $N \cap M_i$ and D have nonzero isomorphic submodules, because M_i is atomic. But since $N + D = N \oplus D$, it follows that $N \perp D$. But then $(N \cap M_i) \perp D$ contradicts that M_i is atomic. □

It is rare that one is able to completely and explicitly identify all type submodules of a module as completely as in the next proposition. Note that the hypothesis below implies that for all $j \neq i \in I$, $E(M_i) \leq E(M)$ is fully invariant and that $\mathrm{Hom}_R(EM_i, EM_j) = 0$, and that the latter two properties are equivalent.

5.2.26. PROPOSITION. Suppose that $M = \oplus_{i\in I} M_i$ where all the M_i, $i \in I$, are atomic. For any $j \in I$, define $K_j = \oplus\{M_i : i \in I, i \neq j\}$, and assume that $\text{Hom}_R(EK_j, EM_j) = 0$. Then every nonzero type submodule N of M is given by

$$N = \oplus \{M_i : i \in I, M_i \subseteq N\} = \oplus \{M_i : i \in I, M_i \cap N \neq 0\}.$$

PROOF. Denote by $\pi_i : E(M) = E(M_i) \oplus E(K_i) \longrightarrow E(M_i)$ the natural projection that extends the canonical projection from M onto M_i along K_i, and define $I(0) = \{i \in I : M_i \subseteq N\} \subseteq I(1) = \{i \in I : \pi_i N \neq 0\}$. It suffices to show that for any $j \in I(1)$, also $j \in I(0)$. Choose the notation so that $j = 1$, $\pi = \pi_1$, $K = K_1$, and $p : E(M) = E(M_1) \oplus E(K) \longrightarrow E(K)$ is the other natural projection that extends the canonical projection from M onto K along M_1. By (4.3.7), $(N \cap M_1) \oplus (N \cap K) \leq_e N$.

Case 1. $N \cap M_1 = 0$. Then also $N \cap E(M_1) = 0$, and $N \cap K \leq_e N$. If $p\big|_N$ is the restriction $p\big|_N : N \longrightarrow K$ of p to N, then the kernel of $p\big|_N$ is $Ker(p\big|_N) = N \cap E(M_1) = N \cap M_1 = 0$. Consequently the map $p\big|_N : N \longrightarrow K$, $n \longmapsto pn$ is monic. Let g be the monic inverse map $g : pN \longrightarrow N$, where $g(pn) = n$. Define $\varphi = \pi g : pN \longrightarrow M_1$. Since $pN \subseteq K \subseteq E(K)$, extend φ to $\widehat{\varphi} \in \text{Hom}_R(EK, EM_1)$. Note that $gpN = N \neq 0$. If $\pi\big|_N : N \longrightarrow E(M_1)$ is not zero, then also $\widehat{\varphi} pN = \varphi pN = \pi gpN = \pi N \neq 0$. Then $0 \neq \widehat{\varphi} \in \text{Hom}_R(K, EM_1)$ contradicts the hypothesis of this proposition. Therefore, $\pi N = 0$, which contradicts that $1 \in I(1)$.

Case 2. $N \cap M_1 \neq 0$. So far the fact that the M_i are atomic has not been used, but now it is needed to conclude that $N \cap M_1 \leq_e M_1$ by use of (5.2.25). It will be shown that $M_1 \subseteq N$. Form $N \subseteq E(N) \subseteq E(M)$, and since $N \cap M_1 \subseteq N$, choose some injective hull of $N \cap M_1$ inside $E(N)$, i.e., $E(M_1) \cong E(N \cap M_1) \subseteq E(N)$. But by hypothesis $E(M_1) \leq E(M)$ is fully invariant, and hence $E(M)$ contains a unique injective hull of M_1, i.e., $E(M_1) \subseteq E(N)$. Since $N \leq M$ is a complement submodule, and since $N \leq_e E(N) \cap M$, necessarily $N = E(N) \cap M$. Also $E(M_1) \cap M = M_1$. Consequently

$$E(M_1) \subseteq E(N) \Longrightarrow M \cap E(M_1) \subseteq M \cap E(N) \Longleftrightarrow M_1 \subseteq N.$$

Thus, $1 \in I(0)$ and in all cases $I(0) = I(1)$. □

5.2.27. COROLLARY. Under the hypotheses and with the notation of the last proposition, suppose that $P \leq \oplus_{i \in I} M_i = M$ is any submodule. Then $N = \oplus \{M_i : i \in I, P \cap M_i \neq 0\}$ is the unique type closure of P in M.

PROOF. If Q is a type closure of P in M, then by (5.2.26)

$$Q = \oplus \{M_i : i \in I, M_i \cap Q \neq 0\}.$$

But, because $M_i \leq_t M$, $M_i \cap Q \neq 0$ iff $M_i \cap P \neq 0$. Then

$$Q = \oplus \{M_i : i \in I, M_i \cap Q \neq 0\} = \oplus \{M_i : i \in I, M_i \cap P \neq 0\} = N.$$

□

5.2.28. COROLLARY. If the module $M = \oplus_{i\in I} M_i$ satisfies the hypotheses of (5.2.26), then so does also the module $\oplus_{i\in I} E(M_i)$; and hence the conclusions of (5.2.26) and (5.2.27) also hold for $\oplus_{i\in I} E(M_i)$. □

5.2.29. At the time of this writing, we do not know if every 2-decomposable module is 3-decomposable (or finitely decomposable), if every countably decomposable module is fully decomposable, and if every type summand of a 2-decomposable module is again 2-decomposable.

5.2.30. REFERENCES. Dauns [32]; Dauns-Zhou [43]; Goodearl [66]; Goodearl-Boyle [68]; Mohamed-Müller [87]; Müller-Rizvi [89,90]; Zhou [141].

5.3 Unique Type Closure Modules

In the previous two sections decompositions of modules as orthogonal type direct sums were unique only up to superspectivity. For nonsingular modules, all of these decompositions are unique. In fact, the whole theory of natural classes at first for many years was done for nonsingular modules. Thus we now develop a bigger class of modules, called unique type closure modules (abbreviation: UTC), in which any type direct sum decomposition will be unique.

Clearly, a sufficient condition which will guarantee that all type direct sum decompositions of a module M are unique is that M has a unique type submodule of type $\mathcal{K}$ for every natural class $\mathcal{K}$. The next proposition shows that two other conditions are equivalent to this one. The next theorem will give three more counter-intuitive and harder to prove equivalent conditions.

5.3.1. PROPOSITION. For a module M, the following conditions are equivalent:

1. For every natural class $\mathcal{K}$, M has a unique type submodule of type $\mathcal{K}$.
2. Every submodule of M has a unique type closure in M.
3. For every natural class $\mathcal{K}$, $\Sigma\{X : X \leq M, X \in \mathcal{K}\} \in \mathcal{K}$, i.e., M has a largest submodule in $\mathcal{K}$.

PROOF. (2) $\Longrightarrow$ (1). For a natural class $\mathcal{K}$, let P, Q both be type submodules of type $\mathcal{K}$ of M. If $(P\cap Q)\oplus D \leq P$ with $0 \neq D \leq M$, then $D\cap Q = D\cap P\cap Q = 0$. Hence $Q + D = Q \oplus D \in \mathcal{K}$ since $D \subseteq P \in \mathcal{K}$, a contradiction. Hence $P \cap Q \leq_e P$, $P \cap Q \leq_e Q$, and both P and Q are type closures in M of the same submodule $P \cap Q$. By (2), $P = Q$.

(3) $\Longrightarrow$ (2). For any $N \leq M$, set $\mathcal{K} = d(N)$. Any type closure N^{tc} of N is a type submodule of type $\mathcal{K}$ of M, and hence N^{tc} is uniquely given by (3).

(1) $\Longrightarrow$ (3). It suffices to show that for $A, B \in \mathcal{K}$, also $A + B \in \mathcal{K}$. By Zorn's Lemma, there exists a maximal submodule X in the set

$$\{X : X \leq M, A \subseteq X \in \mathcal{K}\}.$$

Such an X is a type submodule of type $\mathcal{K}$. Similarly, there exists a type submodule Y of type $\mathcal{K}$ with $B \subseteq Y$. By (1), $X = Y$. But then $A \subseteq X$, $B \subseteq X$, hence $A + B \subseteq X \in \mathcal{K}$, and $A + B \in \mathcal{K}$. □

5.3.2. DEFINITION. A **unique type closure module** (abbreviation: UTC) is any module M satisfying the equivalent conditions in (5.3.1). A module M each of whose submodules has a unique complement closure in M

is called a **unique complement closure module** (abbreviation: UC). (See Smith [109, p.302].)

A **partial homomorphism** from a module M to a module N is a homomorphism from a submodule of M to N. A partial endomorphism of M is a partial homomorphism from M to M.

5.3.3. THEOREM. For a module M, the following are all equivalent:

1. For any nonzero submodule N of M, if $C_1 \neq C_2$ are two complement closures of N in M, then there exists a $0 \neq X \leq C_1 + C_2$ such that $C_1 \cap X = 0$ and $X \hookrightarrow N$.

2. There does not exist an R-module X and a proper essential submodule $Y <_e X$ such that $X \perp (X/Y)$ and $X \oplus (X/Y) \hookrightarrow M$.

3. For every partial endomorphism $f : A \longrightarrow M$, $A \leq M$ with $fA \perp A$, the kernel $Ker(f) \leq A$ is a complement submodule of A.

4. M is a UTC-module.

PROOF. (2) $\Longrightarrow$ (5.3.1)(1). If (5.3.1)(1) fails, then for some $\mathcal{K} \in \mathcal{N}(R)$, M has type submodules T_1 and T_2, both of type $\mathcal{K}$ with $T_1 \neq T_2$. As in the proof of "(2) $\Longrightarrow$ (1)" in (5.3.1), $0 \neq T_1 \cap T_2 \leq_e T_i$ for $i = 1, 2$. Since $T_1 + T_2 \notin \mathcal{K}$, $T_1 \cap T_2$ is not essential in $T_1 + T_2 \notin \mathcal{K}$ (or since the T_i are complement submodules of M), $T_1 \cap T_2$ is not essential in $T_1 + T_2$. There exists a $0 \neq A \leq T_1 + T_2$, $T_1 \cap T_2 \cap A = 0$. Since $T_1 \cap T_2 \leq_e T_1$, also $T_1 \cap A = 0$. Similarly, $T_2 \cap A = 0$. Since $T_i \leq_t M$, $T_i \perp A$. Thus

$$A \cong A/(T_1 \cap A) \cong (A + T_1)/T_1 \subseteq (T_2 + T_1)/T_1 \cong T_2/(T_1 \cap T_2).$$

But then $A \cong X/(T_1 \cap T_2)$ for some X with $T_1 \cap T_2 \leq_e X \leq_e T_2$. Since $A \perp T_2$ while $X \subseteq T_2$, $A \perp X$. The latter implies that $X + A = X \oplus A \subseteq M$. Set $Y = T_1 \cap T_2 \leq_e X$. Since $0 \neq A \cong X/Y$, $Y \neq X$. And again, $(X/Y) \perp X$. This contradicts (2). Thus $T_1 = T_2$.

(5.3.1)(2) $\Longrightarrow$ (3). If not, there exists a nonzero module X and a nonzero homomorphism $\alpha : X \longrightarrow M$ such that $Y := Ker\alpha <_e X$ and $\alpha(X) \perp X$. Fix an $x \in X \backslash Y$. Write $x^{-1}Y = \{r \in R \mid xr \in Y\} \leq R$. Define β to be the restriction of α to xR, i.e., $\beta : xR \longrightarrow M$. Then $Ker(\beta) = x(x^{-1}Y)$. Let L be a type closure of $Ker(\beta)$ in xR. Define a monic map $f : xR \longrightarrow xR \oplus \beta(xR)$ by $f(xr) = xr + \beta(xr)$, $r \in R$.

Note that the restriction of f to $Ker(\beta)$ is the identity and $Ker(\beta) \subseteq L \cap fL$. Let L^{tc} and $(fL)^{tc}$ denote the type closures of L and fL in M, respectively. Then we have the following implications:

$$Ker(\beta) \parallel L \ , L \parallel L^{tc} \Longrightarrow Ker(\beta) \parallel L^{tc};$$
$$Ker(\beta) \subseteq fL \ , L \cong fL \Longrightarrow Ker(\beta) \parallel fL;$$
$$Ker(\beta) \parallel fL \ , fL \| (fL)^{tc} \Longrightarrow Ker(\beta) \parallel (fL)^{tc}.$$

From the latter it now follows that both L^{tc} and $(fL)^{tc}$ are type closures of $Ker(\beta)$ in M. By hypothesis (5.3.1)(2), $L^{tc} = (fL)^{tc}$. Hence $L^{tc} = L^{tc}+(fL)^{tc} \supseteq L+fL$. Then $L \parallel (L+fL)$. Since $L \leq_t xR$, and $xR \perp \beta xR$, L is a type submodule of $xR \oplus \beta xR$ with $L \perp \beta L$. But $L \parallel (L+fL) \leq xR \oplus \beta xR$, and $\beta L \subseteq L + fL$. It follows that $L \| \beta L$ and $L \perp \beta L$. It must be that $\beta L = 0$. Thus $Ker\beta = L \leq_t xR$. Since $xR/Ker(\beta) \cong \beta xR$, we have that $(xR/Ker(\beta)) \perp xR$. If $Ker\beta \oplus D \leq xR$ with $D \neq 0$, then $D \perp xR$ is a contradiction. So $Ker\beta = xR$ and $0 = \beta x = \alpha x$ is a contradiction of $x \notin Y$.

(1) $\Longrightarrow$ (5.3.1)(3). It suffices to show that for any $\mathcal{K} \in \mathcal{N}(R)$, and $Y, Z \leq M$ with $Y \in \mathcal{K}$ and $Z \in \mathcal{K}$, also $Y + Z \in \mathcal{K}$. By Zorn's Lemma there exist submodules P and Q of M maximal with respect to $Y \subseteq P \in \mathcal{K}$ and $Z \subseteq Q \in \mathcal{K}$. As in the proof of (2) $\Longrightarrow$ (1) in (5.3.1), $P \cap Q \leq_e P$ and $P \cap Q \leq_e Q$, and P and Q being type submodules of type $\mathcal{K}$ of M are, in particular, complement submodules of M. If $P \neq Q$, use (1) to get a submodule $0 \neq X \subseteq P+Q$ with $P \cap X = 0$, such that $X \hookrightarrow P \cap Q \in \mathcal{K}$. Then $P \subset P \oplus X \in \mathcal{K}$ is a contradiction. Thus $P = Q$, and hence $X + Y \subseteq P \in \mathcal{K}$.

(5.3.1)(2) $\Longrightarrow$ (1). For $0 \neq N \leq M$, let $C_1 \neq C_2$ be two complement closures of N in M. Set $\mathcal{K} = d(N)$. Since $N \leq_e C_1$ and $N \leq_e C_2$, the type closures C_1^{tc} and C_2^{tc} of C_1 and C_2 are also two type closures of N in M. By (5.3.1)(2), $C_1^{tc} = C_2^{tc}$. Hence $C_1 + C_2 \subseteq C_1^{tc} \in \mathcal{K}$. Since $C_1 \neq C_2$, $C_1 \cap A = 0$ for some $0 \neq A \subseteq C_1 + C_2$. But then $A \in \mathcal{K}$, and $C_1 \oplus A \subseteq C_1^{tc} \in d(N)$. Thus, there exists a nonzero submodule $0 \neq X \leq A$ such that $X \hookrightarrow N$. Since $C_1 \cap A = 0$, also $C_1 \cap X = 0$ as required in (1).

(3) $\Longrightarrow$ (2). If (2) fails, there exists an embedding $h : X \oplus (X/Y) \hookrightarrow M$, where $Y \leq_e X$, $X \neq Y$, and $X \perp (X/Y)$. Let $\pi : X \longrightarrow X/Y$ be the natural epimorphism. Set $A = hX$ and define $f = h\pi h^{-1} : A \longrightarrow M$. Then $fA = h(X/Y)$. Each $a \in A$ is uniquely of the form $a = hx$ with $x \in X$, and $fa = h\pi h^{-1}(hx) = h(x + Y)$. Consequently, $Ker(f) = hY$. Since $A \cong X$, $fA \cong X/Y$, we have $A \perp fA$. From $Y \leq_e X$ and because h is an embedding, it follows that $Ker(f) = hY \leq_e hX = A$, and $Ker(f)$ is not a complement submodule of A, which is a contradiction.

(2) $\Longrightarrow$ (3). Assume that $f : A \longrightarrow M$, $fA \perp A$, but $Ker(f) \leq A$ is not a complement submodule, where $A \leq M$. Then $A + fA = A \oplus fA \leq M$. Let X be a complement closure of $Y = Ker(f)$ in A, where $Y \leq_e X$, and $Y \neq X$. Then $fX \cong X/Y$, and $X \oplus (X/Y) \cong X \oplus fX \leq A \oplus fA \leq M$. So (2) does not hold, which is a contradiction. □

5.3.4. COROLLARY. A module M is not a UTC-module if and only if there exists an $A \leq M$, and a nonzero homomorphism $f : A \longrightarrow M$ where $Ker(f) \leq_e A$ and $Ker(f) \perp Im(f)$. □

5.3.5. EXAMPLES. The following statements hold:

1. Any UC-module is UTC. In particular, any semisimple module is UTC.

2. In particular, all nonsingular modules are UTC.

3. All atomic modules are UTC.

4. Let M be any module satisfying the hypotheses of (5.2.26). By (5.2.27), M is a UTC module.

5. For the ring $R = \mathbb{Z}$, an Abelian group M is UTC if and only if M is torsion free, or M is torsion.

PROOF. (1) Suppose that $N \leq M$ has type closures A and B in M. Then $A \cap B \leq_e B$ (as well as $A \cap B \leq_e A$). For if not, then for some $0 \neq U \leq B$, $A \cap U = 0$. But then $A \,\|\, (A \oplus U)$ is a contradiction. Thus A and B are complement closures of $A \cap B$, and $A = B$.

(5) If M is torsion free, then it is nonsingular atomic, and hence UTC by either (1) or (3). If M is mixed, it contains an element $y \in M$ of prime order p, $yp = 0$, and another element $x \in M$ of infinite order. Then $x\mathbb{Z}+y\mathbb{Z} = x\mathbb{Z}\oplus y\mathbb{Z}$, and $y\mathbb{Z} \cong x\mathbb{Z}/xp\mathbb{Z}$. By (5.3.3)(2), M is not UTC.

Let M be a torsion Abelian group and let p_i be the ith prime number. Then $M = \oplus_{i\in I} M_i$ where M_i is the p_i-component of M. Since $E(M_i)$ is a p_i-torsion Abelian group, $\text{Hom}_{\mathbb{Z}}(EM_i, EM_j) = 0$ if $i \neq j$, and hence

$$\text{Hom}_R(EK_j, EM_j) = 0 \quad \text{where } K_j = \oplus\{M_i : i \geq 1, i \neq j\}.$$

Thus by (5.2.27), M is UTC. □

Theorem (5.3.3)(2) shows that submodules of UTC-modules are UTC. For any modules $X \leq M$, X is called a UTC-submodule of M if X itself is a UTC-module. Conditions will be found guaranteeing that an essential extension of a UTC-module is UTC, and that a type direct sum of UTC-modules is UTC.

The following proposition is in general false if in it "UTC-submodule of M" is replaced with "direct summand of M". Note that the zero module is UTC.

5.3.6. PROPOSITION. Suppose that $\{M_i \leq M : i \in I\}$ is an ascending chain of UTC-submodules of M. Then $\bigcup\{M_i : i \in I\}$ is a UTC-submodule of M. Hence every module contains maximal UTC-submodules.

PROOF. If $\bigcup_{i\in I} M_i$ is not UTC, then by (5.3.3)(2), for some $X \neq 0$ and some $Y <_e X$ with $X \perp (X/Y)$, there exists a monomorphism

$$\theta : X \oplus X/Y \longrightarrow \cup M_i.$$

Take $x \in X\backslash Y$. Then $xR \cap Y <_e xR$ and $xR \perp (xR/xR \cap Y)$, and

$$\theta : xR \oplus (xR + Y)/Y \;(\cong xR \oplus xR/xR \cap Y) \longrightarrow M_i$$

is monic for some i. This contradicts that M_i is UTC. □

Under what types of operations is the class of all UTC-modules closed? For $R = \mathbb{Z}$, both $\mathbb{Q}$ and $\mathbb{Z}_{3^\infty}$ are UTC-modules. However, $\mathbb{Q} \oplus \mathbb{Z}_{3^\infty}$ is not.

The next example shows that the class of UTC-modules is not closed under essential extensions.

5.3.7. EXAMPLE. Let $M = \oplus_{i=1}^{\infty}\mathbb{Z}_{p_i}$ where p_i is the ith prime number. Let $R = \left\{\begin{pmatrix} n & x \\ 0 & n \end{pmatrix} : n \in \mathbb{Z}, x \in M\right\}$. Then R is a ring under the usual addition and multiplication of matrices, and $Soc(R) = \left\{\begin{pmatrix} 0 & x \\ 0 & 0 \end{pmatrix} : x \in M\right\}$ is essential in R_R. Since $Soc(R_R)$ is semisimple, it is clearly UTC. To see R_R is not UTC, let $N = \oplus_{i\geq 2}\mathbb{Z}_{p_i}$, $A = \left\{\begin{pmatrix} n & x \\ 0 & n \end{pmatrix} : n \in 2\mathbb{Z}, x \in N\right\}$, and $B = \left\{\begin{pmatrix} n & x \\ 0 & n \end{pmatrix} : n \in 4\mathbb{Z}, x \in N\right\}$. Then A and B are right ideals of R. Moreover, $A \leq_e B$, $A \perp (A/B)$, and $A \oplus (A/B)$ embeds in R_R. Thus R_R is not UTC. □

For a UTC-module M, and for any essential submodule $N \leq_e M$, every UTC-essential extension of N will be shown to be given by $N \leq X \leq E(M)$, for $X \in \varphi_t(M)$ where the latter set is defined below.

5.3.8. DEFINITION. For a module M, define

$$\begin{aligned}\varphi_t(M) = \{X \leq E(M) : \forall Y \leq X \;\text{ and } f \in \text{End}_R(EM) \text{ with } Y \perp f(Y),\\ f(Y \cap M) = 0 \;\text{ implies } fY = 0\}.\end{aligned}$$

Define almost as in [18, p.253] the set $\varphi(M)$ by

$$\begin{aligned}\varphi(M) = \{X \leq E(M) : \forall Y \leq X \;\text{ and } f \in \text{End}_R(EM) \text{ with } Y \cap f(Y) = 0,\\ f(Y \cap M) = 0 \text{ implies } f(Y) = 0\}.\end{aligned}$$

5.3.9. THEOREM. For a module M, the following hold:

1. $M \in \varphi(M) \subseteq \varphi_t(M)$.
2. $\varphi_t(M)$ has maximal elements.
3. If M is UTC, then every $X \in \varphi_t(M)$ is UTC.
4. If $X \leq_e E(M)$ and $X \notin \varphi_t(M)$, then X is not UTC.

PROOF. (1) Clearly, $\varphi(M) \subseteq \varphi_t(M)$. In the definition of $\varphi(M)$, take $X = M$. Then $Y \leq X = M$, and hence $Y \cap M = Y$. But then the condition $f(Y \cap M) = 0$ simply means that $f(Y) = 0$. The latter guarantees that $M \in \varphi(M)$.

(2) Let $\{X_i\}_i$ be a chain in $\varphi_t(M)$. Let $X = \bigcup_i X_i$, $Y \leq X$, and $f \in \text{End}_R(EM)$ with $f(Y \cap M) = 0$ and $Y \perp f(Y)$. We need to show that $f(Y) = 0$. If $Y_i = Y \cap X_i$, then $Y = \bigcup_i Y_i$, and $f(Y_i \cap M) = 0$. From

$Y_i \perp f(Y_i)$, and $Y_i \leq X_i \in \varphi_t(M)$, we conclude that all $f(Y_i) = 0$, and hence $f(Y) = 0$ as required.

(3) Suppose that $X \in \varphi_t(M)$ is not a UTC-module, while M is one. By (5.3.4), there exists a submodule $B \leq X$ and a homomorphism $0 \neq g : B \longrightarrow X$ with $B \perp g(B)$ and $Ker(g) \leq_e B$. We will use this information to construct a partial endomorphism $f : A \longrightarrow M$ of M which violates Theorem (5.3.3)(3).

Since $M \leq_e E(M)$, $g(B) \cap M \neq 0$; so $A := g^{-1}(g(B) \cap M)$ is a nonzero submodule of B. Let $f : A \longrightarrow M$ be the restriction of g to A. Thus, $A \perp f(A)$, because $B \perp g(B)$ and $f \neq 0$. Moreover, $Kerf = Kerg \cap A \leq_e A$, because $Kerg \leq_e B$. This shows that M is not UTC by (5.3.4).

(4) For X as in (4), since $X \notin \varphi_t(M)$, there exists a $g \in \mathrm{End}_R(EM)$ and $Y \leq X$ with $Y \perp g(Y)$, $g(M \cap Y) = 0$, but $g(Y) \neq 0$. Our objective will be to show that X satisfies (5.3.4). Since $M \leq_e E(M)$ and $X \leq_e E(M)$, we have $X \cap M \cap g(Y) \neq 0$. This means that there exists an element $y_0 \in Y$ such that $0 \neq g(y_0) \in X \cap M$. Then $y_0 \in Y \cap g^{-1}(X \cap M \cap gY) = A$. Thus $g(A) \neq 0$, and $g(A) \subseteq g(g^{-1}X) \subseteq X$, and $M \cap Y \subseteq Ker(g) \cap Y \subseteq A \subseteq Y$, and $M \cap Y \leq_e Y$. Since $g(A) \subseteq X$, let $f : A \longrightarrow X$ be the restriction of g. Then $Ker(f) = Ker(g) \cap A \leq_e A$, and $A \perp f(A)$, which is a consequence of $Y \perp f(Y)$ and $A \subseteq Y$. By (5.3.4), X is not UTC. □

The next corollary is a consequence of (5.3.9) above.

5.3.10. COROLLARY. Every UTC-module has a maximal UTC-essential extension. □

5.3.11. REMARK. The previous proof of Theorem (5.3.9) used some ideas from Camillo and Zelmanowitz [18]. They define M to be a **dimension module** if for any $A, B \leq M$,

$$u.\dim(A + B) + u.\dim(A \cap B) = u.\dim(A) + u.\dim(B),$$

where "$u.\dim$" denotes the finite uniform dimension, or $+\infty$ if it does not exist. One can combine our previous Theorem (5.3.9) with [18, Theorem 8, p.253] and [109, Proposition 14, p.311] to deduce the following.

Assume that every finitely generated submodule of M has finite uniform dimension, and that M is a UC-module and hence automatically a UTC-module. Then $\varphi(M)$ are precisely all the essential submodules $X \leq_e E(M)$ such that X is a dimension module. Furthermore, all of these modules in $\varphi(M)$ are UC (and hence UTC). □

The above remark is never needed or used later on. However, it does explain why our next theorem below is a type analogue of a theorem about dimension modules (Camillo and Zelmanowitz [18, Theorem 13, p.256]). Note that the partial homomorphism hypothesis below is equivalent to $\mathrm{Hom}_R(EM_i, EM_j) = 0$ for all $i, j \in I$ with $i \neq j$.

5.3.12. THEOREM. Let $M = \oplus_{i \in I} M_i$, where $M_i \perp M_j$ for all $i \neq j$ in I. Then M is UTC if and only if each M_i is UTC, and every partial homomorphism between two distinct summands M_i is zero.

PROOF. $\Longrightarrow$. For any homomorphism $h : A \longrightarrow M_j$ with $A \leq M_i$ and $i \neq j$, it follows that $A \perp h(A)$. Take any B such that $Ker(h) \oplus B \leq_e A$. Since $Ker(h)$ is a complement submodule of A by (5.3.3)(3), B embeds as an essential submodule in $A/Ker(h) \cong hA$. This contradicts that $M_i \perp M_j$ unless $h(A) = 0$, or $h = 0$.

$\Longleftarrow$. For any subset J of I, write $M(J) = \oplus_{i \in J} M_i$, and let

$$S = \{J \subseteq I : M(J) \text{ is UTC}\}.$$

By (5.3.6), S is an inductive set; so Zorn's Lemma shows that S has a maximal element, say J. We next show that $J = I$. If $I \neq J$, take $i \in I \backslash J$. We show that this will lead to a contradiction by proving that $J \cup \{i\} \in S$. The latter will follow from the next claim.

Claim. If M_1, M_2 are UTC with $M_1 \perp M_2$ such that every partial homomorphism between M_1 and M_2 is zero, then $M = M_1 \oplus M_2$ is UTC.

Proof of Claim. By contradiction, assume M is not UTC, and hence (5.3.3)(3) fails. So for some $A \leq M$, and homomorphism $0 \neq f : A \longrightarrow M$ with $A \perp fA$, $Ker(f) \leq_e A$ by (5.3.4). This implies $f(A) \neq 0$.

Case 1: $f(A) \cap M_1 \neq 0$. Let $B = f^{-1}(f(A) \cap M_1) \subseteq A$ and consider $g = f|_B : B \longrightarrow M_1$. Then $B \neq 0$ and $g \neq 0$. Moreover, $B \perp g(B)$ since $A \perp f(A)$. Now $Ker(g) = B \cap Ker(f) \leq_e B$, because $Ker(f) \leq_e A$. Therefore, M_1 is not UTC by (5.3.4), a contradiction.

Case 2: $f(A) \cap M_2 \neq 0$. This will lead to a contradiction as in case 1.

Case 3: $f(A) \cap M_1 = 0$ and $f(A) \cap M_2 = 0$. Let $\pi_i : M \longrightarrow M_i$ $(i = 1, 2)$ be the natural projection. Then $\pi_1 f(A) \neq 0$ and $\pi_2 f(A) \neq 0$, and

$$\pi_1 f(A) \longrightarrow \pi_2 f(A),\ \pi_1 f(a) \longmapsto \pi_2 f(a),\ a \in A,$$

gives an isomorphism. This contradicts the assumption that every partial homomorphism between M_1 and M_2 is zero. Therefore, the claim is proved. □

5.3.13. OBSERVATIONS. For modules $N \leq M$, the following hold:

1. Suppose that there exists a unique injective hull $E(N)$ of N in $E(M)$. Then N has a unique complement closure N^c in M, and N^c is a type submodule of M.

2. If $E(N) \leq E(M)$ is fully invariant, then N has a unique injective hull $E(N)$ in $E(M)$, and the previous observation (1) applies.

3. If M is UTC and $N \leq E(M)$ is fully invariant, then N has a unique complement closure N^c in M which also is the unique type closure of N in M.

PROOF. (1) If $N \subseteq C_i \leq M$, $i = 1, 2$, are two complement closures of N, then $E(C_1) = E(C_2) = E(N)$, and $N \leq_e C_1 + C_2 \leq_e E(N)$ implies that $C_1 = C_2 = N^c$. If $N^c \leq M$ is not a type submodule, then there exist submodules $A, V_1, V_2 \leq M$ such that $A \oplus V_1 \leq_e N$, $N \oplus V_2 \leq M$, and $0 \neq V_1 \cong V_2$. Then $E(N) = E(A) \oplus E(V_1) \cong E(A) \oplus E(V_2)$ is a contradiction.

(3) If $C_1 \neq C_2$ are two complement closures of N in M, then by (5.3.3)(1), there exists a $0 \neq X \subseteq C_1 + C_2$ such that $C_1 \cap X = 0$ and $X \hookrightarrow N$. The latter contradicts the full invariance of N in $E(M)$. Thus $C_1 = C_2 = N^c$. If T is any type closure of N in M with $T \neq N^c$, then a complement closure of N in T is a complement closure of N in M (since T is a complement submodule of M); so $N^c \subseteq T$, and hence $N^c \oplus V_2 \leq T$, and $0 \neq V_2 \cong V_1 \leq N$ for some submodules $V_i \leq M$. If $f : V_1 \longrightarrow V_2$ is an isomorphism, then any extension $\widehat{f} : E(M) \longrightarrow E(M)$ of f contradicts the full invariance of N in $E(M)$. Thus N^c is the unique type closure of N in M. □

Note that $\mathbb{Z} \oplus \mathbb{Z}_2$ is not UTC though $\mathbb{Z}, \mathbb{Z}_2$ are UTC and $\mathbb{Z} \perp \mathbb{Z}_2$ (see 5.3.5(5)). For a UTC-module M, if M is 2-decomposable, then M is TS because any type submodule N of M is the unique type submodule of type $d(N)$, and hence is a direct summand of M. Thus far we know that every TS-module is finitely decomposable (by 5.2.21), and if it also satisfies (T_3), then any finite type decomposition is unique up to superspectivity (by 5.2.6).

5.3.14. PROPOSITION. Let the module M be both TS and UTC and let $\{\mathcal{K}_1, \cdots, \mathcal{K}_n\}$ be any maximal set of pairwise orthogonal types. Then M has a unique decomposition $M = M_1 \oplus \cdots \oplus M_n$ where $M_i \in \mathcal{K}_i$. If in addition, M is also nonsingular, then all M_i are fully invariant submodules.

PROOF. The existence follows because M is finitely decomposable. For the uniqueness, suppose that $M = M_1 \oplus \cdots \oplus M_n = N_1 \oplus \cdots \oplus N_n$ with $M_i, N_i \in \mathcal{K}_i$. Then both M_i and N_i are type submodules of type $\mathcal{K}_i$ of M. Therefore, $M_i = N_i$ by (5.3.1)(1).

If M is nonsingular, then for any map $f : M_i \longrightarrow M_j$ where $i \neq j$, $Ker(f)$ is a complement submodule of M_i. Hence $M_i/Ker(f) \in \mathcal{K}_i \cap \mathcal{K}_j = \{0\}$, and $f = 0$. Therefore $M_i \leq M$ is fully invariant. □

5.3.15. COROLLARY. A TS-module M has type direct sum decompositions with respect to the following finite maximal sets of pairwise orthogonal types.

1. C, D (5.1.8).
2. $C \cap A$, $D \cap A$, B.
3. $\mathcal{I}_1, \mathcal{I}_2, \mathcal{I}_3$.
4. $\mathcal{I}_1 \cap \mathfrak{G}, \mathcal{I}_2 \cap \mathfrak{G}, \mathcal{I}_3 \cap \mathfrak{G}, \mathcal{I}_1 \cap \mathfrak{F}, \mathcal{I}_2 \cap \mathfrak{F}, \mathcal{I}_3 \cap \mathfrak{F}$.

The decompositions are unique if M is in addition UTC. □

5.3.16. EXAMPLES. (1) For $R = \mathbb{Z}$, and p a prime, $M = \mathbb{Z}/p\mathbb{Z} \oplus \mathbb{Z}/p^2\mathbb{Z}$ is atomic and hence UTC, but not UC. The submodule $(0, p + p^2\mathbb{Z}) \cdot \mathbb{Z}$ has the complement closures $(1 + p\mathbb{Z}, 1 + p^2\mathbb{Z}) \cdot \mathbb{Z}$ and $(0) \oplus \mathbb{Z}/p^2\mathbb{Z}$.

(2) The next example illustrates how Theorem (5.3.3)(2) can be used to show that a module M is not UTC. Let $R = F\{x, y\}$ be the free algebra on two non-commuting indeterminates x and y over any field F. Set $L = xR + yR \triangleleft R$, $\overline{r} = r + L \in R/L$, $r \in R$. Then $R/L \cong \overline{F}$, where $\overline{f}x = \overline{f}y = 0$ for $f \in F$. Let $M = R_R \oplus \overline{F}$. Since $L \leq_e R$, and $R_R \perp \overline{F}$, by (5.3.3)(2), M is not UTC. Since $x^2R \oplus (0) \leq (x, \overline{1})R$, $x^2R \cong (x, \overline{1})R \cong R_R$, there exists a type closure K of x^2R in $x^2R \subseteq (x, \overline{1})R \subseteq K \leq M$. But $K \neq R_R \oplus (0)$ is also a type closure of $x^2R \oplus (0)$. □

We next consider only one fixed natural class $\mathcal{K}$, and determine conditions under which type submodules of type $\mathcal{K}$ are unique either for a single given module, or for all modules.

5.3.17. THEOREM. Let $\mathcal{K}$ be a natural class of R-modules, and A a right R-module. Then the following are all equivalent:

1. A has a unique type submodule of type $\mathcal{K}$.
2. For any type submodule $N \leq_t A$ of type $\mathcal{K}$ and for any $\phi \in \mathrm{Hom}_R(K, A)$ where $K \leq N$, necessarily $\phi(K) \subseteq N$.
3. For any type submodule $N \leq_t A$ with N being of type $\mathcal{K}$ and submodules $L \leq N$ and $W \leq A$ with $W \in c(\mathcal{K})$, $\mathrm{Hom}_R(L, W) = 0$.

PROOF. (3) $\Longrightarrow$ (2). If not, then for ϕ as in (2), $\phi(K) \nsubseteq N$. Since $N \leq N + \phi(K)$ is not essential, there exists a $0 \neq W \leq A$ with $N \oplus W \subseteq N + \phi(K)$. Form the composite f of the following surjective homomorphisms

$$K \longrightarrow \phi(K) \longrightarrow \phi(K)/(\phi K \cap N) \longrightarrow (\phi(K) + N)/N.$$

Since $(N \oplus W)/N \leq (\phi(K) + N)/N$, define $L = f^{-1}[(N \oplus W)/N] \leq K$. Then the nonzero map $f : L \longrightarrow f(L) = (N \oplus W)/N \cong W$ contradicts (3) that $\mathrm{Hom}_R(L, W) = 0$. Trivially, (2) $\Longrightarrow$ (3), and so far (2) $\Longleftrightarrow$ (3).

(3) $\Longrightarrow$ (1). Suppose that $A_1 \leq_t A$ and $A_2 \leq_t A$ are two distinct type submodules both of type $\mathcal{K}$. Then $A_2 < A_1 + A_2$ is not essential, and again for some $0 \neq W \in c(\mathcal{K})$, $W \oplus A_2 \leq A_1 + A_2$. Let $\pi : A_1 \longrightarrow (A_1 + A_2)/A_2$ be the quotient map, and define $L = \pi^{-1}[(W \oplus A_2)/A_2]$. Then $0 \neq L \leq A_1$. Thus $\pi(L) = (W \oplus A_2)/A_2$, and let ϕ be the composite $L \xrightarrow{\pi} (W \oplus A_2)/A_2 \longrightarrow W$ where the last map is the natural isomorphism. Hence $\phi(L) = W \neq 0$ gives the contradiction that $0 \neq \phi \in \mathrm{Hom}_R(L, W) = 0$.

(1) $\Longrightarrow$ (3). If not, let $0 \neq \phi \in \mathrm{Hom}_R(L, W)$ where $L \leq N$, $W \leq A$, and $W \in c(\mathcal{K})$. There exist elements $0 \neq b = \phi(a), a \in L$. Note that $L \leq N \leq A$ and $W \leq A$. From $a^\perp \subseteq b^\perp$ and $aR \cap bR = 0$, it follows that $(a - b)^\perp = a^\perp \cap b^\perp = a^\perp$. From this it follows that $aR \cong (a - \phi(a))R$. By Zorn's Lemma, there exist type submodules $A_1, A_2 \leq A$, both of type $\mathcal{K}$ with $aR \subseteq A_1$ and $(a - \phi(a))R \subseteq A_2$. By hypothesis (1), $A_1 = A_2 = N$. From $a \in A_1 = N$ and $a - \phi(a) \in A_2 = N$ it follows that also $\phi(a) \in N$, or $\phi(a)R \subseteq N \in \mathcal{K}$. This contradicts that $0 \neq \phi(a)R \leq W \in c(\mathcal{K})$. □

Our next objectives will be to determine when a natural class is closed under quotient modules and when it is closed under direct products.

5.3.18. THEOREM. For a natural class $\mathcal{K}$, the following five conditions are all equivalent:

1. $\mathcal{K}$ is closed under quotient modules.
2. Any one of (1), (2), or (3) of (5.3.17) holds for all R-modules A.
3. For any R-module A, every type submodule $N \leq_t A$ of type $\mathcal{K}$ is fully invariant in A.

PROOF. Let $(1^{'}), (2^{'}), (3^{'})$ be the statements of (5.3.17) holding for every R-module A. Then by the previous theorem, $(1^{'}) \Longleftrightarrow (2^{'}) \Longleftrightarrow (3^{'}) \Longleftrightarrow (2)$.

$(2^{'}) \Longrightarrow (3)$. This follows from the fact that statement (5.3.17)(2) with $K = N$ implies that N is fully invariant in A.

$(3) \Longrightarrow (3^{'})$. By way of contradiction, assume that $(3^{'})$ is false for some $L \leq N \leq_t A$ and $W \leq A$, $W \in c(\mathcal{K})$. Then verify that we may take all of these modules to be injective, and that there is a nonzero homomorphism $0 \neq \phi : L \longrightarrow W$. Extend ϕ to $\overline{\phi} : A \longrightarrow A$, with $\overline{\phi}(L) \cap N = 0$, and hence $\overline{\phi}(N) \nsubseteq N$, contradicting (3). Clearly, $(1) \Longrightarrow (3^{'})$.

$(3^{'}) \Longrightarrow (1)$. If not, then $N/V \notin \mathcal{K}$ for some $V < N \in \mathcal{K}$. Take complement submodules $P/V, Q/V$ of N/V such that $P/V \oplus Q/V \leq_e N/V$ with $P/V \in \mathcal{K}$ and $0 \neq Q/V \in c(\mathcal{K})$. Thus Q/V embeds in N/P as an essential submodule, and consequently $0 \neq N/P \in c(\mathcal{K})$. Let $W = N/P$ and let $A = N \oplus W$. Then $N <_t A$ and $\mathrm{Hom}_R(N, W) \neq 0$ contradicts (3). □

5.3.19. COROLLARY. If a natural class $\mathcal{K}$ is closed under quotient modules, then $c(\mathcal{K})$ is closed under products.

PROOF. If not, then $N = \Pi\{N_i : i \in I\} \notin c(\mathcal{K})$ although all $N_i \in c(\mathcal{K})$. But then N contains a cyclic submodule $0 \neq xR \in \mathcal{K}$ where $x = (x_i)$. For any $x_i \neq 0$, there is an epimorphism $xR \longrightarrow x_iR \in c(\mathcal{K}) \cap \mathcal{K}$, a contradiction. □

In the next two examples, the last corollary can be directly verified.

5.3.20. EXAMPLES. (1) Over any ring R the class $\mathfrak{G}$ of Goldie torsion modules is closed under quotient modules. Then the class $c(\mathfrak{G}) = \mathfrak{F}$ is closed under products.

(2) For $R = \mathbb{Z}$ and a prime p, the class $\mathcal{K}$ of all p-torsion modules is closed under quotients, and the class $c(\mathcal{K}) = \{N : N \text{ is } p\text{-torsion free}\}$ is closed under products. □

5.3.21. REFERENCES. Dauns [32,37]; Camillo-Zelmanowitz [18]; Dauns and Zhou [43]; Goodearl and Boyle [68]; Smith [109].

5.4 TS-Modules

The concept of a TS-module was invented and their theory developed in Zhou [141]. Most of the results in this section first appeared and come from the latter. By the very definition of TS-modules, their use necessarily involves direct sum decompositions of modules. The TS-modules are one of the more easily accessible examples of 2-decomposable modules. In the previous section, TS-modules inevitably came up in the study of 2-decomposable modules. (See 5.2.1, 5.2.12, 5.2.20, 5.2.21(1), 5.2.24(3).)

Let us begin by proving a converse of (5.2.20). The question arises whether the nonsingular hypothesis below can be weakened to UTC.

5.4.1. PROPOSITION. For a nonsingular TS-module M, the following are all equivalent:

1. M is a direct sum of atomic modules.
2. For any ascending chain $m_1^\perp \subseteq m_2^\perp \subseteq \cdots \subseteq m_i^\perp \subseteq \cdots$ where $m_i \in M$, $t.\dim(\oplus_{i=1}^\infty R/m_i^\perp) < \infty$.
3. Every finitely generated submodule of M has finite type dimension.
4. Every cyclic submodule of M has finite type dimension.

PROOF. (2) $\Longrightarrow$ (1). By (5.2.20) this holds even if $Z(M) \neq 0$.

(1) $\Longrightarrow$ (4). Every finitely generated submodule of M is contained in a finite direct sum of atomic modules, and hence is of finite dimension.

(4) $\Longrightarrow$ (3). A technique of Camillo [16, p.338] will be used. Let $M = x_1R + \cdots + x_nR$, $E(M) = E(x_1R) \oplus K_1$, $x_2 = a_1 + k_1$, $a_1 \in E(x_1R)$, $k_1 \in K_1$, and $K_1 = E(k_1R) \oplus K_2$. Then $E(M) = E(x_1R) \oplus E(k_1R) \oplus K_2$, where $E(x_1R + x_2R) \subseteq E(x_1R) \oplus E(k_1R)$. It is not difficult to continue this way and to express $E(M)$ as a finite direct sum of injective hulls of cyclics. Hence $E(M)$ and M are of finite type dimension by (4.1.10)(2).

(3) $\Longrightarrow$ (2). Suppose that $t.\dim\left(\oplus_{i=1}^\infty R/m_i^\perp\right) = \infty$ for a strictly increasing countable chain of right annihilator ideals $m_1^\perp \subset m_2^\perp \subset \cdots \subset R$ where $m_i \in M$. By (3) each $R/m_i^\perp \cong m_iR$ is of finite type dimension. There exists a type direct sum $\oplus_{k=1}^\infty P_k \subseteq \oplus_{k=1}^\infty R/m_k^\perp$ with all $P_k \neq 0$ and $P_i \perp P_j$ for all $i \neq j$. By use of the projection argument, it may be assumed that each $0 \neq P_k \subseteq R/m_{i(k)}^\perp$ for some $i(k)$. By renumbering it may be assumed that $i(k) = k$, and that $\oplus_{i=1}^\infty X_i/m_i^\perp$ is a type direct sum where $m_i^\perp \subset X_i \leq R$. Each $m_i^\perp \leq R_R$ is a complement submodule. (For if $a \in R\backslash m_i^\perp$ is in the complement closure of $m_i^\perp$, then $a^{-1}m_i^\perp \leq_e R$, $m_ia\left(a^{-1}m_i^\perp\right) \subseteq m_im_i^\perp = 0$,

shows that $0 \neq m_i a \in Z(m_i R) = 0$.) Let $m_i^\perp \oplus Y_i \leq_e X_i \leq R$ for some Y_i. Then

$$\frac{m_i^\perp}{m_1^\perp} \oplus \frac{Y_i \oplus m_1^\perp}{m_1^\perp} \leq_e \frac{X_i}{m_1^\perp} \leq \frac{R}{m_1^\perp} .$$

There is an embedding $Y_i \hookrightarrow X_i/m_i^\perp$. Here the hypothesis that M is nonsingular has to be invoked to conclude that, first $Y_i \neq 0$, secondly that $m_i^\perp/m_1^\perp \leq R/m_1^\perp$ is a complement, and thirdly that the above embedding

$$0 \neq Y_i \hookrightarrow X_i/m_i^\perp \neq 0$$

is essential and nonzero. Since $0 \neq Y_i \cong (Y_i + m_1^\perp)/m_1^\perp \leq R/m_1^\perp$ and $Y_i \perp Y_j$, if $i \neq j$, we get that

$$\oplus_{i=1}^\infty Y_i \cong \oplus_{i=1}^\infty (Y_i + m_1^\perp)/m_1^\perp \leq R/m_1^\perp,$$

contradicting that $t.\dim R/m_1^\perp < \infty$. □

Note that a direct sum of nonsingular CS-modules need not be CS.

5.4.2. PROPOSITION. Every direct sum of nonsingular TS-modules is TS. In particular, every direct sum of nonsingular atomic modules is a TS-module.

PROOF. For nonsingular modules, the TS and 2-decomposable conditions coincide. Any direct sum of 2-decomposable modules is 2-decomposable by (5.2.3). All atomic modules are TS. □

5.4.3. LEMMA. Suppose that $M = M_1 \oplus M_2$ where $M_1 \perp M_2$ are two orthogonal TS-modules. If every type submodule $K \leq_t M$ with $K \cap M_1 = 0$ or $K \cap M_2 = 0$ is a direct summand of M, then M is a TS-module.

PROOF. For $L \leq_t M$, let A be a type closure of $L \cap M_2$ in L, where $L \cap M_2 \subseteq A \leq_t L$. By (4.1.5), $A \leq_t M$ is also a type submodule. It is asserted that since $(L \cap M_2) \perp M_1$, then also $A \perp M_1$. Hence $A \cap M_1 = 0$. By our hypothesis, $M = A \oplus B$ for some $B \leq M$. Then $L = A \oplus (L \cap B)$. Since $A \leq_t M$, $A \perp (L \cap B)$. But then $L \cap B$ is a type submodule of L, and hence $L \cap B \leq_t M$. Since $A + (L \cap B) = A \oplus (L \cap B)$, $L \cap M_2 \subseteq A$, we have

$$(L \cap B) \cap M_2 \subseteq A \cap (L \cap B) = 0.$$

By our assumption again, $L \cap B$ is a direct summand of M and hence of B. That is, $M = (L \cap B) \oplus C$, $B = (L \cap B) \oplus (C \cap B)$ for some $C \leq M$, and hence $M = A \oplus B = [A \oplus (L \cap B)] \oplus (C \cap B) = L \oplus (C \cap B)$. □

5.4.4. PROPOSITION. Suppose that $M = M_1 \oplus \ldots \oplus M_n$ where $M_i \perp M_j$ and M_i is M_j-injective whenever $i \neq j$. Then M is a TS-module if and only if every M_i is a TS-module.

PROOF. $\Longrightarrow$. Type submodules $M_i \leq_t M$ of a TS-module M are TS.

$\Longleftarrow$. It suffices to take $n = 2$. Let $K \leq_t M$ with $K \cap M_1 = 0$ or $K \cap M_2 = 0$. Without loss of generality, assume $K \cap M_2 = 0$. By (4.3.1) there exists a submodule $K \subseteq N \leq M$ with $M = N \oplus M_2$. Then $N \cong M_1$ is a TS-module. Since $K \leq_t M$, also $K \leq_t N$. But then K is a direct summand of N, and hence one of also M. By (5.4.3), M is a TS-module. □

In most of the remainder of this section, the module M will be TS with (T_3)

5.4.5. PROPOSITION. The following are all equivalent for a module M:

1. M is TS with (T_3).
2. If X and Y are type complements of each other, then $M = X \oplus Y$.
3. If X and Y are complement submodules of M with $X \oplus Y \leq_e M$ and $X \perp Y$, then $M = X \oplus Y$.
4. For any $f^2 = f \in \text{End}_R(EM)$ with $Im(f) \perp Ker(f)$, $f(M) \subseteq M$.
5. If $E(M) = F_1 \oplus F_2$ with $F_1 \perp F_2$, then $M = (F_1 \cap M) \oplus (F_2 \cap M)$.

PROOF. (1) $\Longrightarrow$ (2). Let X, Y be given as in (2). Then X and Y are both type submodules of M. Hence $X \oplus Y = X^{tc} \oplus Y \leq_e M$. Since M is TS, X and Y are both summands of M. Thus, $M = X \oplus Y$ because M satisfies (T_3).

(2) $\Longrightarrow$ (3). Let X, Y be given as in (3). Then, by (4.1.2), X and Y are both type submodules of M. Since $X \oplus Y \leq_e M$, it follows that X and Y are type complements of each other; so $M = X \oplus Y$.

(3) $\Longrightarrow$ (1). To see that M is TS, let $X \leq_t M$. Let Y be a complement to X in M. By (3), $M = X \oplus Y$ and hence M is TS. The hypothesis (3) automatically implies (T_3).

(3) $\Longrightarrow$ (4). Set $X_1 = M \cap f(EM)$ and $X_2 = M \cap (1 - f)(EM)$. Then $X_1 \oplus X_2 \leq_e M$ and $X_1 \perp X_2$. Choose complement submodules $Y_i \leq M$, $i = 1, 2$, such that

$$X_1 \subseteq Y_1,\ X_2 \subseteq Y_2,\ Y_1 \oplus X_2 \leq_e M,\ X_1 \oplus Y_2 \leq_e M.$$

Then $X_1 \leq_e Y_1, X_2 \leq_e Y_2$, and $Y_1 \oplus Y_2 \leq_e M$. Thus $Y_1 \perp Y_2$. By (3), $M = Y_1 \oplus Y_2$. Let $\pi : M = Y_1 \oplus Y_2 \longrightarrow Y_1$ be the projection. For $x, y \in M$, suppose that $(f - \pi)(x) = y \in M \cap (f - \pi)(M)$. Then $f(x) = y + \pi(x) \in M \cap f(M) \subseteq X_1 \subseteq Y_1$. Hence $\pi f(x) = f(x)$. Next, $(1 - f)(x) \in M \cap (1 - f)(M) \subseteq X_2$, and $\pi(1 - f)(x) = 0$, or $0 = \pi(x) - \pi f(x) = \pi(x) - f(x)$. Thus $y = 0$, $M \cap (f - \pi)(M) = 0$, and since $M \leq_e E(M)$, $(f - \pi)(M) = 0$. Thus $f(M) = \pi(M) \subseteq M$.

(4) $\Longrightarrow$ (5). Let $\pi_i : F_1 \oplus F_2 \longrightarrow F_i$ be the projections. By (4),

$$M \subseteq \pi_1(M) \oplus \pi_2(M) \subseteq (F_1 \cap M) \oplus (F_2 \cap M) \subseteq M,$$

or $M = (F_1 \cap M) \oplus (F_2 \cap M)$.

(5) $\Longrightarrow$ (3). For complement submodules $X, Y \leq M$ with $X \perp Y$ and $X \oplus Y \leq_e M$, we have $E(X) \perp E(Y)$, $X = M \cap E(X)$, and $Y = M \cap E(Y)$. By (5), $M = X \oplus Y$. □

5.4.6. LEMMA. For modules $M_1 \perp M_2$, if $M = M_1 \oplus M_2$ is TS with (T_3), then M_1 and M_2 are relatively injective.

PROOF. Suppose that $X \leq M_1$ and $f : X \longrightarrow M_2$ is a homomorphism. Set $N = \{x - f(x) : x \in X\} \leq M$. Then $N \cap M_2 = 0$. Let $N \subseteq P$, $P \oplus M_2 \leq_e M$, where $P \supset N$ is any complement of M_2 in M. Because $M_2 \leq_t M$, we have $P \perp M_2$. By (5.4.5)(3), $M = P \oplus M_2$. Now let $\pi : P \oplus M_2 \longrightarrow M_2$ be the projection. For $x \in X$,

$$0 = \pi(x - f(x)) = \pi(x) - \pi f(x) = \pi(x) - f(x).$$

Thus the restriction $\pi|_{M_1}$ of π to M_1 extends f. □

5.4.7. EXAMPLES. For two uniform (or atomic) TS-modules M_1 and M_2 with $M_1 \perp M_2$, $M = M_1 \oplus M_2$ may not be TS.

(1) For $R = \mathbb{Z}$, $M = \mathbb{Z} \oplus \mathbb{Z}_2$ is not TS, since $Z_2M = ZM = \mathbb{Z}_2$ is not $M_1(= \mathbb{Z})$-injective, by (5.2.12). The complement submodule $\mathbb{Z}(2, \overline{1})$ of M is a type submodule because $\mathbb{Z}(2, \overline{1}) \oplus \mathbb{Z}_2 \leq_e M$, yet it is not a summand of M.

(2) For $R = F\{x, y\}$, the free algebra on two noncommuting variables x and y over a field F, $I = xR + yR$, $M = R \oplus R/I$ is an orthogonal sum of two atomic modules, $R \perp R/I$. Since $Z_2(M) = Z(M) = R/I$ is not M_1-injective $(= R$-injective), again M is not TS by (5.2.12).

We state without proof that there exists a type closure of $(x, 1 + I)R$ in M which is not a summand of M. □

5.4.8. For a module $M = \oplus_{\alpha \in I} M_\alpha$, a countable chain condition (A_2) due to [87, p.4] is defined below. It holds if for all possible choices of $\alpha \in I$, $x \in M_\alpha$, distinct $\{\alpha(i) : i = 1, 2, \cdots\} \subseteq I$, $m_i \in M_{\alpha(i)}$ such that $x^\perp \subseteq m_i^\perp$ for all i, the ascending sequence of right ideals $\bigcap_{i \geq n} m_i^\perp$ for $n = 1, 2, \cdots$ becomes stationary, i.e., is finite.

For $J \subseteq I$ where $M = \oplus_{i \in I} M_i$, $M(J)$ denotes $M(J) = \oplus_{j \in J} M_j$. For $\alpha \in I$, we write $I - \alpha = I \backslash \{\alpha\}$ and $M(I - \alpha)$ in place of $M(I \backslash \{\alpha\})$.

5.4.9. THEOREM. For a module $M = \oplus\{M_\alpha : \alpha \in I\}$ where the M_α are pairwise orthogonal, the following are equivalent:

1. M is a TS-module with (T_3).
2. For any $\alpha \in I$, M_α is TS with (T_3), and $M(I - \alpha)$ is M_α-injective.
3. M_α is TS with (T_3), and M_β-injective for all $\alpha \neq \beta \in I$, and (A_2) holds.

PROOF. (1) $\Longrightarrow$ (2). Since $M = M_\alpha \oplus M(I - \alpha)$ and $M_\alpha \perp M(I - \alpha)$, both modules M_α and $M(I - \alpha)$ are type submodules of the TS-module M, and hence both are TS-modules. By (5.4.6), these two modules must also be mutually self-injective. Lastly, in view of (5.4.5), each M_α satisfies (T_3).

(2) $\Longrightarrow$ (1). Let $N \leq_t M$. For each $\alpha \in I$, define X_α to be any submodule of M_α maximal with respect to $M_\alpha \cap N \cap X_\alpha = N \cap X_\alpha = 0$. Thus $N \perp X_\alpha$, and $(M_\alpha \cap N) \perp X_\alpha$. The latter makes X_α a type submodule of M_α. Let Y_α be a complement closure of $M_\alpha \cap N$ in M_α. Then $X_\alpha \oplus Y_\alpha \leq_e M_\alpha$, $X_\alpha \perp Y_\alpha$, and by (5.4.5), $M_\alpha = X_\alpha \oplus Y_\alpha$. If $0 \neq \xi \in N \cap (\oplus_{\alpha \in I} X_\alpha) \neq 0$, then (4.3.7) shows that for some $r_0 \in R$, $0 \neq \xi r_0 \in N \cap X_\alpha$ for some $\alpha \in I$, a contradiction. Thus $(\oplus_{\alpha \in I} X_\alpha) \oplus N \leq M$ is a direct sum, and it will be shown next that the latter is an equality.

Since M_α has (T_3), by (5.4.6) the modules X_α and Y_α are relatively injective. By hypothesis (2), $M(I-\alpha)$ is M_α-injective, and hence Y_α-injective. But then as a summand of $M(I-\alpha)$, $\oplus_{\alpha \neq \beta \in I} X_\beta$ is Y_α-injective. Combine this with the fact that X_α is Y_α-injective, to get that $\oplus_{\alpha \in I} X_\alpha = X_\alpha \oplus (\oplus_{\alpha \neq \beta} X_\beta)$ is Y_α-injective. But then $\oplus_{\alpha \in I} X_\alpha$ is also $\oplus_{\alpha \in I} Y_\alpha$-injective by [87, Prop.1.5]. Similarly, $\oplus_{\alpha \in I} Y_\alpha$ is $\oplus_{\alpha \in I} X_\alpha$-injective, which will later be used to prove that M satisfies (T_3). Thus, so far,

$$M = (\oplus_{\alpha \in I} X_\alpha) \oplus (\oplus_{\alpha \in I} Y_\alpha), \quad (\oplus_{\alpha \in I} X_\alpha) \cap N = 0,$$

and hence, by (4.3.1), for some $N \subseteq N^* \leq M$, $M = (\oplus_{\alpha \in I} X_\alpha) \oplus N^*$. We show that $N = N^*$.

From $M_\alpha \cap N \leq_e Y_\alpha$, it follows that $(\oplus_{\alpha \in I} X_\alpha) \oplus (\oplus_{\alpha \in I} M_\alpha \cap N) \leq_e M$. But $\oplus_{\alpha \in I} M_\alpha \cap N \subseteq N \subseteq N^*$ shows that all of these inclusions are essential. Since $N \leq_t M$, $N = N^*$. Thus M is TS.

To verify (T_3), let $N, P \leq M$ be complement submodules of each other which are also orthogonal. Thus $N \leq_t M$. Then the above previous construction can be used to construct

$$M = (\oplus_{\alpha \in I} X_\alpha) \oplus N = (\oplus_{\alpha \in I} X_\alpha) \oplus (\oplus_{\alpha \in I} Y_\alpha),$$

where $N \cong \oplus_{\alpha \in I} Y_\alpha$ is $\oplus_{\alpha \in I} X_\alpha$-injective. By (4.3.1), since $P \cap N = 0$, $M = P^* \oplus N$ with $P \subseteq P^* \leq M$. Then $P \oplus N \leq_e M$ and also $P \leq_e P^*$. But M is TS and so P is a summand of M, and hence P is a complement in M. Thus $P = P^*$, and $M = P \oplus N$ as required.

(2) $\Longleftrightarrow$ (3). If in both (2) and (3), the hypothesis "M_α is TS" is removed, then the equivalence of the so-modified (2) and (3) is just [87, Prop.1.9]. $\square$

5.4.10. PROPOSITION. Suppose that M_i, $i \in I$, is a set of pairwise orthogonal modules such that their direct sum $M = \oplus_{i \in I} M_i$ complements type summands. Then the following hold:

1. For each $i \in I$, M_i is t-indecomposable.
2. If $M = \oplus_{\gamma \in \Lambda} X_\gamma$ is a type direct sum, then I has a partition $I = \bigcup \{I_\gamma : \gamma \in \Lambda\}$ such that $X_\gamma \cong M(I_\gamma)$.
3. If $M = \oplus_{j \in J} N_j$ where each $N_j \leq_t M$ is t-indecomposable, then there is a bijection $\sigma : I \longrightarrow J$ such that $M_i \cong N_{\sigma(i)}$ for all $i \in I$.

4. For any type direct summand N of M, N has a decomposition that complements type summands.

PROOF. (1) Suppose that for some $k \in I$, $M_k = X \oplus Y$ with $X \perp Y$ and $X \neq 0$. We will show that $Y = 0$. Since $X \leq_t M_k \leq_t M$, X is a type summand of M. By hypothesis, $M = X \oplus M(J)$ for some $J \subseteq I$, with $X \cong M(I \backslash J)$ and $X \perp M(I - k)$. Now if $|I \backslash J| \geq 2$, then for some $i(1) \neq i(2) \in I \backslash J$, $M_{i(1)} \oplus M_{i(2)} \hookrightarrow X$. Since $\{i(1), i(2)\} \cap I \backslash \{k\} \neq \emptyset$, this is a contradiction. Hence $I \backslash J = \{k\}$ and $X \cong M_k$. But then $Y \hookrightarrow X$ and $Y \perp X$ implies that $Y = 0$.

(2) For $\gamma \in \Lambda$, since X_γ is a type summand of M, and since $M = \oplus_{i \in I} M_i$ complements type summands, $X_\gamma \cong M(I_\gamma)$ for some $I_\gamma \subseteq I$. For $\alpha \neq \beta \in \Lambda$, since $X_\alpha \perp X_\beta$, also $M(I_\alpha) \perp M(I_\beta)$, and hence $I_\alpha \cap I_\beta = \emptyset$. For any $i \in I$, $M_i \subseteq \oplus_{\gamma \in \Lambda} X_\gamma \cong \oplus_{\gamma \in \Lambda} M(I_\gamma)$. By (4.3.7), $i \in I_\gamma$ for some $\gamma \in \Lambda$ and $I = \bigcup \{I_\gamma : \gamma \in \Lambda\}$ is the required partition of I. (3) This is an immediate consequence of (2).

(4) If N is a type summand of M, then $N \cong M(J)$ for some $J \subseteq I$. So assume $N = M(J)$, and we show that this is the decomposition of N that complements type summands of N. Let X be a type summand of N. Then X is a type summand of M. Hence by hypothesis, $M = X \oplus M(K)$ for some $K \subseteq I$. Thus $N = M(J) = X \oplus [M(J) \cap M(K)]$. Clearly, $M(J \cap K) \subseteq M(J) \cap M(K)$. Suppose $0 \neq \xi \in M(J) \cap M(K)$. By a double application of (4.3.7), we get that $0 \neq \xi r(1) R \subseteq M_{j(1)}$ for some $j(1) \in J$, and $0 \neq \xi r(1) r(2) R \subseteq M_{k(2)}$ for some $k(2) \in K$ and $r(1), r(2) \in R$. Then $j(1) = k(2)$ and hence $0 \neq \xi r(1) r(2) R \subseteq M(J \cap K)$. Hence $M(J \cap K) \leq_e M(J) \cap M(K)$. Because $M(J \cap K)$ is a direct summand of M, it must be that $M(J \cap K) = M(J) \cap M(K)$. Therefore $N = X \oplus M(J \cap K)$ as required. □

The next result extends (4.3.10).

5.4.11. PROPOSITION. For a TS-module M with (T_3), the following are all equivalent:

1. M is a direct sum of atomic modules.
2. M has a decomposition that complements type summands.
3. Every local type summand of M is a summand.

PROOF. Similar to the proof of (4.3.10). □

It was claimed in [141, Example 23(d)] that there exists a TS-module without (T_3). Unfortunately, the module M in that example is not TS. So no TS-modules without (T_3) are known so far.

5.4.12. REFERENCES. Mohamed and Müller [87]; Dauns and Zhou [43]; Zhou [141].

Chapter 6

Lattices of Module Classes

The collection $\mathcal{N}_r^p(R)$ of pre-natural classes of R-modules is a complete lattice, the study of which was initiated in [142]. It contains as subsets almost all the important lattices of module classes associated with the ring R. For example, $\mathcal{N}_r^p(R)$ contains a complete sublattice isomorphic to the complete lattice of all linear topologies of R and a sublattice anti-isomorphic to the frame of all hereditary torsion theories of R. The complete Boolean lattice of all natural classes of R-modules is also a sublattice of $\mathcal{N}_r^p(R)$. The lattice $\mathcal{N}_r^p(R)$ and some of its sublattices are introduced and discussed in sections 6.1 and 6.2.

In section 6.3, several properties of the lattice $\mathcal{N}_r^p(R)$ and of some of its sublattices are discussed and related to properties of the ring R and the category Mod-R. In section 6.4, the rings R for which every hereditary pretorsion class is hereditary torsion are investigated as applications. Section 6.5 explores the functoriality of $\mathcal{N}_r(\cdot)$, and some of its subfunctors. Finally in section 6.6, it is shown that every ring R contains a lattice of ideals isomorphic to $\mathcal{N}_f(R)$, the lattice of all natural classes of nonsingular R-modules.

6.1 Lattice of Pre-Natural Classes

In this section, we introduce the lattice of pre-natural classes and several sublattice structures. Recall that the collection $\mathcal{N}_r^p(R)$ of pre-natural classes is a set.

6.1.1. LEMMA. Let $\{\mathcal{K}_i : i \in I\}$ be a set of pre-natural classes. Then $\cap_{i\in I}\mathcal{K}_i$ is a pre-natural class.

PROOF. Write $\mathcal{K}_i = d(M_{\mathcal{K}_i}) \cap \sigma[M_{\mathcal{K}_i}]$ for each $i \in I$ by (2.5.4). Then

$$\cap_{i\in I}\mathcal{K}_i = \cap_i(d(M_{\mathcal{K}_i}) \cap \sigma[M_{\mathcal{K}_i}]) = (\cap_i d(M_{\mathcal{K}_i})) \cap (\cap_i \sigma[M_{\mathcal{K}_i}])$$

is an intersection of a natural class and a hereditary pretorsion class, so it is a pre-natural class by (2.5.5). □

6.1.2. THEOREM. $\mathcal{N}_r^p(R)$ is a complete lattice with the least element $\mathbf{0} = \{0\}$ and the greatest element $\mathbf{1} =$ Mod-R under the following partial ordering and lattice operations:

1. For $\mathcal{K}_1, \mathcal{K}_2 \in \mathcal{N}_r^p(R)$, $\mathcal{K}_1 \leq \mathcal{K}_2 \Longleftrightarrow \mathcal{K}_1 \subseteq \mathcal{K}_2$.

2. For a set of pre-natural classes $\mathcal{K}_i$, $\wedge\mathcal{K}_i = \cap\mathcal{K}_i$ and $\vee\mathcal{K}_i = d(\mathcal{K}) \cap \sigma[M_{\mathcal{K}}]$ where $\mathcal{K} = \cup\mathcal{K}_i$.

PROOF. In view of (6.1.1), we only need to verify that $d(\mathcal{K}) \cap \sigma[M_{\mathcal{K}}]$ is the smallest pre-natural class containing $\cup_i\mathcal{K}_i$, but this is certainly true by (2.5.4) since $d(\mathcal{K}) = d(M_{\mathcal{K}})$ (see 2.3.6). □

6.1.3. COROLLARY. If $\mathcal{K}_i = \sigma[X_i] \cap d(X_i), i \in I$, then

$$\bigvee_{i\in I} \mathcal{K}_i = \sigma[\oplus_{i\in I} X_i] \cap d(\oplus_{i\in I} X_i).$$

PROOF. By (2.5.4), $\mathcal{K}_i$ is the smallest pre-natural class containing X_i. Thus, $\vee_{i\in I}\mathcal{K}_i$ is the smallest pre-natural class containing $\oplus_{i\in I}X_i$. So, by (2.5.4), $\vee_{i\in I}\mathcal{K}_i = \sigma[\oplus_{i\in I}X_i] \cap d(\oplus_{i\in I}X_i)$. □

Next, we study several sublattice structures of $\mathcal{N}_r^p(R)$. For a pre-natural class $\mathcal{K}$ and a module M, $\mathcal{K} \subseteq \sigma[M]$ does not imply that $\mathcal{K}$ is an M-natural class in general. In fact, every pre-natural class is a subclass of $\sigma[R_R]$, but many pre-natural classes are not natural classes. This suggests that the following result is not trivial.

6.1.4. THEOREM. For any module M, $\mathcal{N}(R, M)$ is a sublattice of $\mathcal{N}_r^p(R)$.

PROOF. Let $\mathcal{K}_1, \mathcal{K}_2 \in \mathcal{N}(R, M)$. Clearly, $\mathcal{K}_1 \wedge \mathcal{K}_2 \in \mathcal{N}(R, M)$. We want to show that $\mathcal{K}_1 \vee \mathcal{K}_2 \in \mathcal{N}(R, M)$. Since $\mathcal{K}_1 \vee \mathcal{K}_2$ is a pre-natural class, we only need to show that it is closed under M-injective hulls. For $N \in \mathcal{K}_1 \vee \mathcal{K}_2$, let $\{X_t : t \in I\}$ be a maximal independent set of submodules of N in $\mathcal{K}_1$ and let $X = \oplus X_t$. Then X is in $\mathcal{K}_1$. Let P be a submodule of N which is maximal with respect to $X \cap P = 0$. Then $X \oplus P$ is essential in N. Let $\{Y_s : s \in J\}$ be a maximal independent set of submodules of P in $\mathcal{K}_2$ and $Y = \oplus Y_s$. Then Y is in $\mathcal{K}_2$. If $Y \cap Q = 0$ for some $0 \neq Q \subseteq P$, then, by (6.1.3), Q contains a nonzero submodule which is in $\mathcal{K}_1$ or $\mathcal{K}_2$. But the choices of X and Y show that this is impossible. So Y is essential in P and hence $X \oplus Y$ is essential in N. Then $E(N) = E(X) \oplus E(Y)$ and

$$\begin{aligned} E_M(N) &= \Sigma\{f(M) : f \in \text{Hom}(M, EN)\} \\ &\subseteq \Sigma\{f_1(M) : f_1 \in \text{Hom}(M, EX)\} + \Sigma\{f_2(M) : f_2 \in \text{Hom}(M, EY)\} \\ &= E_M(X) \oplus E_M(Y). \end{aligned}$$

Since $\mathcal{K}_1$ and $\mathcal{K}_2$ are closed under M-injective hulls, $E_M(X)$ is in $\mathcal{K}_1$ and $E_M(Y)$ is in $\mathcal{K}_2$. Thus, $E_M(N) = E_M(X) \oplus E_M(Y)$ is in $\mathcal{K}_1 \vee \mathcal{K}_2$. □

6.1.5. COROLLARY. $\mathcal{N}_r(R)$ is a sublattice of $\mathcal{N}_r^p(R)$. □

In general, $\mathcal{N}(R,M)$ is not a complete sublattice of $\mathcal{N}_r^p(R)$ even though $\mathcal{N}(R,M)$ itself is a complete lattice (see 6.2.16). In fact there exists a ring R such that $\mathcal{N}_r(R)$ is not a complete sublattice of $\mathcal{N}_r^p(R)$.

6.1.6. EXAMPLE. Let $Q = \prod_{i=1}^{\infty} F_i$, where each $F_i = \mathbb{Z}_2$, be the direct product of rings $\mathbb{Z}_2$. Let R be the subring of Q generated by $\oplus_{i=1}^{\infty} F_i$ and 1_Q. Then $Soc(R) = \oplus_i F_i$ is the only essential right ideal of R. For each i and each index set I, $F_i^{(I)}$ is an injective R-module. To see this, let $f : Soc(R) \longrightarrow F_i^{(I)}$ be an R-homomorphism. Then $f(F_j) = 0$ for all $j \neq i$. Note that F_i is a direct summand of R. Write $R = F_i \oplus A_i$ and let π be the projection of R onto F_i. Then $f\pi : R \longrightarrow F_i^{(I)}$ extends f. So $F_i^{(I)}$ is R-injective. For each i, let $\mathcal{K}_i = d(F_i) \cap \sigma[F_i]$. Then $\mathcal{K}_i = \sigma[F_i]$ is closed under injective hulls and so $\mathcal{K}_i \in \mathcal{N}_r(R)$. Let $\mathcal{K} = \vee_i \mathcal{K}_i$. Then, by (6.1.3),

$$\mathcal{K} = d(Soc(R)) \cap \sigma[Soc(R)] = \sigma[Soc(R)].$$

Since R is not semisimple, R is not in $\sigma[Soc(R)]$. But $Soc(R)$ is essential in R, so $\sigma[Soc(R)]$ is not closed under injective hulls. Thus, $\mathcal{K}$ is not a natural class, showing that $\mathcal{N}_r(R)$ is not a complete sublattice of $\mathcal{N}_r^p(R)$. □

6.1.7. COROLLARY. If $\mathcal{K}, \mathcal{L} \in \mathcal{N}_r(R)$, then

$$\mathcal{K} \vee \mathcal{L} = \{M \in \text{Mod-}R : X \in \mathcal{K} \text{ and } M/X \in \mathcal{L} \\ \text{for some } X \subseteq M\}.$$

PROOF. By (6.1.5), $\mathcal{K} \vee \mathcal{L} \in \mathcal{N}_r(R)$, so $\mathcal{K} \vee \mathcal{L}$ is closed under extensions of modules by (2.4.5). Hence the inclusion in one direction is clear.

Suppose $M \in \mathcal{K} \vee \mathcal{L}$. Let N be a submodule of M maximal with respect to $N \in \mathcal{L}$ and let P be a submodule of M maximal with respect to $N \cap P = 0$ (N and P exist by Zorn's Lemma). Then N is essentially embeddable in M/P, implying $M/P \in \mathcal{L}$. By the maximality of N, we have $P \in c(\mathcal{L})$. Therefore, by (5.1.5),

$$\begin{aligned} P \in c(\mathcal{L}) \wedge (\mathcal{K} \vee \mathcal{L}) &= (c(\mathcal{L}) \wedge \mathcal{K}) \vee (c(\mathcal{L}) \wedge \mathcal{L}) \\ &= (c(\mathcal{L}) \wedge \mathcal{K}) \vee 0 \\ &= c(\mathcal{L}) \wedge \mathcal{K} \leq \mathcal{K}. \end{aligned}$$

□

Let $\mathcal{F}_r(R)$ be the set of all hereditary torsion free classes, and let $\mathcal{T}_r^p(R)$ (respectively, $\mathcal{T}_r(R)$) be the set of all hereditary pretorsion classes (respectively, all hereditary torsion classes).

6.1.8. COROLLARY. $\mathcal{F}_r(R)$ is a sublattice of $\mathcal{N}_r^p(R)$.

PROOF. Let $\mathcal{K}_1, \mathcal{K}_2 \in \mathcal{F}_r(R)$. Clearly we have $\mathcal{K}_1 \wedge \mathcal{K}_2 \in \mathcal{F}_r(R)$. Let $\mathcal{L} = \mathcal{K}_1 \vee \mathcal{K}_2$. We prove that $\mathcal{L} \in \mathcal{F}_r(R)$. By (6.1.7),

$$\mathcal{L} = \{M \in \text{Mod-}R : \exists X \leq M \text{ such that } X \in \mathcal{K}_1 \text{ and } M/X \in \mathcal{K}_2\}.$$

We only need to prove that $\mathcal{L}$ is closed under direct products. Let M_i be in $\mathcal{L}$ $(i \in I)$. Then for each i, there exists an X_i in $\mathcal{K}_1$ such that M_i/X_i is in $\mathcal{K}_2$. Thus, we have ΠX_i is in $\mathcal{K}_1$ and $(\Pi M_i)/(\Pi X_i) \cong \Pi(M_i/X_i)$ is in $\mathcal{K}_2$. So ΠM_i is in $\mathcal{L}$. □

There is a well-known frame structure on the set **Tor**-R of all hereditary torsion theories of R (see Golan [59]). For $\tau \in$ **Tor**-R, let $\mathcal{F}_\tau$ be the class of all τ-torsion free modules. It can be checked that the correspondence $\tau \longmapsto \mathcal{F}_\tau$ gives a lattice anti-isomorphism of **Tor**-R onto $\mathcal{F}_r(R)$.

6.1.9. COROLLARY. $\mathcal{T}_r^p(R)$ is a complete sublattice of $\mathcal{N}_r^p(R)$.

PROOF. Let $\{\mathcal{K}_i : i \in I\} \subseteq \mathcal{T}_r^p(R)$. Clearly, we have $\wedge_{i\in I}\mathcal{K}_i \in \mathcal{T}_r^p(R)$. Let $\{X_t : t \in J\}$ be a complete set of representatives of the isomorphic classes of cyclic modules in $\cup_{i\in I}\mathcal{K}_i$ and $M = \oplus_{t\in J}X_t$. By (6.1.2), $\vee_{i\in I}\mathcal{K}_i = d(\mathcal{K})\cap\sigma[M]$ where $\mathcal{K} = \cup_{i\in I}\mathcal{K}_i$. If $\vee_{i\in I}\mathcal{K}_i \neq \sigma[M]$, then there exists an $N \in \sigma[M]$ but $N \notin d(\mathcal{K})$. Thus, there exists a nonzero cyclic submodule Y of N such that Y has no nonzero submodules in $\cup_{i\in I}\mathcal{K}_i$. Since $Y \in \sigma[M]$ is cyclic, there exists some X_t such that Y has a nonzero submodule which is isomorphic to a subfactor of X_t. But X_t is in some $\mathcal{K}_i$, so Y has a nonzero submodule which is in $\mathcal{K}_i$. This is a contradiction. □

There is a complete lattice structure on **fil**-R, the set of all right linear topologies of R (see Golan [60]). For $\mathsf{A} \in$ **fil**-R, let $\mathcal{T}_\mathsf{A}$ be the class of all M in Mod-R such that $x^\perp \in \mathsf{A}$ for all $x \in M$. Then $\mathsf{A} \longmapsto \mathcal{T}_\mathsf{A}$ gives a lattice isomorphism of **fil**-R onto $\mathcal{T}_r^p(R)$.

6.1.10. COROLLARY. The following hold:

1. For modules X and Y, $d(X) \vee d(Y) = d(X \oplus Y)$.
2. For a family $\{X_i : i \in I\}$ of modules, $\vee_{i\in I}\sigma[X_i] = \sigma[\oplus_{i\in I}X_i]$.

PROOF. (1) By (6.1.5), $d(X) \vee d(Y)$ must be the smallest natural class containing X and Y, so $d(X) \vee d(Y) = d(X \oplus Y)$.

(2) By (6.1.9), $\vee_{i\in I}\sigma[X_i]$ must be the smallest hereditary pretorsion class containing all X_i, so $\vee_{i\in I}\sigma[X_i] = \sigma[\oplus_{i\in I}X_i]$. □

6.1.11. REFERENCES. Dauns [31,32]; Dauns and Zhou [40]; Golan [59,60]; Zhou [136,142].

6.2 More Sublattice Structures

There is a rich supply of sublattices of the lattice $\mathcal{N}_r^p(R)$. Some of these sublattices will be described in this section.

For a pre-natural class $\mathcal{K}$, let

$$\mathcal{T}_r^p(\mathcal{K}, R) = \{\mathcal{K} \cap \mathcal{T} : \mathcal{T} \in \mathcal{T}_r^p(R)\}.$$

We will show next that $\mathcal{T}_r^p(\mathcal{K}, R)$ is a complete sublattice of $\mathcal{N}_r^p(R)$, and that the lattice $\mathcal{T}_r^p(\mathcal{K}, R)$ is algebraic, upper continuous, and modular. Note that when $\mathcal{K} = $ Mod-R, $\mathcal{T}_r^p(\mathcal{K}, R) = \mathcal{T}_r^p(R)$ is the set of all hereditary pretorsion classes. When $\mathcal{K} = \sigma[M]$, $\mathcal{T}_r^p(\mathcal{K}, R)$ is the set of all hereditary pretorsion classes contained in $\sigma[M]$, which was denoted by *M-ptors* in [104] and is the subject of [104], [124], and [125].

6.2.1. LEMMA. The following are equivalent for a pre-natural class $\mathcal{K}$:

1. $\mathcal{L} \in \mathcal{T}_r^p(\mathcal{K}, R)$.

2. $\mathcal{L}$ is a subclass of $\mathcal{K}$ such that $\mathcal{L}$ is closed under submodules, direct sums, and whenever $A \longrightarrow B \longrightarrow 0$ with $A \in \mathcal{L}$ and $B \in \mathcal{K}$ we have $B \in \mathcal{L}$.

PROOF. (1) $\Longrightarrow$ (2). It is obvious.

(2) $\Longrightarrow$ (1). We claim that $\mathcal{L} = \mathcal{K} \cap \sigma[M_{\mathcal{L}}]$. Clearly, $\mathcal{L} \subseteq \mathcal{K} \cap \sigma[M_{\mathcal{L}}]$. Let $X \in \mathcal{K} \cap \sigma[M_{\mathcal{L}}]$. There exists an index set I and a submodule C of $M_{\mathcal{L}}^{(I)}$ such that X is embeddable in $M_{\mathcal{L}}^{(I)}/C$. Write $X \cong A/C$ where A is a submodule of $M_{\mathcal{L}}^{(I)}$. Since $M_{\mathcal{L}}^{(I)}$ is in $\mathcal{L}$, A is in $\mathcal{L}$. So, by (2), X is in $\mathcal{L}$. Thus, the claim is proved. □

6.2.2. LEMMA. For any $\mathcal{K} \in \mathcal{N}_r(R)$ and $\mathcal{K}_i \in \mathcal{T}_r^p(R)$ $(i \in I)$,

$$\vee_{i \in I}(\mathcal{K} \wedge \mathcal{K}_i) = \mathcal{K} \wedge (\vee_{i \in I}\mathcal{K}_i).$$

PROOF. Obviously, we have $\vee_{i \in I}(\mathcal{K} \wedge \mathcal{K}_i) \subseteq \mathcal{K} \wedge (\vee_{i \in I}\mathcal{K}_i)$. In view of (2.5.6), it suffices to show that every cyclic module $N \in \mathcal{K} \wedge (\vee_{i \in I}\mathcal{K}_i)$ is in $\vee_{i \in I}(\mathcal{K} \wedge \mathcal{K}_i)$. Write $\mathcal{K}_i = \sigma[M_i]$ for each $i \in I$. Then

$$N \in \vee_{i \in I}\mathcal{K}_i = \vee_{i \in I}\sigma[M_i] = \sigma[\oplus_{i \in I} M_i]$$

by (6.1.10). Since N is cyclic, $N \in \sigma[\oplus_{i \in I_0} M_i] = \vee_{i \in I_0}\mathcal{K}_i$, where I_0 is a finite subset of I. So we may assume that I is a finite set. Since any join of elements in $\mathcal{T}_r^p(R)$ is still in $\mathcal{T}_r^p(R)$ (by 6.1.9), we may assume that $I = \{1, 2\}$. Let $M = M_1 \oplus M_2$. Now $N \in \mathcal{K} \wedge \sigma[M]$. Then $N \subseteq E_M(N) \in \mathcal{K} \wedge \sigma[M]$ and $E_M(N) = E_{M_1}(N) + E_{M_2}(N)$. It follows that $E_{M_1}(N) \in \mathcal{K} \wedge \sigma[M_1]$ and $E_{M_2}(N) \in \mathcal{K} \wedge \sigma[M_2]$. We actually proved that every nonzero module in $\mathcal{K} \wedge \sigma[M_1 \oplus M_2]$ has a nonzero submodule in $\mathcal{K} \wedge \sigma[M_1]$ or in $\mathcal{K} \wedge \sigma[M_2]$. So we have $E_M(N) \in d(M_{\mathcal{L}})$, where $\mathcal{L} = (\mathcal{K} \wedge \mathcal{K}_1) \vee (\mathcal{K} \wedge \mathcal{K}_2)$. Since $\mathcal{L} = \sigma[M_{\mathcal{L}}] \cap d(M_{\mathcal{L}})$,

$$E_M(N) = E_{M_1}(N) + E_{M_2}(N) \in \sigma[M_{\mathcal{L}}].$$

Thus, $E_M(N) \in \mathcal{L}$, which shows $N \in \mathcal{L}$. □

6.2.3. THEOREM. For every pre-natural class $\mathcal{K}$, $\mathcal{T}_r^p(\mathcal{K}, R)$ is a complete sublattice of $\mathcal{N}_r^p(R)$.

PROOF. Let $\mathcal{K}_i = \mathcal{K} \cap \mathcal{T}_i \in \mathcal{T}_r^p(\mathcal{K}, R)$ $(i \in I)$, where all $\mathcal{T}_i \in \mathcal{T}_r^p(R)$. Clearly,

$$\wedge_{i \in I} \mathcal{K}_i = \mathcal{K} \cap (\cap_{i \in I} \mathcal{T}_i) \in \mathcal{T}_r^p(\mathcal{K}, R).$$

Write $\mathcal{K} = \mathcal{L}_0 \cap \mathcal{T}_0$, where $\mathcal{L}_0 \in \mathcal{N}_r(R)$ and $\mathcal{T}_0 \in \mathcal{T}_r^p(R)$. Note that $\vee_{i \in I} \mathcal{T}_i \in \mathcal{T}_r^p(R)$ by (6.1.9). In view of (6.2.2), we have

$$\begin{aligned} \vee_{i \in I} \mathcal{K}_i &= \vee_{i \in I}(\mathcal{L}_0 \cap \mathcal{T}_0 \cap \mathcal{T}_i) \\ &= \mathcal{L}_0 \wedge [\vee_{i \in I}(\mathcal{T}_0 \cap \mathcal{T}_i)] \\ &= \mathcal{L}_0 \wedge \{\mathcal{T}_0 \wedge [\vee_{i \in I}(\mathcal{T}_0 \cap \mathcal{T}_i)]\} \\ &= \mathcal{K} \cap [\vee_{i \in I}(\mathcal{T}_0 \cap \mathcal{T}_i)] \in \mathcal{T}_r^p(\mathcal{K}, R). \end{aligned}$$

Therefore, $\mathcal{T}_r^p(\mathcal{K}, R)$ is a complete sublattice of $\mathcal{N}_r^p(R)$. □

6.2.4. LEMMA. Let $\mathcal{K}$ be a pre-natural class. If $M = \Sigma_{i \in I} M_i \in \mathcal{K}$, then $\mathcal{K} \cap \sigma[M] = \vee_{i \in I}(\mathcal{K} \cap \sigma[M_i])$.

PROOF. Write $\mathcal{K} = \mathcal{L}_0 \cap \mathcal{T}_0$, where $\mathcal{L}_0 \in \mathcal{N}_r(R)$ and $\mathcal{T}_0 \in \mathcal{T}_r^p(R)$. Since $M \in \mathcal{K}$, $\mathcal{T}_0 \cap \sigma[M] = \sigma[M]$. Thus,

$$\begin{aligned} \mathcal{K} \cap \sigma[M] &= \mathcal{L}_0 \cap \sigma[M] \\ &= \mathcal{L}_0 \wedge (\vee_{i \in I} \sigma[M_i]) \\ &= \vee_{i \in I}(\mathcal{L}_0 \wedge \sigma[M_i]) \\ &= \vee_{i \in I}(\mathcal{K} \wedge \sigma[M_i]). \end{aligned}$$

The second to the last equality is by (6.2.2). □

6.2.5. LEMMA. Let $\mathcal{K}$ be a pre-natural class. If $\mathcal{L} \in \mathcal{T}_r^p(\mathcal{K}, R)$, then $\mathcal{L} = \mathcal{K} \cap \sigma[M_{\mathcal{L}}]$.

PROOF. Write $\mathcal{L} = \mathcal{K} \cap \mathcal{T}$, where $\mathcal{T} \in \mathcal{T}_r^p(R)$. Thus $\mathcal{L} \subseteq \sigma[M_{\mathcal{L}}] \subseteq \mathcal{T}$. It follows that $\mathcal{L} = \mathcal{K} \cap \sigma[M_{\mathcal{L}}]$. □

An element c in a complete lattice L is said to be **compact** if for any subset Y of L with $c \leq \vee\{a : a \in Y\}$, we have $c \leq \vee\{a : a \in F\}$ for a finite subset F of Y. A complete lattice is said to be **algebraic** if each of its elements is the join of compact elements.

6.2.6. PROPOSITION. Let $\mathcal{K}$ be a pre-natural class. $\mathcal{L} \in \mathcal{T}_r^p(\mathcal{K}, R)$ is compact iff $\mathcal{L} = \mathcal{K} \cap \sigma[M]$ for some finitely generated module $M \in \mathcal{K}$.

PROOF. Suppose $\mathcal{L} \in \mathcal{T}_r^p(\mathcal{K}, R)$ is compact. Write $\mathcal{L} = \mathcal{K} \cap \mathcal{T}$ for some $\mathcal{T} \in \mathcal{T}_r^p(R)$. By (6.2.5), we may write $\mathcal{L} = \mathcal{K} \cap \sigma[X]$, where $X = \oplus_{t \in I} X_t \in \mathcal{K}$ with all X_t cyclic. By (6.2.4), $\mathcal{L} = \mathcal{K} \cap \sigma[X] = \vee_{i \in I}(\mathcal{K} \cap \sigma[X_i])$. Since $\mathcal{L}$ is compact in $\mathcal{T}_r^p(\mathcal{K}, R)$, there exists a finite subset F of I such that $\mathcal{L} = \vee_{i \in F}(\mathcal{K} \wedge \sigma[X_i])$, which is, by (6.2.4), equal to $\mathcal{K} \cap \sigma[N]$, where $N = \oplus_{i \in F} X_i \in \mathcal{K}$.

Suppose that $\mathcal{L} = \mathcal{K} \cap \sigma[M]$ for some finitely generated module $M \in \mathcal{K}$ and $\mathcal{L} \leq \vee_{i \in I} \mathcal{K}_i$ where $\mathcal{K}_i \in \mathcal{T}_r^p(\mathcal{K}, R)$. By (6.2.5), we may write $\mathcal{K}_i = \mathcal{K} \cap \sigma[X_i]$ with $X_i \in \mathcal{K}$. Then

$$\mathcal{L} \leq \vee_{i \in I}(\mathcal{K} \cap \sigma[X_i]) \leq \mathcal{K} \cap (\vee_{i \in I} \sigma[X_i]) = \mathcal{K} \cap \sigma[\oplus_{i \in I} X_i].$$

Since $M \in \mathcal{L}$ is finitely generated, $M \in \sigma[\oplus_{i \in F} X_i]$ for a finite subset F of I. Thus

$$\mathcal{L} = \mathcal{K} \cap \sigma[M] \subseteq \mathcal{K} \cap \sigma[\oplus_{i \in F} X_i] = \vee_{i \in F}(\mathcal{K} \cap \sigma[X_i]) = \vee_{i \in F} \mathcal{K}_i.$$

The second last equality is by (6.2.4). Thus $\mathcal{L}$ is compact in $\mathcal{T}_r^p(\mathcal{K}, R)$. □

6.2.7. THEOREM. For any pre-natural class $\mathcal{K}$, the lattice $\mathcal{T}_r^p(\mathcal{K}, R)$ is algebraic.

PROOF. Let $\mathcal{L} \in \mathcal{T}_r^p(\mathcal{K}, R)$. By (6.2.5), there exists a module X such that $X = \oplus_{i \in I} X_i$ with all X_i cyclic, $X \in \mathcal{K}$, and $\mathcal{L} = \mathcal{K} \cap \sigma[X]$. By (6.2.4), $\mathcal{L} = \vee_{i \in I}(\mathcal{K} \cap \sigma[X_i])$. By (6.2.6), each $\mathcal{K} \cap \sigma[X_i]$ is compact in $\mathcal{T}_r^p(\mathcal{K}, R)$. Thus $\mathcal{T}_r^p(\mathcal{K}, R)$ is algebraic. □

6.2.8. COROLLARY. [60] The lattice $\mathcal{T}_r^p(R)$ is algebraic. □

The next proposition shows that every $\mathcal{T}_r^p(\mathcal{K}, R)$ can be regarded as a sublattice of $\mathcal{T}_r^p(R)$.

6.2.9. PROPOSITION. For $\mathcal{K} = \mathcal{L}_0 \wedge \mathcal{T}_0 \in \mathcal{N}_r^p(R)$ where $\mathcal{L}_0 \in \mathcal{N}_r(R)$ and $\mathcal{T}_0 \in \mathcal{T}_r^p(R)$, define $f : \mathcal{T}_r^p(R) \longrightarrow \mathcal{T}_r^p(\mathcal{K}, R)$ by $f(\mathcal{T}) = \mathcal{K} \cap \mathcal{T}$ and let f_0 be the restriction of f to $\mathcal{T}_r^p(\mathcal{T}_0, R)$. Then the following statements hold:

1. f is onto, $f(\mathbf{0}) = \mathbf{0}$, and $f(\mathbf{1}) = \mathcal{K}$ the greatest element of $\mathcal{T}_r^p(\mathcal{K}, R)$.
2. f preserves arbitrary infima.
3. $\mathcal{T}_r^p(\mathcal{T}_0, R)$ is a complete sublattice of $\mathcal{T}_r^p(R)$ and
$$f_0 : \mathcal{T}_r^p(\mathcal{T}_0, R) \longrightarrow \mathcal{T}_r^p(\mathcal{K}, R)$$
is a lattice isomorphism whose inverse sends $\mathcal{F}$ to $\sigma[M_{\mathcal{F}}]$.
4. If $\mathcal{K} \in \mathcal{N}_r(R)$, then f preserves arbitrary suprema.

PROOF. (1) and (2) are obvious. (4) is by (6.2.2). For (3), clearly $\mathcal{T}_r^p(\mathcal{T}_0, R)$ is a complete sublattice of $\mathcal{T}_r^p(R)$, and f_0 is onto by (6.2.5). It is easy to see that f_0 is one-to-one. Also f_0 preserves arbitrary suprema by (6.2.4). It follows that f_0 is a lattice isomorphism whose inverse sends $\mathcal{F}$ to $\sigma[M_{\mathcal{F}}]$. □

6.2.10. THEOREM. [104] The lattice $\mathcal{T}_r^p(R)$ is modular.

PROOF. Let $\mathcal{K}, \mathcal{L}, \mathcal{J} \in \mathcal{T}_r^p(R)$ with $\mathcal{K} \subseteq \mathcal{L}$. We need to show that

$$\mathcal{K} \vee (\mathcal{L} \wedge \mathcal{J}) = \mathcal{L} \wedge (\mathcal{K} \vee \mathcal{J}).$$

It is clear that $\mathcal{K} \vee (\mathcal{L} \wedge \mathcal{J}) \leq \mathcal{L} \wedge (\mathcal{K} \vee \mathcal{J})$. By (2.5.6), it suffices to show that every cyclic module $mR \in \mathcal{L} \wedge (\mathcal{K} \vee \mathcal{J})$ is in $\mathcal{K} \vee (\mathcal{L} \wedge \mathcal{J})$. Write $\mathcal{K} = \sigma[X]$ and $\mathcal{J} = \sigma[Y]$. Then $mR \in \mathcal{K} \vee \mathcal{J} = \sigma[X \oplus Y]$. By (2.2.5), there exist $x_i + y_i \in X \oplus Y$ $(i = 1, \cdots, n)$ such that $\cap_i(x_i^\perp \cap y_i^\perp) = \cap_i(x_i + y_i)^\perp \subseteq m^\perp$. Let $A = \cap_i x_i^\perp$ and $B = \cap_i y_i^\perp$. Then $A \cap B \subseteq m^\perp$, $R/A \in \mathcal{K}$ and $R/B \in \mathcal{J}$. Since $\mathcal{K} \leq \mathcal{L}$, it follows that $R/(A \cap m^\perp) \hookrightarrow R/A \oplus mR \in \mathcal{L}$, and hence $R/(A \cap m^\perp + B) \in \mathcal{L} \wedge \mathcal{J}$. Thus,

$$R/[A \cap (A \cap m^\perp + B)] \hookrightarrow R/A \oplus R/(A \cap m^\perp + B) \in \mathcal{K} \vee (\mathcal{L} \wedge \mathcal{J}).$$

But

$$A \cap (A \cap m^\perp + B) = (A \cap m^\perp) + (A \cap B) \subseteq m^\perp.$$

So $mR \cong R/m^\perp \in \mathcal{K} \vee (\mathcal{L} \wedge \mathcal{J})$. □

6.2.11. COROLLARY. $\mathcal{T}_r^p(\mathcal{K}, R)$ is modular for every $\mathcal{K} \in \mathcal{N}_r^p(R)$.

PROOF. This is by (6.2.9)(3) and (6.2.10). □

A partially ordered set I is called **directed** if for every couple $i, j \in I$ there exists a $k \in I$ such that $i \leq k$ and $j \leq k$. A lattice L is **upper continuous** if for every directed subset D of L and every $a \in L$, one has $(\vee_{d \in D} d) \wedge a = \vee_{d \in D}(d \wedge a)$.

6.2.12. THEOREM. For every $\mathcal{K} \in \mathcal{N}_r^p(R)$, the lattice $\mathcal{T}_r^p(\mathcal{K}, R)$ is upper continuous.

PROOF. We first prove that $\mathcal{T}_r^p(R)$ is upper continuous. Let $\{\sigma[X_i] : i \in D\}$ be a directed subset of $\mathcal{T}_r^p(R)$ and Y be any module. For any cyclic module X in $(\vee_{i \in D}\sigma[X_i]) \wedge \sigma[Y]$, $X \in \vee_{i \in D}\sigma[X_i] = \sigma[\oplus_{i \in D} X_i]$. It follows that there exist $i_1, \cdots, i_n \in D$ such that $X \in \sigma[\oplus_{k=1}^n X_{i_k}] = \vee_{k=1}^n \sigma[X_{i_k}]$. Since $\{\sigma[X_i] : i \in D\}$ is directed, there exists some $j \in D$ such that $\sigma[X_{i_k}] \leq \sigma[X_j]$ for $k = 1, \cdots, n$. So $X \in \sigma[X_j] \wedge \sigma[Y] \leq \vee_{i \in D}(\sigma[X_i] \wedge \sigma[Y])$. This shows that $(\vee_{i \in D}\sigma[X_i]) \wedge \sigma[Y] \leq \vee_{i \in D}(\sigma[X_i] \wedge \sigma[Y])$. It is obvious that

$$\vee_{i \in D}(\sigma[X_i] \wedge \sigma[Y]) \leq (\vee_{i \in D}\sigma[X_i]) \wedge \sigma[Y].$$

So $\vee_{i \in D}(\sigma[X_i] \wedge \sigma[Y]) = (\vee_{i \in D}\sigma[X_i]) \wedge \sigma[Y]$ and $\mathcal{T}_r^p(R)$ is upper continuous.

Now let $\{\mathcal{K}_i : i \in D\}$ be a directed subset of $\mathcal{T}_r^p(\mathcal{K}, R)$ and $\mathcal{L} \in \mathcal{T}_r^p(\mathcal{K}, R)$. By (6.2.5), $\mathcal{K}_i = \mathcal{K} \cap \mathcal{T}_i$, where $\mathcal{T}_i = \sigma[M_{\mathcal{K}_i}]$. Write $\mathcal{L} = \mathcal{K} \cap \mathcal{T}$ and $\mathcal{K} = \mathcal{N} \cap \mathcal{T}_0$, where $\mathcal{N} \in \mathcal{N}_r(R)$ and $\mathcal{T}, \mathcal{T}_0 \in \mathcal{T}_r^p(R)$. Then $\{\mathcal{T}_i : i \in D\}$ and hence $\{\mathcal{T}_i \cap \mathcal{T}_0 : i \in D\}$ is a directed subset of $\mathcal{T}_r^p(R)$. Then

$$
\begin{aligned}
(\vee_{i\in D}\mathcal{K}_i)\wedge\mathcal{L} &= \{\vee_{i\in D}(\mathcal{N}\cap\mathcal{T}_0\cap\mathcal{T}_i)\}\wedge(\mathcal{N}\cap\mathcal{T}_0\cap\mathcal{T}) \\
&= \mathcal{N}\wedge\{\vee_{i\in D}(\mathcal{T}_0\cap\mathcal{T}_i)\}\wedge(\mathcal{N}\cap\mathcal{T}_0\cap\mathcal{T}) \text{ (by 6.2.2)} \\
&= \{[\vee_{i\in D}(\mathcal{T}_0\cap\mathcal{T}_i)]\wedge(\mathcal{T}_0\cap\mathcal{T})\}\wedge\mathcal{N} \\
&= [\vee_{i\in D}(\mathcal{T}_0\cap\mathcal{T}_i\wedge\mathcal{T}_0\cap\mathcal{T})]\wedge\mathcal{N} \text{ ($\mathcal{T}_r^p(R)$ is upper continuous)} \\
&= \vee_{i\in D}(\mathcal{T}_i\wedge\mathcal{T}_0\wedge\mathcal{T}\wedge\mathcal{N}) \text{ (by 6.2.2)} \\
&= \vee_{i\in D}(\mathcal{K}_i\wedge\mathcal{L}).
\end{aligned}
$$

□

In the last part of this section, we discuss another family of sublattices of $\mathcal{N}_r^p(R)$. For a pre-natural class $\mathcal{K}$, let

$$\mathcal{N}(R,\mathcal{K}) = \{\mathcal{L}\cap\mathcal{K} : \mathcal{L}\in\mathcal{N}_r(R)\}.$$

It will be proved that $\mathcal{N}(R,\mathcal{K})$ is a sublattice of $\mathcal{N}_r^p(R)$, $\mathcal{N}(R,\mathcal{K})$ is a complete Boolean lattice, and $\mathcal{N}(R,\mathcal{K})$ is an algebraic lattice iff every nonzero module in $\mathcal{K}$ contains an atomic submodule. Finally, lattice decompositions of $\mathcal{N}_r^p(R)$ are discussed. Note that when $\mathcal{K} = \text{Mod-}R$, $\mathcal{N}(R,\mathcal{K}) = \mathcal{N}_r(R)$. When $\mathcal{K} = \sigma[M]$, $\mathcal{N}(R,\mathcal{K}) = \mathcal{N}(R,M)$ is the set of all M-natural classes.

6.2.13. LEMMA. The following are equivalent for a pre-natural class $\mathcal{K}$:

1. $\mathcal{F}\in\mathcal{N}(R,\mathcal{K})$.

2. $\mathcal{F}$ is a subclass of $\mathcal{K}$ closed under submodules, direct sums, and whenever A is an essential submodule of B with $A\in\mathcal{F}$ and $B\in\mathcal{K}$, we have $B\in\mathcal{F}$.

PROOF.(1) $\Longrightarrow$ (2). This can easily be verified by noting that $\mathcal{K} = \mathcal{K}_0\cap\mathcal{T}_0$, where $\mathcal{K}_0\in\mathcal{N}_r(R)$ and $\mathcal{T}_0\in\mathcal{T}_r^p(R)$.

(2) $\Longrightarrow$ (1). Suppose (2) holds. We show that $\mathcal{F} = d(M_{\mathcal{F}})\cap\mathcal{K}$. Clearly, $\mathcal{F}\subseteq d(M_{\mathcal{F}})\cap\mathcal{K}$. Let $X\in d(M_{\mathcal{F}})\cap\mathcal{K}$. Since $X\in d(M_{\mathcal{F}})$, X contains $\oplus_{i\in I}X_i$ as an essential submodule with all $X_i\in\mathcal{F}$. Thus $\oplus_{i\in I}X_i\in\mathcal{F}$. By (2), $X\in\mathcal{F}$. □

6.2.14. LEMMA. For any $\mathcal{K}_i\in\mathcal{N}_r(R)$ $(i = 1, 2)$ and any pre-natural class $\mathcal{L}$, $(\mathcal{K}_1\vee\mathcal{K}_2)\wedge\mathcal{L} = (\mathcal{K}_1\wedge\mathcal{L})\vee(\mathcal{K}_2\wedge\mathcal{L})$.

PROOF. We first assume $\mathcal{L} = \sigma[M]$ is a hereditary pretorsion class. One inclusion is obvious. Let $A\in(\mathcal{K}_1\vee\mathcal{K}_2)\wedge\sigma[M]$. Then $A\in\sigma[M]$ and, by (6.1.7), there exists an exact short sequence $0\longrightarrow B\longrightarrow A\longrightarrow C\longrightarrow 0$, where $B\in\mathcal{K}_1$ and $C\in\mathcal{K}_2$. Thus, $B\in\mathcal{K}_1\wedge\sigma[M]$ and $C\in\mathcal{K}_2\wedge\sigma[M]$. By (6.1.4), $(\mathcal{K}_1\wedge\sigma[M])\vee(\mathcal{K}_2\wedge\sigma[M])$ is an M-natural class. It follows from (2.4.5)(2) that $A\in(\mathcal{K}_1\wedge\sigma[M])\vee(\mathcal{K}_2\wedge\sigma[M])$. Therefore, the lemma holds in this case.

For a pre-natural class $\mathcal{L}$, write $\mathcal{L} = \mathcal{L}_0 \cap \sigma[M]$ for some module M, where $\mathcal{L}_0$ is a natural class. Then

$$\begin{aligned}(\mathcal{K}_1 \vee \mathcal{K}_2) \wedge \mathcal{L} &= \{(\mathcal{K}_1 \vee \mathcal{K}_2) \wedge \mathcal{L}_0\} \wedge \sigma[M] \\ &= \{(\mathcal{K}_1 \wedge \mathcal{L}_0) \vee (\mathcal{K}_2 \wedge \mathcal{L}_0)\} \wedge \sigma[M] \ (\mathcal{N}_r(R) \text{ is distributive}) \\ &= (\mathcal{K}_1 \wedge \mathcal{L}_0 \wedge \sigma[M]) \vee (\mathcal{K}_2 \wedge \mathcal{L}_0 \wedge \sigma[M]) \ (\text{as above}) \\ &= (\mathcal{K}_1 \wedge \mathcal{L}) \vee (\mathcal{K}_2 \wedge \mathcal{L}).\end{aligned}$$

□

6.2.15. THEOREM. For any pre-natural class $\mathcal{K}$, $\mathcal{N}(R, \mathcal{K})$ is a sublattice of $\mathcal{N}_r^p(R)$.

PROOF. Let $\mathcal{K}_i \in \mathcal{N}(R, \mathcal{K})$ $(i = 1, 2)$. Write $\mathcal{K}_i = \mathcal{L}_i \cap \mathcal{K}$, where $\mathcal{L}_i \in \mathcal{N}_r(R)$. It is obvious that $\mathcal{K}_1 \wedge \mathcal{K}_2 \in \mathcal{N}(R, \mathcal{K})$. And

$$\begin{aligned}\mathcal{K}_1 \vee \mathcal{K}_2 &= (\mathcal{L}_1 \wedge \mathcal{K}) \vee (\mathcal{L}_2 \wedge \mathcal{K}) \\ &= (\mathcal{L}_1 \vee \mathcal{L}_2) \wedge \mathcal{K} \in \mathcal{N}(R, \mathcal{K}) \ (\text{by } 6.2.14).\end{aligned}$$

□

It was proved by Dauns [31] that $\mathcal{N}(R, \mathcal{K})$ is a complete Boolean lattice for every hereditary pretorsion class $\mathcal{K}$. The following result extends this to any pre-natural class $\mathcal{K}$.

6.2.16. THEOREM. For any pre-natural class $\mathcal{K}$, $\mathcal{N}(R, \mathcal{K})$ is a complete Boolean lattice.

PROOF. It is easy to show that the intersection of any family of elements in $\mathcal{N}(R, \mathcal{K})$ is still in $\mathcal{N}(R, \mathcal{K})$. So $\mathcal{N}(R, \mathcal{K})$ is a complete lattice.

$\mathcal{N}(R, \mathcal{K})$ is complemented: For $\mathcal{L} \cap \mathcal{K} \in \mathcal{N}(R, \mathcal{K})$ with $\mathcal{L} \in \mathcal{N}_r(R)$, let $\mathcal{L}' = c(\mathcal{L})$. Then $\mathcal{L}' \cap \mathcal{K} \in \mathcal{N}(R, \mathcal{K})$. We have

$$(\mathcal{L}' \cap \mathcal{K}) \wedge (\mathcal{L} \cap \mathcal{K}) = (\mathcal{L}' \wedge \mathcal{L}) \wedge \mathcal{K} = \mathbf{0} \wedge \mathcal{K} = \mathbf{0},$$

and

$$(\mathcal{L}' \cap \mathcal{K}) \vee (\mathcal{L} \cap \mathcal{K}) = (\mathcal{L}' \vee \mathcal{L}) \wedge \mathcal{K} \ (\text{by } 6.2.14) = \mathbf{1} \wedge \mathcal{K} = \mathcal{K}.$$

$\mathcal{N}(R, \mathcal{K})$ is distributive: Let $\mathcal{L}_i \cap \mathcal{K} \in \mathcal{N}(R, \mathcal{K})$ with $\mathcal{L}_i \in \mathcal{N}_r(R)$ for $i = 1, 2, 3$. Then

$$\begin{aligned}&(\mathcal{L}_1 \cap \mathcal{K}) \wedge \{(\mathcal{L}_2 \cap \mathcal{K}) \vee (\mathcal{L}_3 \cap \mathcal{K})\} \\ &= (\mathcal{L}_1 \cap \mathcal{K}) \wedge \{(\mathcal{L}_2 \vee \mathcal{L}_3) \wedge \mathcal{K}\} \ (\text{by } 6.2.14) \\ &= \mathcal{L}_1 \wedge (\mathcal{L}_2 \vee \mathcal{L}_3) \wedge \mathcal{K} \\ &= \{(\mathcal{L}_1 \wedge \mathcal{L}_2) \vee (\mathcal{L}_1 \wedge \mathcal{L}_3)\} \wedge \mathcal{K} \ (\mathcal{N}_r(R) \text{ is distributive}) \\ &= (\mathcal{L}_1 \wedge \mathcal{L}_2 \wedge \mathcal{K}) \vee (\mathcal{L}_1 \wedge \mathcal{L}_3 \wedge \mathcal{K}) \ (\text{by } 6.2.14) \\ &= \{(\mathcal{L}_1 \cap \mathcal{K}) \wedge (\mathcal{L}_2 \cap \mathcal{K})\} \vee \{(\mathcal{L}_1 \cap \mathcal{K}) \wedge (\mathcal{L}_3 \cap \mathcal{K})\}.\end{aligned}$$

□

Next, we consider when the lattice $\mathcal{N}(R,\mathcal{K})$ is algebraic for $\mathcal{K} \in \mathcal{N}_r^p(R)$.

6.2.17. LEMMA. Let $\mathcal{K}$ be a pre-natural class. If $\mathcal{L} \in \mathcal{N}(R,\mathcal{K})$, then $\mathcal{L} = d(M_{\mathcal{L}}) \cap \mathcal{K}$.

PROOF. Write $\mathcal{L} = \mathcal{N} \cap \mathcal{K}$, where $\mathcal{N} \in \mathcal{N}_r(R)$. Since $\mathcal{L} = d(M_{\mathcal{L}}) \cap \sigma[M_{\mathcal{L}}]$, $\mathcal{L} \subseteq d(M_{\mathcal{L}}) \subseteq \mathcal{N}$. It follows that $\mathcal{L} = d(M_{\mathcal{L}}) \cap \mathcal{K}$. □

We note that, for a pre-natural class $\mathcal{K}$ and $X \in \mathcal{K}$, $d(X) \cap \mathcal{K}$ is the smallest element in $\mathcal{N}(R,\mathcal{K})$ that contains X.

6.2.18. LEMMA. Let $\mathcal{K}$ be a pre-natural class and $X \in \mathcal{K}$. Then $d(X) \cap \mathcal{K}$ is a compact element of $\mathcal{N}(R,\mathcal{K})$ iff X has finite type dimension.

PROOF. Suppose X is of finite type dimension. Then X contains an essential submodule Y such that $Y = Y_1 \oplus \cdots \oplus Y_n$, where each Y_i is an atomic submodule and $Y_i \perp Y_j$ for all $i \neq j$. Suppose that $d(X) \cap \mathcal{K} \leq \vee_{t \in I}\mathcal{K}_t$, where each $\mathcal{K}_t$ is in $\mathcal{N}(R,\mathcal{K})$ and the infinite supremum is taken in $\mathcal{N}(R,\mathcal{K})$ (not in $\mathcal{N}_r^p(R)$). By (6.2.17), for each $t \in I$ there exists an $X_t \in \mathcal{K}$ such that $\mathcal{K}_t = d(X_t) \cap \mathcal{K}$. By the remark before (6.2.18),

$$\vee_{t \in I}\mathcal{K}_t = \vee_{t \in I}[d(X_t) \cap \mathcal{K}] = d(\oplus_{t \in I} X_t) \cap \mathcal{K}.$$

Thus $Y \in d(\oplus_{t \in I} X_t)$. For each Y_i, there exists a $t_i \in I$ such that Y_i and X_{t_i} have nonzero isomorphic submodules. Thus, $Y \in d(\oplus_{t \in J} X_t)$, where $J = \{t_1, \cdots, t_n\}$. It follows that

$$\begin{aligned} d(X) \cap \mathcal{K} = d(Y) \cap \mathcal{K} &\leq d(\oplus_{t \in J} X_t) \cap \mathcal{K} \\ &= [\vee_{t \in J} d(X_t)] \cap \mathcal{K} \\ &= \vee_{t \in J}[d(X_t) \cap \mathcal{K}] = \vee_{t \in J}\mathcal{K}_t. \end{aligned}$$

Note that the second to the last equality is by (6.2.14). So $d(X) \cap \mathcal{K}$ is compact in $\mathcal{N}(R,\mathcal{K})$.

If X is not of finite type dimension, then X contains an essential submodule Y such that $Y = \oplus_{i=1}^{\infty} Y_i$, where each Y_i is nonzero and $Y_i \perp Y_j$ for all $i \neq j$. Thus

$$d(X) \cap \mathcal{K} = d(\oplus_{i=1}^{\infty} Y_i) \cap \mathcal{K} = \vee_{i=1}^{\infty}[d(Y_i) \cap \mathcal{K}],$$

where the infinite supremum is taken in $\mathcal{N}(R,\mathcal{K})$. If $d(X) \cap \mathcal{K}$ is compact in $\mathcal{N}(R,\mathcal{K})$, then there exists an $m > 0$ such that

$$d(X) \cap \mathcal{K} = \vee_{i=1}^{m}[d(Y_i) \cap \mathcal{K}] = [\vee_{i=1}^{m} d(Y_i)] \cap \mathcal{K} = d(\oplus_{i=1}^{m} Y_i) \cap \mathcal{K}.$$

The second to the last equality is by (6.2.14). This shows that X and hence Y_{m+1} is in $d(\oplus_{i=1}^{m} Y_i)$. This is impossible. So $d(X) \cap \mathcal{K}$ is not compact in $\mathcal{N}(R,\mathcal{K})$. □

6.2.19. THEOREM. Let $\mathcal{K}$ be a pre-natural class. Then the lattice $\mathcal{N}(R, \mathcal{K})$ is algebraic iff every R-module in $\mathcal{K}$ contains an atomic submodule.

PROOF. "$\Longleftarrow$". For any $\mathcal{L} \in \mathcal{N}(R, \mathcal{K})$, we can write $\mathcal{L} = d(X) \cap \mathcal{K}$ for some $X \in \mathcal{K}$ by (6.2.17). By our assumption, X contains an essential submodule N such that $N = \oplus_{t \in I} N_t$ is a direct sum of pairwise orthogonal atomic submodules. Then

$$\mathcal{L} = d(N) \cap \mathcal{K} = \vee_{t \in I}[d(N_t) \cap \mathcal{K}],$$

which is a join of compact elements in $\mathcal{N}(R, \mathcal{K})$ by (6.2.18) (Note that the infinite supremum is taken in $\mathcal{N}(R, \mathcal{K})$). So $\mathcal{N}(R, \mathcal{K})$ is algebraic.

"$\Longrightarrow$". For any module $X \in \mathcal{K}$, by our assumption $d(X) \cap \mathcal{K}$ is a join of compact elements in $\mathcal{N}(R, \mathcal{K})$. So $d(X) \cap \mathcal{K} \supseteq d(Y) \cap \mathcal{K}$ for some $Y \in \mathcal{K}$ such that $d(Y) \cap \mathcal{K}$ is a compact element of $\mathcal{N}(R, \mathcal{K})$. By (6.2.18), Y is of finite type dimension. So Y contains an atomic submodule N. Since $N \in d(X)$, X must contain an atomic submodule. □

6.2.20. COROLLARY. For any module M, the lattice $\mathcal{N}(R, M)$ is algebraic iff every module in $\sigma[M]$ contains an atomic submodule. □

6.2.21. COROLLARY. The lattice $\mathcal{N}_r(R)$ is algebraic iff every R-module contains an atomic submodule. □

A sublattice K of the lattice L is called **convex** if $a, b \in K, c \in L$, and $a \leq c \leq b$ implies that $c \in K$.

6.2.22. THEOREM. For $\mathcal{K} \in \mathcal{N}_r^p(R)$, define $g : \mathcal{N}_r(R) \longrightarrow \mathcal{N}(R, \mathcal{K})$ given by $g(\mathcal{F}) = \mathcal{F} \cap \mathcal{K}$. Let $\mathcal{N}_1 = \mathcal{N}(R, c(\mathcal{K}))$ and $\mathcal{N}_2 = \mathcal{N}(R, d(\mathcal{K}))$. Then

1. $g(\mathbf{0}) = \mathbf{0}$ and $g(\mathbf{1}) = \mathcal{K}$, the greatest element in $\mathcal{N}(R, \mathcal{K})$.
2. g preserves suprema of finitely many elements and arbitrary infima.
3. $\mathcal{N}_1$ and $\mathcal{N}_2$ are complete convex sublattices of $\mathcal{N}_r(R)$.
4. $\mathcal{N}_r(R) = \mathcal{N}_1 \oplus \mathcal{N}_2$ is a lattice direct sum.
5. $Ker(g) = \mathcal{N}_1$ and $g|_{\mathcal{N}_2} : \mathcal{N}_2 \longrightarrow \mathcal{N}(R, \mathcal{K})$ is a lattice isomorphism whose inverse sends $\mathcal{L}$ to $d(M_{\mathcal{L}})$.

PROOF. Parts (1) and (3) are obvious. Part (2) follows from (6.2.14). Part (4) holds because for any $\mathcal{K}_1 \in \mathcal{N}_1$ and $\mathcal{K}_2 \in \mathcal{N}_2$, $\mathcal{K}_1 \wedge \mathcal{K}_2 = \mathbf{0}$ and for any $\mathcal{L} \in \mathcal{N}_r(R)$,

$$\mathcal{L} = \mathcal{L} \wedge \mathbf{1} = \mathcal{L} \wedge [c(\mathcal{K}) \vee d(\mathcal{K})] = [\mathcal{L} \wedge c(\mathcal{K})] \vee [\mathcal{L} \wedge d(\mathcal{K})]$$

(by 6.2.14). For (5), $g|_{\mathcal{N}_2}$ is onto by (6.2.17) and the proof of the rest is straightforward. □

Finally, we extend g in (6.2.22) to a similar lattice homomorphism

$$\psi : \mathcal{N}_r^p(R) = \mathcal{N}_{d(\mathcal{K})}^p \oplus \mathcal{N}_{c(\mathcal{K})}^p \longrightarrow \mathcal{N}(R, \mathcal{K}),$$

where $\mathcal{K} \in \mathcal{N}_r^p(R)$,

$$\mathcal{N}_{d(\mathcal{K})}^p = \{\mathcal{F} \wedge d(\mathcal{K}) : \mathcal{F} \in \mathcal{N}_r^p(R)\},$$
$$\mathcal{N}_{c(\mathcal{K})}^p = \{\mathcal{F} \wedge c(\mathcal{K}) : \mathcal{F} \in \mathcal{N}_r^p(R)\}.$$

We identify the lattice congruence that plays the role of the kernel of ψ and show that $\psi|_{\mathcal{N}_{d(\mathcal{K})}^p} : \mathcal{N}_{d(\mathcal{K})}^p \longrightarrow \mathcal{N}(R, \mathcal{K})$ is a retract.

Let $\rho : \mathcal{N}_r^p(R) \longrightarrow \mathcal{N}_r(R)$ be the lattice homomorphism and retract defined by $\rho(\mathcal{F}) = d(\mathcal{F}) = c(c(\mathcal{F}))$ for $\mathcal{F} \in \mathcal{N}_r^p(R)$ (see [32, Thm.2.6]). It is interesting to note that although $\mathcal{N}_r^p(R)$ is not known to be "completely pseudo-complemented distributive," it behaves as if it were [10, p.148]. Define

$$\psi : \mathcal{N}_r^p(R) \longrightarrow \mathcal{N}(R, \mathcal{K}) \ \text{ by } \psi = g\rho$$

for g as in (6.2.22), where $\mathcal{K} \in \mathcal{N}_r^p(R)$. Thus for $\mathcal{F} \in \mathcal{N}_r^p(R)$, $\psi(\mathcal{F}) = g(\rho(\mathcal{F})) = g(d(\mathcal{F})) = d(\mathcal{F}) \cap \mathcal{K}$. For any $\mathcal{N} \subseteq \mathcal{N}_r^p(R)$, write

$$\mathcal{N}^{\perp} = \{\mathcal{L} \in \mathcal{N}_r^p(R) : \mathcal{L} \wedge \mathcal{K} = \mathbf{0} \ \text{ for all } \mathcal{K} \in \mathcal{N}\}.$$

6.2.23. LEMMA. For $\mathcal{K} \in \mathcal{N}_r^p(R)$,

1. $\mathcal{N}_r^p(R) = \mathcal{N}_{d(\mathcal{K})}^p \oplus \mathcal{N}_{c(\mathcal{K})}^p$ is a lattice direct sum of convex and complete sublattices of $\mathcal{N}_r^p(R)$.
2. $(\mathcal{N}_{d(\mathcal{K})}^p)^{\perp} = \mathcal{N}_{c(\mathcal{K})}^p$ and $(\mathcal{N}_{c(\mathcal{K})}^p)^{\perp} = \mathcal{N}_{d(\mathcal{K})}^p$.

PROOF. (1) Note that $\mathcal{N}_{d(\mathcal{K})}^p$ is the collection of all pre-natural classes $\mathcal{F}$ such that $\mathcal{F} \leq d(\mathcal{K})$. So $\mathcal{N}_{d(\mathcal{K})}^p$ is clearly a convex and complete sublattice of $\mathcal{N}_r^p(R)$. The same reason shows that $\mathcal{N}_{c(\mathcal{K})}^p$ is a convex and complete sublattice of $\mathcal{N}_r^p(R)$. Obviously, $\mathcal{F}_1 \wedge \mathcal{F}_2 = \mathbf{0}$ for all $\mathcal{F}_1 \in \mathcal{N}_{d(\mathcal{K})}^p$ and $\mathcal{F}_2 \in \mathcal{N}_{c(\mathcal{K})}^p$. For $\mathcal{F} \in \mathcal{N}_r^p(R)$, since $\mathbf{1} = d(\mathcal{K}) \vee c(\mathcal{K})$, in view of (6.2.14),

$$\mathcal{F} = \mathcal{F} \wedge \mathbf{1} = [\mathcal{F} \wedge d(\mathcal{K})] \vee [\mathcal{F} \wedge c(\mathcal{K})].$$

So (1) is proved.

(2) It follows from (1). □

6.2.24. THEOREM. For $\mathcal{K} \in \mathcal{N}_r^p(R)$, let

$$\psi : \mathcal{N}_r^p(R) = \mathcal{N}_{d(\mathcal{K})}^p \oplus \mathcal{N}_{c(\mathcal{K})}^p \longrightarrow \mathcal{N}(R, \mathcal{K})$$

as above. Then the following statements hold:

1. ψ is a lattice homomorphism preserving suprema and infima of finitely many elements such that $\psi(\mathbf{0}) = \mathbf{0}$ and $\psi(\mathbf{1}) = \mathcal{K}$.

2. $Ker(\psi) = \mathcal{N}^p_{c(\mathcal{K})}$.

3. $\psi|_{\mathcal{N}_r(R)} = g$ as in (6.2.22).

4. $\psi|_{\mathcal{N}^p_{d(\mathcal{K})}} : \mathcal{N}^p_{d(\mathcal{K})} \longrightarrow \mathcal{N}(R, \mathcal{K})$ is a retract.

PROOF. (1) Both g and ρ are lattice homomorphisms (6.2.22 and [32, Thm.2.6]); hence so is $\psi = g\rho$. (Note that g preserves arbitrary infima, while ρ preserves arbitrary suprema.)

(2) For any $c(\mathcal{K}) \wedge \mathcal{F} \in \mathcal{N}^p_{c(\mathcal{K})}$,

$$\psi(c(\mathcal{K}) \wedge \mathcal{F}) = \psi(c(\mathcal{K})) \wedge \psi(\mathcal{F}) = c(\mathcal{K}) \wedge \mathcal{K} \wedge d(\mathcal{F}) = \mathbf{0}.$$

Thus, $\psi(\mathcal{N}^p_{c(\mathcal{K})}) = \{\mathbf{0}\}$.

Conversely, for any $\mathcal{F} \in \mathcal{N}^p_r(R)$, $\mathcal{F} = [\mathcal{F} \wedge d(\mathcal{K})] \vee [\mathcal{F} \wedge c(\mathcal{K})]$. If $\mathcal{F} \notin \mathcal{N}^p_{c(\mathcal{K})}$, then $\mathcal{F} \wedge d(\mathcal{K}) \neq \mathbf{0}$. Choose $0 \neq V \in \mathcal{F} \wedge d(\mathcal{K})$. Then V has a nonzero submodule W in $\mathcal{K}$. Thus, $W \in \mathcal{F} \wedge \mathcal{K} \leq d(\mathcal{F}) \wedge \mathcal{K} = \psi(\mathcal{F})$. So $\psi(\mathcal{F}) \neq \mathbf{0}$.

(3) For $\mathcal{F} \in \mathcal{N}_r(R)$, $\mathcal{F} = d(\mathcal{F})$, and always $\psi(\mathcal{F}) = d(\mathcal{F}) \cap \mathcal{K} = \mathcal{F} \cap \mathcal{K} = g(\mathcal{F})$. Thus, $\psi|_{\mathcal{N}_r(R)} = g$.

(4) By (2) and (3) above, we have that $\psi(\mathcal{N}^p_r(R)) = \psi(\mathcal{N}^p_{d(\mathcal{K})}) = \mathcal{N}(R, \mathcal{K})$. For $\mathcal{F} \in \mathcal{N}(R, \mathcal{K})$, $\mathcal{F} = \mathcal{L} \wedge \mathcal{K}$, $\mathcal{L} \in \mathcal{N}_r(R)$. Then

$$\psi(\mathcal{F}) = \psi(\mathcal{L}) \wedge \psi(\mathcal{K}) = \mathcal{L} \wedge d(\mathcal{K}) \wedge \mathcal{K} = \mathcal{L} \wedge \mathcal{K} = \mathcal{F}.$$

Thus, $\psi|_{\mathcal{N}(R,\mathcal{K})} = Id$. □

6.2.25. REFERENCES. Dauns [31,32,34]; Golan [60]; Dauns and Zhou [40]; Raggi, Montes and Wisbauer [104]; Zhou [141,142].

6.3 Lattice Properties of $\mathcal{N}_r^p(R)$

We first describe the atoms and coatoms of $\mathcal{N}_r^p(R)$.

6.3.1. PROPOSITION. The following are equivalent for a class $\mathcal{K}$ of modules:

1. $\mathcal{K}$ is an atom of $\mathcal{N}_r^p(R)$.

2. $\mathcal{K} = d(M) \cap \sigma[M]$ where M is an atomic module and $M \in \sigma[N]$ for any $0 \neq N \subseteq M$.

PROOF. (1) $\Longrightarrow$ (2). Write $\mathcal{K} = d(M) \cap \sigma[M]$ for some module M (see 2.5.4). If $X \perp Y$ for nonzero submodules X and Y of M, then $\mathcal{K}_1 := d(X) \cap \sigma[X] \leq \mathcal{K}$ and $Y \in \mathcal{K} \backslash \mathcal{K}_1$. This is a contradiction. So M is atomic. For a nonzero submodule N of M, (1) implies $d(N) \cap \sigma[N] = \mathcal{K}$. Thus M is in $\sigma[N]$.

(2) $\Longrightarrow$ (1). Suppose $\mathcal{L} \leq \mathcal{K}$ where $\mathbf{0} \neq \mathcal{L} \in \mathcal{N}_r^p(R)$. Then there exists a $0 \neq X \leq M$ such that $X \in \mathcal{L}$. Since M is atomic, $d(X) = d(M)$. So $d(X) \cap \sigma[X] = \mathcal{K}$ since $\sigma[X] = \sigma[M]$ by (2). It follows that $\mathcal{L} = \mathcal{K}$. □

Homogeneous semisimple modules and the $\mathbb{Z}$-modules $\mathbb{Z}$, $\mathbb{Z}_{p^n}$ ($n \geq 2$ and p is prime), and $\mathbb{Z}_{p^\infty}$ are atomic, but only the first two satisfy (6.3.1)(2).

Let L be a lattice with the greatest element $\mathbf{1}$. A **coatom** of L is an element $c \in L$ such that $c \neq \mathbf{1}$ and $c \not\lneq d$ for any $\mathbf{1} \neq d \in L$. An R-module M is called a **large module** if, for any nonzero R-module N, M and N have nonzero isomorphic submodules. Thus a module N is large iff $d(N) = \text{Mod-}R$.

6.3.2. PROPOSITION. The following are equivalent for a class $\mathcal{K}$ of modules:

1. $\mathcal{K}$ is a coatom of $\mathcal{N}_r^p(R)$.

2. (a) $\mathcal{K} = \sigma[M]$ is a coatom of $\mathcal{T}_r^p(R)$ for a large module M or

   (b) $\mathcal{K} = d(M)$ where M is not large, but for any nonzero module Y with $M \perp Y$, $M \oplus Y$ is large and $X^\perp = 0$ for a finite subset X of $M \oplus Y$.

PROOF. (1) $\Longrightarrow$ (2). Write $\mathcal{K} = d(M) \cap \sigma[M]$ for some module M. If M is large, then $d(M) = \mathbf{1}$. So $\mathcal{K} = \sigma[M]$ is a coatom of $\mathcal{N}_r^p(R)$. Thus $\mathcal{K}$ is a coatom of $\mathcal{T}_r^p(R)$. Thus, (2a) holds. If M is not large, then $d(M) < \mathbf{1}$ and so $\mathcal{K} = d(M)$. Then, for any nonzero module Y with $M \perp Y$, $Y \notin d(M)$. But $\mathcal{K} = d(M) \cap \sigma[M] \leq d(M \oplus Y) \cap \sigma[M \oplus Y]$ and $Y \in d(M \oplus Y) \cap \sigma[M \oplus Y]$, so $d(M \oplus Y) \cap \sigma[M \oplus Y] = \mathbf{1}$. It follows that $M \oplus Y$ is a large module and $R \in \sigma[M \oplus Y]$, showing that $X^\perp = 0$ for a finite subset X of $M \oplus Y$.

(2) $\Longrightarrow$ (1). Clearly (2) implies that $\mathcal{K} \neq \mathbf{1}$. Let $\mathcal{L} = d(N) \cap \sigma[N]$ with $\mathcal{K} < \mathcal{L}$. If (2$a$) holds, then $M \in d(N)$ is large and $\mathcal{K} = \sigma[M]$. Thus, N is large and $\mathcal{L} = \sigma[N]$. Since $\sigma[M]$ is a coatom of $\mathcal{T}_r^p(R)$, $\mathcal{L} = \sigma[N] = \mathbf{1}$.

Suppose (2b) holds. Then $d(M) < d(N)$. So there exists a $0 \neq Y \leq N$ such that $M \perp Y$. By (2b), $M \oplus Y$ is large and $R \in \sigma[M \oplus Y]$. So $d(M \oplus Y) \cap \sigma[M \oplus Y] = \mathbf{1}$. Since $M \oplus Y \in \mathcal{L}$, $\mathcal{L} = \mathbf{1}$. □

6.3.3. PROPOSITION. [60] Every proper element of $\mathcal{T}_r^p(R)$ is contained in a coatom of $\mathcal{T}_r^p(R)$.

PROOF. Let $\mathcal{T} \in \mathcal{T}_r^p(R)$ with $\mathcal{T} < \mathbf{1}$. For any chain $\{\mathcal{T}_\alpha : \alpha \in \Lambda\}$ of elements in $\mathcal{T}_r^p(R)$ with $\mathcal{T} \leq \mathcal{T}_\alpha < \mathbf{1}$, $\vee_\alpha \mathcal{T}_\alpha \in \mathcal{T}_r^p(R)$. We claim $\vee_\alpha \mathcal{T}_\alpha < \mathbf{1}$. To see so, write $\mathcal{T}_\alpha = \sigma[M_\alpha]$ for some module M_α. Then by (6.1.10), $\vee_\alpha \mathcal{T}_\alpha = \sigma[\oplus_\alpha M_\alpha]$. If $\vee_\alpha \mathcal{T}_\alpha = \mathbf{1}$, that is $R \in \sigma[\oplus_\alpha M_\alpha]$, then there exists a $k > 0$ such that $R \in \sigma[\oplus_{j=1}^k M_{\alpha_j}] = \vee_{j=1}^k \mathcal{T}_{\alpha_j} = \mathcal{T}_{\alpha_k}$ (assume that $\mathcal{T}_{\alpha_1} \leq \cdots \leq \mathcal{T}_{\alpha_k}$). Thus, $\mathcal{T}_{\alpha_k} = \mathbf{1}$, a contradiction. By Zorn's Lemma, $\mathcal{T}$ is contained in a coatom of $\mathcal{T}_r^p(R)$. □

We next consider the question of when every proper element of $\mathcal{N}_r^p(R)$ is contained in a coatom of $\mathcal{N}_r^p(R)$.

6.3.4. THEOREM. The following are equivalent for a ring R:

1. Every proper element of $\mathcal{N}_r^p(R)$ is contained in a coatom of $\mathcal{N}_r^p(R)$.
2. There does not exist a pair of nonzero R-modules X and Y such that
   (a) R is in $\sigma[X]$.
   (b) $X \perp Y$.
   (c) Y contains no atomic submodules.

PROOF. (1) $\Longrightarrow$ (2). Suppose that there exist nonzero modules X and Y satisfying (a), (b), and (c). Let $\mathcal{K} = c(Y)$. Then $X \in \mathcal{K} \in \mathcal{N}_r^p(R)$. We claim that $\mathcal{K}$ is not contained in any coatom of $\mathcal{N}_r^p(R)$. To see this, let $\mathcal{K} \leq \mathcal{L} < \mathbf{1}$ for some $\mathcal{L} \in \mathcal{N}_r^p(R)$. Write $\mathcal{L} = d(M) \cap \sigma[M]$ for some module M. Since $X \in \mathcal{K}$, X is in $\sigma[M]$. So $\sigma[M] = \mathbf{1}$ because of 2(a) and hence $\mathcal{L} = d(M)$. Since $\mathcal{L} \neq \mathbf{1}$, there exists a nonzero module A such that $A \perp M$. Then A is not in $\mathcal{K}$. Thus, there exists a $0 \neq B \leq A$ such that $B \hookrightarrow Y$. By (c), B is not atomic, so B contains two nonzero submodules B_1 and B_2 such that $B_1 \perp B_2$. Let $\mathcal{H} = d(\mathcal{F})$ where $\mathcal{F} = \mathcal{L} \cup \{B_1\}$. Then $\mathcal{L} \leq \mathcal{H}$ such that B_1 is in $\mathcal{H} \backslash \mathcal{L}$ and $B_2 \notin \mathcal{H}$.

(2) $\Longrightarrow$ (1). Suppose that (1) does not hold. Then there exists $\mathbf{1} \neq \mathcal{K} \in \mathcal{N}_r^p(R)$ such that $\mathcal{K}$ is not contained in any coatom of $\mathcal{N}_r^p(R)$. If $\sigma[M_\mathcal{K}] \neq \mathbf{1}$, then by (6.3.3), there exists a maximal element $\mathcal{T}$ in $\mathcal{T}_r^p(R)$ such that $\sigma[M_\mathcal{K}] \leq \mathcal{T} \neq \mathbf{1}$. Since $\mathcal{T}$ is not a coatom of $\mathcal{N}_r^p(R)$, we have $\mathcal{T} < \mathcal{H} \neq \mathbf{1}$ for some $\mathcal{H} \in \mathcal{N}_r^p(R)$. Since $\mathcal{K} \leq \mathcal{H}$, $\mathcal{H}$ is not contained in any coatom of $\mathcal{N}_r^p(R)$. Note that $\mathcal{H} \subseteq \sigma[M_\mathcal{H}]$. By the maximality of $\mathcal{T}$, R is in $\sigma[M_\mathcal{H}]$. Therefore, we may assume $R \in \sigma[M_\mathcal{K}]$. It follows that $\mathcal{K} = d(M_\mathcal{K}) \cap \sigma[M_\mathcal{K}] = d(M_\mathcal{K})$.

Since $d(M_{\mathcal{K}}) \neq \mathbf{1}$, there exists a nonzero module Y such that $Y \perp M_{\mathcal{K}}$. Let $X = M_{\mathcal{K}}$. We now prove that Y contains no atomic submodules and hence the pair of X and Y gives a contradiction to (2). If Y has an atomic submodule P, then $\mathcal{K} \leq c(P) \neq \mathbf{1}$. Suppose $c(P) < \mathcal{L} \in \mathcal{N}_r^p(R)$. Since $R \in \sigma[M_{\mathcal{K}}] \leq \sigma[M_{\mathcal{L}}]$, $\mathcal{L} = d(M_{\mathcal{L}}) \cap \sigma[M_{\mathcal{L}}] = d(M_{\mathcal{L}})$. So $c(P) < d(M_{\mathcal{L}})$. It follows that $P \in d(M_{\mathcal{L}})$ since P is atomic. Thus, $\mathbf{1} = c(P) \vee d(P) \leq d(M_{\mathcal{L}}) = \mathcal{L}$, so $\mathcal{L} = \mathbf{1}$. This shows that $c(P)$ is a coatom of $\mathcal{N}_r^p(R)$, a contradiction. □

6.3.5. COROLLARY. If every nonzero R-module contains an atomic submodule, then every proper element of $\mathcal{N}_r^p(R)$ is contained in a coatom of $\mathcal{N}_r^p(R)$.

6.3.6. EXAMPLE. There exists a ring R such that some proper element of $\mathcal{N}_r^p(R)$ is not contained in any coatom of $\mathcal{N}_r^p(R)$.

PROOF. Let $F_1, F_2, \cdots$ be fields and $R = \Pi_{i=1}^{\infty} F_i$. Let $X_R = R$ and $Y_R = (\Pi F_i)/(\oplus F_i)$. Then it can easily be verified that the pair of X and Y satisfies the conditions $(a) - (c)$ of (6.3.4). Therefore, R is the required ring. □

For any ring R, $\mathcal{N}_r^p(R)$ has at least one coatom as shown below.

6.3.7. PROPOSITION. There exists at least one coatom of $\mathcal{N}_r^p(R)$.

PROOF. Take a simple module X and let $\mathcal{K} = c(X)$. If $\mathcal{K}$ is a coatom, then we are done. If not, then $\mathcal{K} < \mathcal{L} \neq \mathbf{1}$ for some $\mathcal{L} \in \mathcal{N}_r^p(R)$. It follows that X is in $\mathcal{L}$. For any module Y, either X is embeddable in Y or Y is in $\mathcal{K}$. So Y has an essential submodule in $d(M_{\mathcal{L}})$, showing that $Y \in d(M_{\mathcal{L}})$. Thus, $d(M_{\mathcal{L}}) = \mathbf{1}$ and so $\mathcal{L} = d(M_{\mathcal{L}}) \cap \sigma[M_{\mathcal{L}}] = \sigma[M_{\mathcal{L}}]$. Since $\sigma[M_{\mathcal{L}}] \neq \mathbf{1}$, (6.3.3) implies that there exists $\mathcal{H}$ which is maximal with respect to $\sigma[M_{\mathcal{L}}] \subseteq \mathcal{H} \in \mathcal{T}_r^p(R)$ and $\mathcal{H} \neq \mathbf{1}$. We claim that $\mathcal{H}$ is a coatom of $\mathcal{N}_r^p(R)$. In fact, if $\mathcal{H} < \mathcal{G}$ for some $\mathcal{G} \in \mathcal{N}_r^p(R)$, then $d(M_{\mathcal{L}}) \leq d(M_{\mathcal{G}})$ and $\sigma[M_{\mathcal{H}}] \leq \sigma[M_{\mathcal{G}}]$. So $d(M_{\mathcal{G}}) = \mathbf{1}$ and hence $\mathcal{G} = \sigma[M_{\mathcal{G}}]$. Thus, $\sigma[M_{\mathcal{H}}] < \sigma[M_{\mathcal{G}}]$. By the maximality of $\mathcal{H}$, $\mathcal{G} = \sigma[M_{\mathcal{G}}] = \mathbf{1}$. □

For a ring R, Nicholson and Sarath [91, Theorem 1] gives a necessary and sufficient condition for $\mathcal{T}_r^p(R)$ to have a unique coatom. That result was used in [142] to characterize the ring R for which $\vee\{\mathcal{K} : \mathbf{1} \neq \mathcal{K} \in \mathcal{N}_r^p(R)\} \neq \mathbf{1}$.

A lattice L with the greatest element $\mathbf{1}$ and the least element $\mathbf{0}$ is called a **complemented lattice** if, for any $a \in L$, there exists $a_c \in L$ such that $a \wedge a_c = \mathbf{0}$ and $a \vee a_c = \mathbf{1}$. If such an a_c is unique for every $a \in L$, then L is called **uniquely complemented**.

6.3.8. PROPOSITION. The following are equivalent for a ring R:

1. $\mathcal{N}_r^p(R)$ is a complemented lattice.

2. For any R-module A, there exists a module B such that $A \perp B$, $A \oplus B$ is a large module, and $X^{\perp} = 0$ for a finite subset X of $A \oplus B$.

PROOF. (1) $\Longrightarrow$ (2). For any module A, let $\mathcal{K} = \sigma[A] \cap d(A)$. By (1), there exists $\mathcal{K}' \in \mathcal{N}_r^p(R)$ such that $\mathcal{K} \wedge \mathcal{K}' = \mathbf{0}$ and $\mathcal{K} \vee \mathcal{K}' = \mathbf{1}$. It follows that $\mathcal{K}' \subseteq c(\mathcal{K})$. Thus, we have $\mathcal{K} \vee c(\mathcal{K}) = \mathbf{1}$. By (6.1.3), $\sigma[A \oplus M_{c(\mathcal{K})}] \cap d(A \oplus M_{c(\mathcal{K})}) = \mathbf{1}$. Let $B = M_{c(\mathcal{K})}$. Then $A \perp B$ and $A \oplus B$ is a large module. Since $R \in \sigma[A \oplus B]$, there exists a finite subset X of $A \oplus B$ such that $X^{\perp} = 0$.

(2) $\Longrightarrow$ (1). Let $\mathcal{K} \in \mathcal{N}_r^p(R)$. Write $\mathcal{K} = \sigma[M_{\mathcal{K}}] \cap d(M_{\mathcal{K}})$. Let $A = M_{\mathcal{K}}$. By (2), there exists a B such that A and B satisfy the properties in (2). Let $\mathcal{L} = \sigma[B] \cap d(B)$. Then $\mathcal{K} \wedge \mathcal{L} = \mathbf{0}$ since $A \perp B$. Because $A \oplus B$ is a large module, $d(A \oplus B) = \mathbf{1}$. By the fact that $X^{\perp} = 0$ for a finite subset X of $A \oplus B$, we see that $\sigma[A \oplus B] = \mathbf{1}$. Therefore, $\mathcal{K} \vee \mathcal{L} = \sigma[A \oplus B] \cap d(A \oplus B) = \mathbf{1} \wedge \mathbf{1} = \mathbf{1}$. □

Thus, $\mathcal{N}_r^p(\mathbb{Z})$ is a complemented lattice but $\mathcal{N}_r^p(\mathbb{Z}_4)$ is not. Next, we will see that several concepts of rings can be characterized by properties of some sublattices of $\mathcal{N}_r^p(R)$. A ring R is called **right QI** if every quasi-injective module is injective. The equivalence "(1) $\Longleftrightarrow$ (3)" of the next result belongs to Gabriel [54], and Golan and López-Permouth [62].

6.3.9. THEOREM. The following are equivalent for a ring R:

1. R is a right QI-ring.
2. $\mathcal{N}_r^p(R) = \mathcal{N}_r(R)$.
3. $\mathcal{T}_r^p(R) \subseteq \mathcal{N}_r(R)$.
4. $\mathcal{N}_r^p(R)$ is a uniquely complemented lattice.

PROOF. (1) $\Longrightarrow$ (2). For $\mathcal{K} \in \mathcal{N}_r^p(R)$, we need to show that $\mathcal{K}$ is closed under injective hulls. Write $\mathcal{K} = d(M) \cap \sigma[M]$ for some module M and let $N \in \mathcal{K}$. Since $N \in \sigma[M]$, we have $N \subseteq E_M(N) \subseteq E(N)$ by (2.2.9). Note $E_M(N)$ is M-injective, so $E_M(N)$ is quasi-injective by (2.2.10). Thus, $E_M(N)$ is injective by (1). This shows that $E(N) = E_M(N)$ is in $\sigma[M]$. But, clearly $E(N)$ is in $d(M)$. So $E(N) \in \mathcal{K}$.

(2) $\Longrightarrow$ (4). This follows from (5.1.5) that $\mathcal{N}_r(R)$ is a complete Boolean lattice.

(4) $\Longrightarrow$ (3). Let $\mathcal{K} \in \mathcal{T}_r^p(R)$. By (4) there exists $\mathcal{L} \in \mathcal{N}_r^p(R)$ such that $\mathcal{K} \wedge \mathcal{L} = \mathbf{0}$ and $\mathcal{K} \vee \mathcal{L} = \mathbf{1}$. Then $\mathcal{K} \leq c(\mathcal{L})$, so $c(\mathcal{L}) \vee \mathcal{L} = \mathbf{1}$ and $\mathcal{L} \wedge c(\mathcal{L}) = \mathbf{0}$. By (4), $\mathcal{K} = c(\mathcal{L})$ is a natural class.

(3) $\Longrightarrow$ (1). Let M be a quasi-injective module. Then $\mathcal{K} = \sigma[M] \in \mathcal{T}_r^p(R)$, so $E(M)$ is in $\sigma[M]$ by (3). Since M is M-injective, M is N-injective for each N in $\sigma[M]$ by [87, 1.3 and 1.5]. Thus, M is $E(M)$-injective. This shows that M is a direct summand of $E(M)$. So $M = E(M)$ is injective. □

A ring R is called **right semiartinian** if every nonzero right R-module contains a nonzero simple submodule.

6.3.10. THEOREM. The following are equivalent for a ring R:

1. R is right semiartinian.

2. $\mathcal{N}_r(R) = \mathcal{F}_r(R)$.

3. The natural class $\mathcal{K} = \{M \in \text{Mod-}R : Soc(M) \leq_e M\}$ is in $\mathcal{F}_r(R)$.

4. $\mathcal{F}_r(R)$ is a Boolean lattice.

PROOF. (1) $\Longrightarrow$ (2). Let R be right semiartinian and let $\mathcal{K}$ be a natural class. Set $\mathcal{F} = \{Soc(M) : M \in \mathcal{K}\}$. Then $\mathcal{K} = d(\mathcal{F})$. Consider $M = \Pi_i M_i$ with $M_i \in \mathcal{K}$. If $M \notin \mathcal{K}$, then there exists a nonzero submodule N of M such that N has no nonzero submodule in $\mathcal{F}$. Since R is right semiartinian, we may assume $N = xR$ is simple. Write $x = (x_i)$ with $x_i \in M_i$ and some $x_k \neq 0$. Then the map $N = xR \longrightarrow x_kR$ by $xr \longmapsto x_kr$ is an isomorphism. Thus, the fact that $N \cong x_kR \subseteq M_k \in \mathcal{K}$ implies $N \in \mathcal{F}$, a contradiction. So M is in $\mathcal{K}$ and hence $\mathcal{K}$ is closed under products.

(2) $\Longrightarrow$ (3). The class $\mathcal{K} = \{M \in \text{Mod-}R : Soc(M) \leq_e M\}$ is a natural class, so is in $\mathcal{F}_r(R)$ by (2).

(3) $\Longrightarrow$ (1). By (3), $\mathcal{K} = \{M : Soc(M) \leq_e M\}$ is a hereditary torsion free class. Since all injective hulls of simple modules are in $\mathcal{K}$, the least cogenerator C, i.e., the direct sum of injective hulls of non-isomorphic simple modules, is in $\mathcal{K}$. Since C cogenerates all modules, it follows that $\mathcal{K} = \mathbf{1}$. So R is right semiartinian.

(2) $\Longrightarrow$ (4) follows from (5.1.5).

(4) $\Longrightarrow$ (1). Suppose (1) does not hold. Let $\mathcal{K}$ be a complete set of representatives of isomorphism classes of simple right R-modules, and let

$$\mathcal{F} = \{F \in \text{Mod-}R : \text{Hom}_R(K, F) = 0, \forall K \in \mathcal{K}\},$$
$$\mathcal{T} = \{T \in \text{Mod-}R : \text{Hom}_R(T, F) = 0, \forall F \in \mathcal{F}\}.$$

Since $\mathcal{F}$ is clearly closed under injective hulls, $(\mathcal{T}, \mathcal{F})$ is a hereditary torsion theory by (2.1.6). By (4), there exists a hereditary torsion theory $(\mathcal{T}^{'}, \mathcal{F}^{'})$ such that $\mathcal{F} \vee \mathcal{F}^{'} = \mathbf{1}$ and $\mathcal{F} \wedge \mathcal{F}^{'} = \mathbf{0}$. If $\mathcal{T}^{'} \neq \mathbf{0}$, then there exists a simple module $N \in \mathcal{T}^{'}$. Thus, $N \notin \mathcal{F}^{'}$ but $N \in \mathcal{F} \vee \mathcal{F}^{'}$. It must be that $N \in \mathcal{F}$. This is a contradiction. Therefore, $\mathcal{T}^{'} = \mathbf{0}$ and so $\mathcal{F}^{'} = \mathbf{1}$. It follows that $\mathcal{F} = \mathbf{0}$. Thus, every nonzero module has a nonzero socle, so (1) holds. □

It was proved in Popescu [103] that R is right semiartinian iff **Tor**-R is a Boolean lattice. This result follows from (6.3.10) because the two lattices $\mathcal{F}_r(R)$ and **Tor**-R are anti-isomorphic. The next result is taken from Alin and Armendariz [6] and Dlab [44] (see Golan [59]). A ring R is called **left perfect** if $R/J(R)$ is a semisimple ring and $J(R)$ is left T-nilpotent.

6.3.11. THEOREM. The following are equivalent for a ring R:

1. R is a left perfect ring.

2. $\mathcal{F}_r(R)$ is a Boolean lattice and every $\mathcal{T} \in \mathcal{T}_r(R)$ is closed under direct products.

PROOF. (1) $\Longrightarrow$ (2). Since R is left perfect, R is right semiartinian and there are only finitely many simple right R-modules, up to isomorphism. So $\mathcal{F}_r(R)$ is a Boolean lattice by (6.3.10). Let $(\mathcal{T}, \mathcal{F})$ be a hereditary torsion theory. We want to show that $\mathcal{T}$ is closed under direct products. We can assume $\mathcal{F} \neq \mathbf{0}$ and assume $\{M_1, \cdots, M_k\}$ is a complete set of representatives of isomorphic classes of simple modules in $\mathcal{F}$. Because R is left perfect, $M_1 \oplus \cdots \oplus M_k$ has a projective cover. That is, there exists a projective module P and an epimorphism $\sigma : P \longrightarrow M_1 \oplus \cdots \oplus M_k$ such that $Ker(\sigma)$ is small in P (i.e., $Ker(\sigma) + X \neq P$ whenever $X \subset P$).

We claim that

$$\mathcal{T} = \{M \in \text{Mod-}R : \text{Hom}_R(P, M) = 0\}.$$

In fact, for $M \in \mathcal{T}$ and $f \in \text{Hom}_R(P, M)$, $P/[Ker(\sigma) + Ker(f)]$ is an image of $P/Ker(\sigma) \cong M_1 \oplus \cdots \oplus M_k$, so $P/[Ker(\sigma) + Ker(f)] \in \mathcal{F}$; but since $P/Ker(f) \in \mathcal{T}$, $P/[Ker(\sigma) + Ker(f)] \in \mathcal{T}$. So $P/[Ker(\sigma) + Ker(f)] = 0$. Thus, $P = Ker(f)$ since $Ker(\sigma)$ is a small submodule of P, and hence $f = 0$. On the other hand, if $N \notin \mathcal{T}$, then $0 \neq N/X \in \mathcal{F}$ for some submodule X of N. Since R is left perfect, N/X contains a simple submodule Y/X. So $Y/X \cong M_j$ for some j with $1 \leq j \leq k$. Thus there exists an epimorphism $P \longrightarrow Y/X$. Since P is projective, there exists a nonzero homomorphism $P \longrightarrow Y$. Hence $\text{Hom}_R(P, N) \neq 0$. So we have proved the claim. It follows from the claim that $\mathcal{T}$ is closed under direct products.

(2) $\Longrightarrow$ (1). By (6.3.10), R is right semiartinian, and hence is left T-nilpotent (see [8, 28.5. Remark]). So it suffices to show that $R/J(R)$ is semisimple Artinian. Since R is right semiartinian,

$$K/J(R) = Soc((R/J(R))_R) \leq_e (R/J(R))_R.$$

We prove next that $K = R$. Suppose that $K \neq R$. Then $K \subsetneq I_0$ for some maximal right ideal I_0 of R. Certainly, $I_0/J(R) \leq_e (R/J(R))_R$.

Let I be a maximal right ideal I with $R/I \cong R/I_0$. If $I/J(R)$ is not essential in $(R/J(R))_R$, then there exists a right ideal L of R properly containing $J(R)$ such that $I \cap L = J(R)$. Thus, $R = I + L$ and so

$$R/J(R) = (I/J(R)) \oplus (L/J(R)).$$

It follows that R/I, and hence R/I_0, is a projective right $R/J(R)$-module. In particular, $I_0/J(R)$ is a summand of $R/J(R)$, contradicting the fact that $I_0/J(R) \leq_e (R/J(R))_R$. Therefore, $I/J(R)$ is indeed essential in $(R/J(R))_R$. Thus, $I/J(R) \supseteq Soc(((R/J(R))_R) = K/J(R)$, i.e., $I \supseteq K$. So we have proved that

$$K \subseteq \cap\{I : I \text{ is a maximal right ideal of } R \text{ with } R/I \cong R/I_0\}.$$

Since $J(R) \subset K$, it follows that

$$J(R) = \cap\{I : I \text{ is a maximal right ideal of } R \text{ with } R/I \not\cong R/I_0\}.$$

Now let $\mathcal{K}$ be a complete set of representatives of isomorphism classes of simple right R-modules not isomorphic to R/I_0, and let

$$\mathcal{F} = \{F \in \text{Mod-}R : \operatorname{Hom}_R(K, F) = 0, \forall K \in \mathcal{K}\},$$
$$\mathcal{T} = \{T \in \text{Mod-}R : \operatorname{Hom}_R(T, F) = 0, \forall F \in \mathcal{F}\}.$$

Since $\mathcal{F}$ is clearly closed under injective hulls, $(\mathcal{T}, \mathcal{F})$ is a hereditary torsion theory by (2.1.6). Since

$R/J(R) \hookrightarrow \Pi\{R/I : I$ is a maximal right ideal of R with $R/I \not\cong R/I_0\}$, $R/J(R) \in \mathcal{T}$ by (2). Thus, $R/I_0 \in \mathcal{T}$ since $J(R) \subseteq I_0$. This is a contradiction since $R/I_0 \in \mathcal{F}$. So $R/J(R) = K/J(R)$. □

Dlab [44] gave a non-left perfect ring with only a finite number of hereditary torsion classes, all of which are closed under direct products. The equivalence "(2) $\Longleftrightarrow$ (3)" of the next corollary is contained in Dlab [45] and Gardner [57].

6.3.12. COROLLARY. The following are equivalent for a ring R:

1. R is left perfect with a unique simple module up to isomorphism.
2. R is right semiartinian with a unique simple module up to isomorphism.
3. $\mathcal{T}_r(R) = \{\mathbf{0}, \mathbf{1}\}$.
4. $\mathcal{N}_r(R) = \{\mathbf{0}, \mathbf{1}\}$.

PROOF. (1) $\Longrightarrow$ (2) is obvious; (2) + (3) $\Longrightarrow$ (1) follows from (6.3.11).

(2) $\Longrightarrow$ (4). Fix a simple module M_0. By (2), M_0 embeds in every nonzero module. It follows that $d(M_0) = \mathbf{1}$ and that $d(M_0) \leq \mathcal{K}$ for every $\mathbf{0} < \mathcal{K} \in \mathcal{N}_r(R)$. So $\mathcal{N}_r(R) = \{\mathbf{0}, \mathbf{1}\}$.

(4) $\Longrightarrow$ (3). (4) clearly implies that $\mathcal{F}_r(R) = \{\mathbf{0}, \mathbf{1}\}$. So (3) follows.

(3) $\Longrightarrow$ (2). Fix a simple right R-module M_0, and let

$$\mathcal{F} = \{F \in \text{Mod-}R : \operatorname{Hom}_R(M_0, F) = 0\},$$
$$\mathcal{T} = \{T \in \text{Mod-}R : \operatorname{Hom}_R(T, F) = 0, \forall F \in \mathcal{F}\}.$$

Since $\mathcal{F}$ is clearly closed under injective hulls, $(\mathcal{T}, \mathcal{F})$ is a hereditary torsion theory by (2.1.6). By (3), $\mathcal{T} = \mathbf{1}$. Thus, M_0 embeds in every nonzero module. So (2) holds. □

The equivalence "(1) $\Longleftrightarrow$ (2)" of the next theorem is obtained in Teply [116].

6.3.13. THEOREM. The following are equivalent for a ring R:

1. R is isomorphic to a finite direct product of left perfect rings, each of which has a unique simple module up to isomorphism.

2. $\mathcal{F}_r(R) \subseteq \mathcal{T}_r(R)$.

3. $\mathcal{N}_r(R) \subseteq \mathcal{T}_r(R)$.

4. $\mathcal{N}_r(R) \subseteq \mathcal{T}_r^p(R)$.

5. $\mathcal{N}_r^p(R) = \mathcal{T}_r^p(R)$.

PROOF. (1) $\Longrightarrow$ (2). Write $R = R_1 \oplus \cdots \oplus R_n$ where each R_i is a left perfect ring with a unique simple module up to isomorphism. Let $(\mathcal{T}, \mathcal{F})$ be a hereditary torsion theory. Suppose that $\mathcal{F}$ is not closed under quotients. Then some nonzero module in $\mathcal{F}$ has a nonzero image in $\mathcal{T}$. This implies that some cyclic module $xR \in \mathcal{F}$ has a simple image Y in $\mathcal{T}$. Let $I = x^{\perp}$. Then $I = (R_1 \cap I) \oplus \cdots \oplus (R_n \cap I)$ and so

$$xR \cong R/I \cong R_1/(R_1 \cap I) \oplus \cdots \oplus R_n/(R_n \cap I).$$

Then Y is a simple image of some $R_i/(R_i \cap I)$. Of course, $0 \neq R_i/(R_i \cap I) \in \mathcal{F}$. Since R_i is right semiartinian with a unique simple module, Y embeds in $R_i/(R_i \cap I)$, so $Y \in \mathcal{F}$. This contradiction shows that $\mathcal{F}$ is closed under quotients.

(2) $\Longrightarrow$ (1). Let $\{X_\alpha : \alpha \in \Lambda\}$ be a complete set of representatives of isomorphism classes of simple R-modules and $M_0 = \oplus\{X_\alpha : \alpha \in \Lambda\}$. Let

$$\mathcal{F} = \{F \in \text{Mod-}R : \text{Hom}_R(M_0, F) = 0\},$$
$$\mathcal{T} = \{T \in \text{Mod-}R : \text{Hom}_R(T, F) = 0, \forall F \in \mathcal{F}\}.$$

Since $\mathcal{F}$ is clearly closed under injective hulls, $(\mathcal{T}, \mathcal{F})$ is a hereditary torsion theory by (2.1.6). Since $\mathcal{F}$ is closed under quotients by hypothesis and since $\mathcal{T}$ contains all simple modules, we have $\mathcal{F} = \mathbf{0}$ and so $\mathcal{T} = \mathbf{1}$. Hence R is right semiartinian. Thus, it follows from (6.3.10) and the hypothesis that every natural class is closed under quotients.

For each $\alpha \in \Lambda$, let I_α be a right ideal of R maximal with respect to $I_\alpha \in d(X_\alpha)$. Since $d(X_\alpha)$ is closed under quotients, I_α is indeed a unique largest right ideal of R in $d(X_\alpha)$. Since $d(X_\alpha)$ is closed under extensions by (2.3.5), the maximality of I_α implies $R/I_\alpha \in c(d(X_\alpha)) = d(\oplus_{\beta \neq \alpha} X_\beta)$. It follows that $R/(\Sigma_{\gamma \in \Lambda} I_\gamma) \in d(\oplus_{\beta \neq \alpha} X_\beta)$ since $d(\oplus_{\beta \neq \alpha} X_\beta)$ is closed under quotients. Thus,

$$R/(\Sigma_{\gamma \in \Lambda} I_\gamma) \in \cap_{\alpha \in \Lambda} d(\oplus_{\beta \neq \alpha} X_\beta) = 0,$$

so $R = \Sigma_{\gamma \in \Lambda} I_\gamma = \Sigma_{i=1}^n I_{\gamma i}$ for some $n > 0$. Thus, $R = \oplus_{i=1}^n I_{\gamma i}$ since

$$I_{\gamma_i} \cap \Sigma_{j \neq i} I_{\gamma_j} \in d(X_{\gamma_i}) \cap d(\oplus_{j \neq i} X_{\gamma_j}) = \mathbf{0}$$

(for every i). Moreover, this direct sum is a ring direct sum, because each $I_{\gamma i}$ is a two-sided ideal of R (since $d(X_{\gamma_i})$ is closed under quotients). Since $I_{\gamma_i} \in d(X_{I_{\gamma_i}})$ and $d(X_{I_{\gamma_i}})$ is closed under quotients, each I_{γ_i} has a unique simple module up to isomorphism. Since R is right semiartinian, it follows that each I_{γ_i} is a right semiartinian ring. Thus, by (6.3.12), each I_{γ_i} is a left perfect ring with one simple module up to isomorphism.

(1) + (2) $\Longrightarrow$ (3). Because of (1), R is left perfect and hence right semiartinian. Thus, $\mathcal{N}_r(R) = \mathcal{F}_r(R)$ by (6.3.10) and hence $\mathcal{N}_r(R) \subseteq \mathcal{T}_r(R)$ by (2).

(3) $\Longrightarrow$ (4) $\Longrightarrow$ (2) and (4) $\Longleftrightarrow$ (5) are obvious. □

6.3.14. COROLLARY. The following are equivalent for a ring R:

1. R is semisimple Artinian.
2. $\mathcal{N}_r^p(R) = \mathcal{F}_r(R)$.
3. $\mathcal{T}_r^p(R) = \mathcal{N}_r(R)$.
4. $\mathcal{T}_r^p(R) = \mathcal{F}_r(R)$.

PROOF. (1) $\Longleftrightarrow$ (2) follows from (6.3.9), (6.3.10), and the fact that a ring is right semiartinian, right QI iff it is semisimple Artinian.

(1) $\Longrightarrow$ (3). Let R be semisimple Artinian. Then $\mathcal{T}_r^p(R) \subseteq \mathcal{N}_r(R)$ by (6.3.9) and, since R satisfies (6.3.13)(1), $\mathcal{N}_r(R) \subseteq \mathcal{T}_r^p(R)$. So $\mathcal{T}_r^p(R) = \mathcal{N}_r(R)$.

(3) or (4) $\Longrightarrow$ (1). Suppose (3) holds. Then R is right QI by (6.3.9) and R is left perfect by (6.3.13). So R is semisimple Artinian. Note that (4) clearly implies that R is right QI by (6.3.9); and (4) also implies $\mathcal{F}_r(R) \subseteq \mathcal{T}_r(R)$, so R is left perfect by (6.3.13). Thus R is semisimple Artinian.

(1) + (3) $\Longrightarrow$ (4). By (6.3.10), $\mathcal{N}_r(R) = \mathcal{F}_r(R)$, so (4) follows from (3). □

6.3.15. THEOREM. [9] The following are equivalent for a ring R:

1. R is right Artinian.
2. Every $\mathcal{T} \in \mathcal{T}_r^p(R)$ is closed under direct products.

PROOF. (1) $\Longrightarrow$ (2). Let $\mathcal{T} \in \mathcal{T}_r^p(R)$ and let $N = \Pi\{N_\alpha : \alpha \in \Lambda\}$ where all $N_\alpha \in \mathcal{T}$. To show $N \in \mathcal{T}$, it suffices to show $xR \in \mathcal{T}$ for every $x \in N$. Write $x = (x_\alpha)$ where $x_\alpha \in N_\alpha$. Then $x^\perp = \cap_\alpha x_\alpha^\perp$. Since R is right Artinian, $x^\perp = \cap_{i=1}^n x_{\alpha_i}^\perp$ for some $n > 0$. Thus,

$$xR \cong R/x^\perp \hookrightarrow \oplus_{i=1}^n R/x_{\alpha_i}^\perp \cong \oplus_{i=1}^n x_{\alpha_i} R \in \mathcal{T}.$$

So $xR \in \mathcal{T}$.

(2) $\Longrightarrow$ (1). Claim: for any module M, $M^\perp = X^\perp$ for a finite subset X of M. In fact, we have

$$R/M^\perp = R/(\cap_{x \in M} x^\perp) \hookrightarrow \Pi_{x \in M} R/x^\perp \cong \Pi_{x \in M} xR \in \sigma[\oplus_{x \in M} xR]$$

by hypothesis. So $R/M^\perp \in \sigma[\oplus_{x \in M} xR]$. Then by (2.2.5), $Z^\perp \subseteq M^\perp$ for a finite subset Z of $\oplus_{x \in M} xR$. It follows that there exists a finite subset X of M such that $M^\perp = X^\perp$.

Let $\{X_\alpha : \alpha \in \Lambda\}$ be a complete set of representatives of isomorphism classes of simple R-modules and $M_0 = \oplus\{X_\alpha : \alpha \in \Lambda\}$. Since $R/J(R)$ embeds in a direct product of simple R-modules, we have $R/J(R) \in \sigma[M_0]$ by hypothesis, so $R/J(R)$ is semisimple. Thus there are only finitely many simple R-modules up to isomorphism and so M_0 is finitely generated. Let $E = E(M_0)$. Since E_R is faithful, it follows from the Claim that $R \hookrightarrow E^n$ for some $n > 0$. So R has a finitely generated essential right socle. Thus, for any faithful module N_R, R embeds in N^k for some $k > 0$ (by the Claim), showing that $Soc(N) \neq 0$. Now for any nonzero module N_R, let $I = N^\perp$. Since R/I inherits the hypothesis and N is faithful over R/I, N has a nonzero socle over R/I and hence over R. So we have proved that every nonzero module over R has a nonzero socle, i.e., R is right semiartinian. Thus, R is left T-nilpotent, and hence is left perfect since $R/J(R)$ is semisimple.

Let $K = \cap\{J(R)^i : i = 1, 2, \cdots\}$. By hypothesis, $R/K \in \sigma[\oplus_{i=1}^\infty R/J(R)^i]$. Thus, $R/K \in \sigma[\oplus_{i=1}^n R/J(R)^i]$ for some $n > 0$. It follows that $J(R)^n = J(R)^{n+1} = \cdots = K$. Since R is left T-nilpotent, it follows that $J(R)^n = 0$. Since $R/J(R)$ is semisimple, $J(R)^i/J(R)^{i+1}$ is semisimple. Since $R/J(R)^{i+1}$ inherits the hypothesis, it has a finitely generated essential right socle as above. Thus, $J(R)^i/J(R)^{i+1}$ is finitely generated and hence has finite length for $i = 1, \cdots, n-1$. Therefore, being of finite length, R is right Artinian. □

6.3.16. PROPOSITION. Suppose that $\mathcal{N}_r(R)$ is a complete sublattice of $\mathcal{N}_r^p(R)$. Then $\mathcal{K} \in \mathcal{N}_r^p(R)$ is compact iff $\mathcal{K} = d(N) \cap \sigma[N]$ for some finitely generated module N of finite type dimension.

PROOF. "$\Longrightarrow$". Recall that $M_\mathcal{K} = \oplus_{t\in I} X_t$ where $\{X_t : t \in I\}$ is a complete set of representatives of the isomorphic classes of cyclic modules in $\mathcal{K}$. By (2.5.4) and (6.1.3), $\mathcal{K} = d(M_\mathcal{K}) \cap \sigma[M_\mathcal{K}] = \vee_t \mathcal{K}_t$ where $\mathcal{K}_t = d(X_t) \cap d(X_t)$. Since $\mathcal{K}$ is compact, there exists a finite subset F of I such that $\mathcal{K} = \vee_{t\in F}\mathcal{K}_t = d(N) \cap \sigma[N]$, where $N = \oplus_{t\in F} X_t$ is finitely generated. Suppose that N is not of finite type dimension. Then N contains an essential submodule $A = \oplus_{j=1}^\infty A_j$ such that all $A_j \neq 0$, $A_i \perp A_j$, whenever $i \neq j$. By hypothesis, $\mathcal{L} = \vee_j d(A_j)$ is a natural class. Then $N \in \mathcal{L}$ since $A \in \mathcal{L}$. This shows that $\mathcal{K} \leq \mathcal{L}$. It follows that $\mathcal{K} \leq \vee_{j=1}^n d(A_j) = d(\oplus_{j=1}^n A_j)$ for some $n > 0$. Since $A_{n+1} \in \mathcal{K}$, there exists a $0 < m \leq n$ such that A_{n+1} and A_m have nonzero isomorphic submodules, a contradiction.

"$\Longleftarrow$". Suppose N is a finitely generated module of finite type dimension and $\mathcal{K} = d(N) \cap \sigma[N]$. Let $\mathcal{K} \leq \vee_{i\in I}\mathcal{K}_i$ with each $\mathcal{K}_i \in \mathcal{N}_r^p(R)$. Write $\mathcal{K}_i = d(M_i) \cap \sigma[M_i]$. Thus, $N \in \vee_{i\in I}\mathcal{K}_i = d(\oplus_i M_i) \cap \sigma[\oplus_i M_i]$. Since N is finitely generated, N is in $\sigma[\oplus_{i\in F_1} M_i]$ for some finite subset F_1 of I. Since N is of finite type dimension, there exists a finite subset F_2 of I such that every nonzero submodule of N has a nonzero submodule embeddable in $\oplus_{i\in F_2} M_i$. Let $F = F_1 \cup F_2$. Then

$$d(N) \cap \sigma[N] \subseteq d(\oplus_{i\in F} M_i) \cap \sigma[\oplus_{i\in F} M_i] = \vee_{i\in F}\mathcal{K}_i.$$

So $\mathcal{K}$ is a compact element in $\mathcal{N}_r^p(R)$. □

6.3.17. COROLLARY. Suppose that $\mathcal{N}(R, M)$ is a complete sublattice of $\mathcal{N}_r^p(R)$ for each module M. Then the following are equivalent:

1. $\mathcal{N}_r^p(R)$ is an algebraic lattice.

2. Every nonzero R-module contains an atomic submodule.

PROOF. (1) $\Longrightarrow$ (2). Let N be a nonzero module. Then, by (6.3.16),

$$d(N) \cap \sigma[N] = \vee_i(d(N_i) \cap \sigma[N_i]) = d(\oplus_i N_i) \cap \sigma[\oplus_i N_i],$$

where each N_i is a finitely generated module of finite type dimension. Thus, $N \in d(\oplus_i N_i)$ and so N contains a nonzero submodule X embeddable in N_i for some i. Since N_i contains an essential submodule, which is a direct sum of atomic modules, X contains an atomic submodule.

(2) $\Longrightarrow$ (1). Let $\mathcal{K} \in \mathcal{N}_r^p(R)$ and write $\mathcal{K} = d(N) \cap \sigma[N]$. By (2), N contains an essential submodule, which is a direct sum $\oplus_{i \in I} N_i$ of atomic submodules. Let $\mathcal{K}_i = d(N_i) \cap \sigma[N]$. Then $\mathcal{K}_i \in \mathcal{N}(R, N)$ by (2.4.4) and hence $\mathcal{L} = \vee \mathcal{K}_i \in \mathcal{N}(R, N)$ by hypothesis. Since $N_i \in \mathcal{K}_i$ for each $i \in I$, $\oplus_i N_i$ is in $\mathcal{L}$ and so N is in $\mathcal{L}$ by (2.4.2). Thus, $\mathcal{K} = \mathcal{L} = \vee_i \mathcal{K}_i$. Now it suffices to show that each $\mathcal{K}_i$ is a join of compact elements of $\mathcal{N}_r^p(R)$. Let $\{X_t^i : t \in I_i\}$ be a complete set of representatives of the isomorphic classes of cyclic modules in $\mathcal{K}_i$ and let $X_i = \oplus_{t \in I_i} X_t^i$. Then, by (2.5.4) and (6.1.3),

$$\mathcal{K}_i = d(X_i) \cap \sigma[X_i] = \vee_t(d(X_t^i) \cap \sigma[X_t^i]).$$

Note that, since N_i is an atomic module, X_t^i is atomic. Thus, $d(X_t^i) \cap \sigma[X_t^i]$ is a compact element of $\mathcal{N}_r^p(R)$ by (6.3.16). □

In the final part of this section, we are concerned with a result of right distributive rings. A ring R is called **right distributive** if the lattice of all right deals of R is a **distributive lattice**, i.e., $I \cap (J + K) = (I \cap J) + (I \cap K)$ for all right ideals I, J, K of R, or equivalently $I + (J \cap K) = (I + J) \cap (I + K)$ for all right ideals I, J, K of R.

6.3.18. THEOREM. [79] If R is right distributive, then **fil**-R is a distributive lattice. Conversely, if **fil**-R is a distributive lattice, then the lattice of ideals of R is distributive. Therefore, a commutative ring R is distributive iff **fil**-R is a distributive lattice.

PROOF. Let R be a right distributive ring. Since **fil**-R is isomorphic to $\mathcal{T}_r^p(R)$, we prove that $\mathcal{T}_r^p(R)$ is a distributive lattice. Let $\mathcal{K}, \mathcal{L}, \mathcal{J} \in \mathcal{T}_r^p(R)$. We need to show that $\mathcal{K} \wedge (\mathcal{L} \vee \mathcal{J}) = (\mathcal{K} \wedge \mathcal{L}) \vee (\mathcal{K} \wedge \mathcal{J})$. It is clear that $\mathcal{K} \wedge (\mathcal{L} \vee \mathcal{J}) \geq (\mathcal{K} \wedge \mathcal{L}) \vee (\mathcal{K} \wedge \mathcal{J})$. By (2.5.6), it suffices to show that every cyclic module $mR \in \mathcal{K} \wedge (\mathcal{L} \vee \mathcal{J})$ is in $(\mathcal{K} \wedge \mathcal{L}) \vee (\mathcal{K} \wedge \mathcal{J})$. Write $\mathcal{L} = \sigma[Y]$ and $\mathcal{J} = \sigma[Z]$. Then $mR \in \mathcal{L} \vee \mathcal{J} = \sigma[Y \oplus Z]$. By (2.2.5), there exist $y_i \in Y$, $z_i \in Z$ $(i = 1, \cdots, n)$ such that $\cap_i(y_i^\perp \cap z_i^\perp) = \cap_i(y_i + z_i)^\perp \subseteq m^\perp$. Let

$B = \cap_i y_i^{\perp}$ and $C = \cap_i z_i^{\perp}$. Then $B \cap C \subseteq m^{\perp}$, $R/B \in \mathcal{L}$, and $R/C \in \mathcal{J}$. Thus, $R/(m^{\perp} + B) \in \mathcal{K} \wedge \mathcal{L}$ and $R/(m^{\perp} + C) \in \mathcal{K} \wedge \mathcal{J}$. It follows that

$$R/[(m^{\perp} + B) \cap (m^{\perp} + C)] \in (\mathcal{K} \wedge \mathcal{L}) \vee (\mathcal{K} \wedge \mathcal{J}).$$

But, by hypothesis,

$$(m^{\perp} + B) \cap (m^{\perp} + C) = m^{\perp} + (B \cap C) = m^{\perp}.$$

So $mR \cong R/m^{\perp} \in (\mathcal{K} \wedge \mathcal{L}) \vee (\mathcal{K} \wedge \mathcal{J})$. Before proving the second part, let us note the following facts: for any ideals I and J of R,

1. $\sigma[R/I] = \sigma[R/J]$ iff $I = J$.
2. $\sigma[R/I] \vee \sigma[R/J] = \sigma[R/(I \cap J)]$.
3. $\sigma[R/I] \wedge \sigma[R/J] = \sigma[R/(I + J)]$.

Facts (1) and (2) are easy to see. Let us verify (3). It is obvious that $\sigma[R/I] \wedge \sigma[R/J] \supseteq \sigma[R/(I+J)]$. To show the other inclusion, we only need to show that every cyclic module $xR \in \sigma[R/I] \wedge \sigma[R/J]$ is in $\sigma[R/(I+J)]$. From $R/x^{\perp} \in \sigma[R/I] \wedge \sigma[R/J]$, it follows that $I \subseteq x^{\perp}$ and $J \subseteq x^{\perp}$. So $I + J \subseteq x^{\perp}$, and thus $xR \cong R/x^{\perp} \in \sigma[R/(I + J)]$.

Now suppose **fil**-R, and hence $\mathcal{T}_r^p(R)$, is a distributive lattice, and let A, B, C be ideals of R. Then

$$\sigma[R/A] \wedge (\sigma[R/B] \vee \sigma[R/C]) = (\sigma[R/A] \wedge \sigma[R/B]) \vee (\sigma[R/A] \wedge \sigma[R/C]).$$

By the facts above, the classes on the two sides are equal to $\sigma[R/(A+(B\cap C))]$ and $\sigma[R/((A + B) \cap (A + C))]$, respectively. Thus, $A + (B \cap C) = (A + B) \cap (A + C)$ by (1). □

6.3.19. EXAMPLE. There exists a ring R such that $\mathcal{N}_r^p(R)$ is a complete Boolean lattice (in particular, **fil**-R is a distributive lattice) but R is not right distributive.

For convenience of references, we give below a ring R such that $\mathcal{N}_l^p(R)$ is a complete Boolean lattice but R is not left distributive. Let F be a field of characteristic 0 and α be the derivation of $F[x]$ defined by $\alpha(\Sigma a_i x^i) = \Sigma i a_i x^{i-1}$. Let Q be the quotient field of $F[x]$. Then α induces a derivation $\overline{\alpha}$ on Q via the usual quotient rule for derivations. Thus, $\overline{\alpha}(Q) \neq 0$ and the characteristic of Q is 0. By Kolchin [81] (or see Faith [48, p.361]), there is a universal differential field $k \supseteq Q$ and a derivation D of k extending $\overline{\alpha}$. Thus, $D(k) \neq 0$. Let $R = k[y, D]$ be the Cozzens domain. As it is well-known, R is a two-sided QI-ring. By (6.3.9), $\mathcal{N}_l^p(R) = \mathcal{N}_l(R)$. Then, by (5.1.5), $\mathcal{N}_l^p(R)$ is a complete Boolean lattice. Consider the maximal left ideal Ry of R. Note that $Ry = \{\Sigma_{i=1}^n a_i y^i : n \geq 1, a_i \in k\}$. Choose $a \in k$ such that $D(a) \neq 0$. Then $ya = D(a) + ay$. If $ya \in Ry$, then $D(a) \in Ry$. This is impossible. So Ry is not a two-sided ideal. By the result of Stephenson [111, Corollary 4] that every maximal left ideal of a left distributive ring is a two-sided ideal, R is not left distributive. □

Our concluding examples illustrate that some of the lattice properties of $\mathcal{N}^p(R)$ as well as $\mathcal{N}(R)$ are not, in general, left-right symmetric.

6.3.20. EXAMPLE. There is a ring R such that $|\mathcal{N}_r^p(R)| = 3$ but $|\mathcal{N}_l^p(R)| > 3$.

We here use an example of A. Viola-Prioli and J. Viola-Prioli [122]. Let K be a field, and $f : K \longrightarrow K$ a field monomorphism which is not onto. Consider the ring S of twisted power series, that is, $S = \{\Sigma x^j a_j : a_j \in K\} = K[[x; f]]$ with the usual addition and $ax = xf(a)$, for every $a \in K$. Let R be the ring S/x^2S. By [122], R has the following properties:

1. R has only three right ideals R, xR, and (0).
2. $|\mathcal{T}_r^p(R)| = 3$.
3. The lattice $\mathcal{T}_l^p(R)$ is not linearly ordered.

Since R is a right Artinian ring with a unique (up to isomorphism) simple R-module, $\mathcal{N}_r^p(R) = \mathcal{T}_r^p(R)$ by (6.3.13).

By (3), there exist left R-modules X and Y such that $\sigma[X] \not\subseteq \sigma[Y]$ and $\sigma[Y] \not\subseteq \sigma[X]$. Thus, $d(X) \cap \sigma[X] \neq d(Y) \cap \sigma[Y]$, for otherwise, $\sigma[X] = \sigma[Y]$. So $\mathbf{0}, d(X) \cap \sigma[X], d(Y) \cap \sigma[Y]$, and $\mathbf{1}$ are distinct pre-natural classes in $\mathcal{N}_l^p(R)$. Therefore, $|\mathcal{N}_l^p(R)| \geq 4$. □

6.3.21. EXAMPLE. There is a ring R such that $|\mathcal{N}_r(R)| = 2$ but $|\mathcal{N}_l(R)| = 4$.

Let R be the ring of all $\mathbb{N}$-square lower triangular matrices over a field K that are constant on the diagonal and have only finitely many nonzero entries off the diagonal. Let J be the set of elements of R with only finitely many nonzero entries, all occurring below the diagonal. Then J is equal to the Jacobson radical of R and is left T-nilpotent ([8, Ex.15.8]) and $R/J \cong K$. So R is a local left perfect ring. Therefore, by (6.3.12), $|\mathcal{N}_r(R)| = 2$.

Since R is local, there is a unique (up to isomorphism) simple left R-module, say ${}_RX$. To show $|\mathcal{N}_l(R)| = 4$, we let $J_k = \{(a_{ij}) \in J : a_{ij} = 0 \text{ if } j \neq k\}$ for $k = 1, 2, \cdots$. It is true that J_k are uniform left ideals of R and $J = \oplus_{k=1}^{\infty} J_k$. Moreover, if $n < m$, we have as left R-modules an embedding $J_m \hookrightarrow J_n$ by $(a_{ij}) \longmapsto (b_{ij})$, where $b_{ij} = a_{i,j+m-n}$. This implies that ${}_RJ$ is an atomic module. For a cyclic socle-free left R-module R/I, it must be true that $I = \oplus_{i \in V} J_i$ for a proper subset V of $\mathbb{N}$. Since J is the only maximal left ideal of R, I is a complement left ideal of R. It follows that $\oplus_{i \in \mathbb{N} \setminus V} J_i$ is essentially embeddable in R/I. Because ${}_RJ$ is atomic, we proved that any two socle-free cyclic left R-modules have isomorphic submodules. Since $Soc({}_RR) = 0$, $X \perp J$. Therefore, $d(X) \wedge d(J) = \mathbf{0}$ and $d(X) \vee d(J) = \mathbf{1}$. So $\mathcal{N}_l(R) = \{\mathbf{0}, d(X), d(J), \mathbf{1}\}$. □

6.3.22. REFERENCES. Alin and Armendariz [6]; Beachy and Blair [9]; Dauns [32]; Dauns and Zhou [40]; Dlab [44,45]; Gabriel [54]; Gardner [57]; Golan [60,61]; Golan and López-Permouth [62]; Katayama [79]; Popescu [103]; Stephenson [111]; Teply [116]; A. Viola-Prioli and J. Viola-Prioli [122]; Zhou [136,142].

6.4 More Lattice Properties of $\mathcal{N}_r^p(R)$

In this section, the interrelatedness of lattice properties of $\mathcal{N}_r^p(R)$ and ring and module theoretic properties of R and Mod-R is pursued further, and applications are given to the rings R for which $\mathcal{T}_r^p(R) = \mathcal{T}_r(R)$. The approach is to relate the join operation of $\mathcal{N}_r^p(R)$ to the **module extension operator** $E(-,-)$. For any module classes $\mathcal{K}_1, \mathcal{K}_2$, let

$$E(\mathcal{K}_1, \mathcal{K}_2) = \{M : X \in \mathcal{K}_1, M/X \in \mathcal{K}_2 \text{ for some } X \subseteq M\}.$$

We have $\mathcal{K}_1 \vee \mathcal{K}_2 = E(\mathcal{K}_1, \mathcal{K}_2)$ for all $\mathcal{K}_1, \mathcal{K}_2 \in \mathcal{N}_r(R)$ by (6.1.7). This fact is an immediate consequence of the following result.

6.4.1. PROPOSITION. The following are equivalent for a module M:

1. $\sigma[M]$ is closed under extensions.

2. $\mathcal{K}_1 \vee \mathcal{K}_2 = E(\mathcal{K}_1, \mathcal{K}_2)$ for all $\mathcal{K}_1, \mathcal{K}_2 \in \mathcal{N}(R, M)$.

PROOF. (2) $\Longrightarrow$ (1). Let $0 \longrightarrow X \longrightarrow N \longrightarrow Y \longrightarrow 0$ be exact with $X, Y \in \sigma[M]$. Let $\mathcal{K}_1 = d(X) \cap \sigma[M]$ and $\mathcal{K}_2 = d(Y) \cap \sigma[M]$. Then $\mathcal{K}_1, \mathcal{K}_2 \in \mathcal{N}(R, M)$. Since $X \in \mathcal{K}_1$ and $Y \in \mathcal{K}_2$, we have $N \in E(\mathcal{K}_1, \mathcal{K}_2)$. So, by (2), $N \in \mathcal{K}_1 \vee \mathcal{K}_2$. Thus, $N \in \sigma[M]$.

(1) $\Longrightarrow$ (2). Let $\mathcal{K}_1, \mathcal{K}_2 \in \mathcal{N}(R, M)$. Write $\mathcal{K}_1 = \mathcal{L}_1 \cap \sigma[M]$ and $\mathcal{K}_2 = \mathcal{L}_2 \cap \sigma[M]$, where $\mathcal{L}_1, \mathcal{L}_2 \in \mathcal{N}_r(R)$. We have that

$$\begin{aligned} \mathcal{K}_1 \vee \mathcal{K}_2 &= (\mathcal{L}_1 \cap \sigma[M]) \vee (\mathcal{L}_2 \cap \sigma[M]) \\ &= (\mathcal{L}_1 \vee \mathcal{L}_2) \wedge \sigma[M] \text{ (by 6.2.14)} \\ &= E(\mathcal{L}_1, \mathcal{L}_2) \wedge \sigma[M] \text{ (by 6.1.7)} \\ &\supseteq E(\mathcal{K}_1, \mathcal{K}_2) \text{ (by (1))}. \end{aligned}$$

It is easy to see that $E(\mathcal{K}_1, \mathcal{K}_2)$ is closed under submodules and direct sums. We now show that $E(\mathcal{K}_1, \mathcal{K}_2)$ is closed under M-injective hulls, and hence is an M-natural class. This then gives $\mathcal{K}_1 \vee \mathcal{K}_2 \subseteq E(\mathcal{K}_1, \mathcal{K}_2)$ since $\mathcal{K}_1 \cup \mathcal{K}_2 \subseteq E(\mathcal{K}_1, \mathcal{K}_2)$. Let $N \in E(\mathcal{K}_1, \mathcal{K}_2)$. Then there exist $A \in \mathcal{K}_1$ and $B \in \mathcal{K}_2$ such that $0 \longrightarrow A \longrightarrow N \longrightarrow B \longrightarrow 0$ is exact. By (2.2.10), we can write $E_M(N) = E_M(A) \oplus C$ for some C. If $C \neq 0$, $C \cap N$ is essential in C and $C \cap N \hookrightarrow B$. So C is always in $\mathcal{K}_2$. Since $E_M(A) \in \mathcal{K}_1$, we see that $E_M(N) \in E(\mathcal{K}_1, \mathcal{K}_2)$. □

6.4.2. LEMMA. Let $\mathcal{T} \in \mathcal{T}_r(R)$ and $\mathcal{K} \in \mathcal{T}_r^p(R)$. Then

1. $E(\mathcal{T}, \mathcal{K}) = E(\mathcal{T}, E(\mathcal{T}, \mathcal{K}))$ and

2. $E(\mathcal{K}, \mathcal{T}) = E(E(\mathcal{K}, \mathcal{T}), \mathcal{T})$.

PROOF. (1) One inclusion is obvious. Let $M \in E(\mathcal{T}, E(\mathcal{T}, \mathcal{K}))$. Let τ and σ be the left exact preradicals corresponding to the hereditary pretorsion classes $E(\mathcal{T}, \mathcal{K})$ and $\mathcal{T}$, respectively. Then $M/\sigma(M) \in E(\mathcal{T}, \mathcal{K})$, and so there exist modules $A \in \mathcal{T}$ and $B \in \mathcal{K}$ such that $0 \longrightarrow A \longrightarrow M/\sigma(M) \longrightarrow B \longrightarrow 0$ is exact. Since σ is a hereditary torsion theory, $M/\sigma(M)$ is σ-torsion free. It follows that $A = 0$. So $M/\sigma(M) \cong B \in \mathcal{K}$. This gives that $M \in E(\mathcal{T}, \mathcal{K})$.

(2) One inclusion is obvious. Let $M \in E(E(\mathcal{K}, \mathcal{T}), \mathcal{T})$. Let τ and σ be the left exact preradicals corresponding to the hereditary pretorsion classes $E(\mathcal{K}, \mathcal{T})$ and $\mathcal{K}$, respectively. Then we have $M/\tau(M) \in \mathcal{T}$. Since $\tau(M) \in E(\mathcal{K}, \mathcal{T})$, $\tau(M)/\sigma(M) = \tau(M)/\sigma(\tau(M)) \in \mathcal{T}$. Because $\mathcal{T}$ is closed under extensions, we have $M/\sigma(M) \in \mathcal{T}$. It follows that $M \in E(\mathcal{K}, \mathcal{T})$. □

6.4.3. THEOREM. The following statements are equivalent for a ring R:

1. $\mathcal{T}_r(R)$ is a sublattice of $\mathcal{T}_r^p(R)$.

2. $\mathcal{K}_1 \vee \mathcal{K}_2 = E(\mathcal{K}_1, \mathcal{K}_2)$ for all $\mathcal{K}_1, \mathcal{K}_2 \in \mathcal{T}_r(R)$.

PROOF. (1) $\Longrightarrow$ (2). For $\mathcal{K}_1, \mathcal{K}_2 \in \mathcal{T}_r(R)$, $\mathcal{K}_1 \cup \mathcal{K}_2 \subseteq \mathcal{K}_1 \vee \mathcal{K}_2$. By (1), $\mathcal{K}_1 \vee \mathcal{K}_2$ is closed under extensions, so $E(\mathcal{K}_1, \mathcal{K}_2) \subseteq \mathcal{K}_1 \vee \mathcal{K}_2$. Since $\mathcal{K}_1 \vee \mathcal{K}_2 \leq E(\mathcal{K}_1, \mathcal{K}_2)$, $\mathcal{K}_1 \vee \mathcal{K}_2 = E(\mathcal{K}_1, \mathcal{K}_2)$.

(2) $\Longrightarrow$ (1). By (2), it is enough to show that $E(\mathcal{K}_1, \mathcal{K}_2)$ is closed under extensions. Let $0 \longrightarrow X \longrightarrow M \longrightarrow M/X \longrightarrow 0$ be exact such that $X, M/X \in E(\mathcal{K}_1, \mathcal{K}_2)$ and let τ, τ_1 be the left exact preradicals corresponding to the hereditary pretorsion classes $E(\mathcal{K}_1, \mathcal{K}_2)$ and $\mathcal{K}_1$, respectively. It follows that $M/\tau(M) \in E(\mathcal{K}_1, \mathcal{K}_2)$. From $\tau(M) \in E(\mathcal{K}_1, \mathcal{K}_2)$, we have $\tau(M)/\tau_1(M) \in \mathcal{K}_2$. Since

$$0 \longrightarrow \tau(M)/\tau_1(M) \longrightarrow M/\tau_1(M) \longrightarrow M/\tau(M) \longrightarrow 0$$

is exact, we see that $M/\tau_1(M) \in E(\mathcal{K}_2, E(\mathcal{K}_1, \mathcal{K}_2))$. Note that (2) implies that $E(\mathcal{K}_1, \mathcal{K}_2) = E(\mathcal{K}_2, \mathcal{K}_1)$. So, by (6.4.2), we have

$$M/\tau_1(M) \in E(\mathcal{K}_2, E(\mathcal{K}_2, \mathcal{K}_1)) = E(\mathcal{K}_2, \mathcal{K}_1) = E(\mathcal{K}_1, \mathcal{K}_2).$$

Thus, $M \in E(\mathcal{K}_1, E(\mathcal{K}_1, \mathcal{K}_2)) = E(\mathcal{K}_1, \mathcal{K}_2)$ by (6.4.2). □

The next example gives a ring R such that $\mathcal{T}_r(R)$ is not a sublattice of $\mathcal{T}_r^p(R)$.

6.4.4. EXAMPLE. Let $Q = \Pi_{i=1}^{\infty} F_i$ be a direct product of rings, where $F_1 = \mathbb{Z}_4$ and $F_i = \mathbb{Z}_2$ for $i > 1$. Let R be the subring of Q generated by $2F_1 \oplus (\oplus_{i=2}^{\infty} F_i)$ and 1_Q. Then $Soc(R) = 2F_1 \oplus (\oplus_{i=2}^{\infty} F_i)$ is the unique proper essential ideal of R. It follows that $Z(R) = 2F_1$ and $K = \oplus_{i=2}^{\infty} F_i$ is nonsingular and hence projective. Note that $Z(R)$ and K are complements of each other in R. So $Z(R)$ is essentially embeddable in R/K and K is essentially embeddable in $R/Z(R)$. It follows that R/K is Goldie torsion and $R/Z(R)$ is nonsingular. So $Z_2(R) = Z(R)$. Let $\mathcal{K}$ be the class of all Goldie torsion modules and $\mathcal{L}$ be the class of all projective semisimple modules. Clearly $\mathcal{K}$ and $\mathcal{L}$ are in $\mathcal{T}_r(R)$. Thus, $R_R \in E(\mathcal{L},\mathcal{K})$. Since $Z_2(R)$ is not a direct summand of R_R, $R/Z_2(R)$ is not projective. So $R/Z_2(R)$ is not in $\mathcal{L}$. This shows that $R_R \notin E(\mathcal{K},\mathcal{L})$. Therefore, $E(\mathcal{L},\mathcal{K}) \neq E(\mathcal{K},\mathcal{L})$. By (6.4.3), $\mathcal{T}_r(R)$ is not a sublattice of $\mathcal{N}_r^p(R)$. □

6.4.5. COROLLARY. The following statements are equivalent for a ring R:

1. $\mathcal{T}_r^p(R) = \mathcal{T}_r(R)$.
2. $\mathcal{K}_1 \vee \mathcal{K}_2 = E(\mathcal{K}_1,\mathcal{K}_2)$ for all $\mathcal{K}_1,\mathcal{K}_2 \in \mathcal{T}_r^p(R)$.

PROOF. (1) $\Longrightarrow$ (2) is by (6.4.3).

(2) $\Longrightarrow$ (1). For any $\mathcal{K} \in \mathcal{T}_r^p(R)$, $\mathcal{K} = \mathcal{K} \vee \mathcal{K} = E(\mathcal{K},\mathcal{K})$. This shows that $\mathcal{K}$ is closed under extensions, and so $\mathcal{K} \in \mathcal{T}_r(R)$. □

It follows clearly that $\mathcal{T}_r^p(R) = \mathcal{T}_r(R)$ implies $E(\mathcal{K}_1,\mathcal{K}_2) = E(\mathcal{K}_2,\mathcal{K}_1)$ for all $\mathcal{K}_1,\mathcal{K}_2 \in \mathcal{T}_r^p(R)$. The study of rings R with $\mathcal{T}_r^p(R) = \mathcal{T}_r(R)$ was initiated by J. Viola-Prioli in [126], and continued by Fenrick [50], Handelman [72], A. Viola-Prioli and J. Viola-Prioli [123], Teply [118], van den Berg [119], and Dauns and Zhou [40]. It is suggested by (6.4.5) that there exist connections between properties of this class of rings and lattice properties of $\mathcal{N}_r^p(R)$.

6.4.6. PROPOSITION. [126] Let R be a ring with $\mathcal{T}_r^p(R) = \mathcal{T}_r(R)$.

1. $Z(R_R) = 0$.
2. $I^2 = I$ for every ideal I of R.

PROOF. (1). Let $\mathcal{Z}$ be the class of all singular right R-modules. By hypothesis, $\mathcal{Z} \in \mathcal{T}_r(R)$. From the exact sequence

$$0 \longrightarrow Z(R_R) \longrightarrow E(Z(R_R)) \longrightarrow E(Z(R_R))/Z(R_R) \longrightarrow 0,$$

it follows that $E(Z(R_R)) \in \mathcal{Z}$. Then $Z(R_R) = 0$ by (3.1.17).

(2). For an ideal I of R, let $M = (I/I^2) \oplus (R/I)$ and let $\mathcal{T} = \sigma[M]$. By hypothesis, $\mathcal{T} \in \mathcal{T}_r(R)$. From the exact sequence

$$0 \longrightarrow I/I^2 \longrightarrow R/I^2 \longrightarrow R/I \longrightarrow 0,$$

it follows that $R/I^2 \in \mathcal{T}$. Since $MI = 0$, $(R/I^2)I = 0$; so $I = I^2$. □

6.4.7. THEOREM. Let R be a ring such that $\mathcal{T}_r^p(R) = \mathcal{T}_r(R)$. Then the following statements hold:

1. For any essential submodule N of a nonsingular module M, $M \in \sigma[N]$.
2. Every nonsingular quasi-injective module is injective.
3. There exists a ring decomposition $R = R_1 \times R_2$, where R_1 is a semisimple Artinian ring and the right socle of R_2 is zero.

4. For any right ideal I of R, $I^\perp$ is a complement right ideal of R and there exists a finitely generated submodule X of I such that $I^\perp = X^\perp$.

5. R_R has finite type dimension.

6. R is a (finite) direct sum of indecomposable rings.

PROOF. (1) Let M be a nonsingular module and $N \leq_e M$. Let $\mathcal{K}$ be the class of all singular R-modules and $\mathcal{L} = \sigma[N]$. Then $M \in E(\mathcal{L}, \mathcal{K})$. By (6.4.5), $M \in E(\mathcal{K}, \mathcal{L})$. Thus, $M \in \mathcal{L} = \sigma[N]$.

(2) Let M be nonsingular quasi-injective. By (1), $E(M) \in \sigma[M]$. It follows that M is $E(M)$-injective. So $M = E(M)$ is injective.

(3) By (6.4.6), $Soc(R_R)$ is nonsingular; so $Soc(R_R)$ is injective by (2). Then $R = Soc(R_R) \oplus I$ for some right ideal I of R. Since R is semiprime (by 6.4.6) and $Soc(R_R)$ is an ideal, $R = Soc(R_R) \oplus I$ is a ring direct sum and so $Soc(I_I) = 0$.

(4) Let I be a right ideal of R. If K is an essential extension of $I^\perp$ in R_R, then $K \in \sigma[I^\perp]$ by (1). Note $I^\perp I = 0$ since R is semiprime. It follows that $KI = 0$. So $IK = 0$ and hence $K = I^\perp$. So $I^\perp$ is a complement right ideal of R. Let J be a complement of I in R_R. Then I is essentially embeddable in R/J. By (1), $R/J \in \sigma[I_R]$. There exists a finitely generated submodule X of I such that $R/J \in \sigma[X]$. Thus, $I \in \sigma[X]$. It follows that $I^\perp = X^\perp$.

(5) If not, then R contains an essential right ideal $\oplus_{i=1}^{\infty} I_i$, where each I_i is nonzero and $I_i \perp I_j$ if $i \neq j$. We claim that $I_i I_j = 0$ for all $i \neq j$. If not, then aI_j for some $i \neq j$. Thus, because R is right nonsingular, $I_j \longrightarrow I_i$, $x \longmapsto ax$, is a nonzero homomorphism whose kernel is a complement submodule of I_j. This contradicts that $I_i \perp I_j$. Now by (1), $R \in \sigma[\oplus_{i=1}^{\infty} I_i]$. It follows that $R \in \sigma[\oplus_{i=1}^{n} I_i]$ for some n. This gives that $I_{n+1} = RI_{n+1} = 0$, which is a contradiction.

(6) Suppose the type dimension of R_R is n. Then R cannot be a direct sum of $n+1$ (non-trivial) rings. So (6) follows. □

6.4.8. COROLLARY. [126] Let R be a commutative ring. Then $\mathcal{T}_r^p(R) = \mathcal{T}_r(R)$ iff R is a finite product of fields.

PROOF. If R is commutative with $\mathcal{T}_r^p(R) = \mathcal{T}_r(R)$, then R is (von Neumann) regular by (6.4.6)(2) and R is a finite product of indecomposable rings by (6.4.7)(6). It follows that R is a finite product of fields. The converse is clear. □

6.4.9. COROLLARY. A ring R is right semiartinian with $\mathcal{T}_r^p(R) = \mathcal{T}_r(R)$ iff R is a semisimple Artinian ring.

PROOF. The implication "$\Longrightarrow$" follows from (6.4.7)(3), and the other implication is clear. □

6.4.10. LEMMA. Let R be a ring direct product $R = \Pi_{i=1}^{n} R_i$. Then $\mathcal{T}_r^p(R) = \mathcal{T}_r(R)$ iff $\mathcal{T}_r^p(R_i) = \mathcal{T}_r(R_i)$ for each i.

PROOF. "$\Longrightarrow$". Let π_i be the projection from R onto R_i. If $N \in$ Mod-R_i, let $N_{(i)} = N \in$ Mod-R denote the induced R-module, where $xr = x\pi_i(r)$ for all $x \in N$ and $r \in R$. For $\mathcal{T} \in \mathcal{T}_r^p(R_i)$, let $\mathcal{K} = \{N_{(i)} : N \in \mathcal{T}\}$. Then $\mathcal{K} \in \mathcal{T}_r^p(R)$, so $\mathcal{K} \in \mathcal{T}_r(R)$ by hypothesis. Thus, $\mathcal{T} \in \mathcal{T}_r(R_i)$.

"$\Longleftarrow$". For $\mathcal{T} \in \mathcal{T}_r^p(R)$, let $\mathcal{T}_i = \{NR_i : N \in \mathcal{T}\}$ for each i. Then $\mathcal{T}_i \in \mathcal{T}_r^p(R_i)$, and so $\mathcal{T}_i \in \mathcal{T}_r(R_i)$ by hypothesis. Let $0 \longrightarrow X \longrightarrow M \longrightarrow Y \longrightarrow 0$ be an exact sequence in Mod-R with $X, Y \in \mathcal{T}$. Then, for each i,

$$0 \longrightarrow XR_i \longrightarrow MR_i \longrightarrow YR_i \longrightarrow 0$$

is an exact sequence in Mod-R_i with $XR_i, YR_i \in \mathcal{T}_i$. So $MR_i \in \mathcal{T}_i$. Thus, $MR_i = NR_i$ for some $N \in \mathcal{T}$. It follows that $MR_i \in \mathcal{T}$. So

$$M = MR_1 \oplus \cdots \oplus MR_n \in \mathcal{T}.$$

□

6.4.11. COROLLARY. $\mathcal{T}_r^p(R) = \mathcal{T}_r(R)$ iff $R = \Pi_{i=1}^n R_i$ where R_i is an indecomposable ring with $\mathcal{T}_r^p(R_i) = \mathcal{T}_r(R_i)$ for each i.

PROOF. It follows from (6.4.7)(6) and (6.4.10). □

The Noetherian condition has attracted much interest in the study of rings R with $\mathcal{T}_r^p(R) = \mathcal{T}_r(R)$. J. Viola-Prioli [126] and Handelman [72] conjectured that any ring R with $\mathcal{T}_r^p(R) = \mathcal{T}_r(R)$ is right Noetherian. Fenrick [50] proved that if R is right Noetherian such that $\mathcal{T}_r^p(R) = \mathcal{T}_r(R)$, then R is a **right V-ring**, i.e., every simple R-module is injective. A. Viola-Prioli and J. Viola-Prioli remarked in [123] that in spite of Fenrick's result, even in the Noetherian case the characterization of the rings R for which $\mathcal{T}_r^p(R) = \mathcal{T}_r(R)$ remains open. A study of rings R with Gabriel dimension such that $\mathcal{T}_r^p(R) = \mathcal{T}_r(R)$ was carried out in Teply [118] where it was proved that these rings are right Noetherian and every $\mathcal{T} \in \mathcal{T}_r^p(R)$ is closed under injective hulls. With the help of Teply's result, it was proved in Dauns and Zhou [40] that a ring R is a right Noetherian ring R with $\mathcal{T}_r^p(R) = \mathcal{T}_r(R)$ iff R is a right QI-ring. In [119], van den Berg gave examples of rings R with $\mathcal{T}_r^p(R) = \mathcal{T}_r(R)$ but which are neither right V-rings nor right Noetherian, thus answering the above question of J. Viola-Prioli and Handelman.

We inductively define the **Gabriel dimension** of a module M, denoted $GdimM$, as follows: We put $Gdim(M) = 0$ iff $M = 0$. Let α be a nonlimit ordinal and assume $Gdim$ is defined for all $\beta < \alpha$. Then we say that a module X is α-**simple** if for every $0 \neq Y \subseteq X$ both $Gdim(Y) \not< \alpha$ and $Gdim(X/Y) < \alpha$. We then say that $GdimM = \alpha$ if $GdimM \not< \alpha$ and for each $N \subset M$, M/N contains a β-simple module for $\beta \leq \alpha$. Note that GdM has been used for the Goldie dimension of M (see 5.1.17).

Several facts on Gabriel dimension needed are stated here as well as in the next lemma: For $N \subseteq M$,

$$GdimM = sup\{Gdim(M/N), GdimN\}$$

if either side exists; if M contains a chain of submodules each of the same dimension, then the union must have the same dimension; for an α-simple module M, $GdimM = \alpha$ and α is a nonlimit ordinal; a nonzero submodule of an α-simple M is α-simple; the 1-simple modules are precisely the usual simple modules; if R has Gabriel dimension, so does every R-module; right Noetherian rings always have Gabriel dimension. For more information on the Gabriel dimension, see Gordon and Robson [69,70].

6.4.12. LEMMA. Let $\mathcal{U} = \{M \in \text{Mod-}R : GdimM \text{ exists}\}$. Then $\mathcal{U} \in \mathcal{T}_r^p(R)$.

In the rest of this section, μ denotes the left exact preradical corresponding to the hereditary pretorsion class $\mathcal{U}$.

6.4.13. LEMMA. Let N be a β-simple submodule of M. Then there exists a β-simple submodule X of M maximal with respect to $N \subseteq X$.

PROOF. Let $\{X_i : i \in I\}$ be a chain of submodules of M where each X_i is β-simple and $N \subseteq X_i$, and let $X = \cup_i X_i$. We prove X is β-simple, and then the claim follows from Zorn's Lemma. Clearly, $GdimX = \beta$. It suffices to show that, for any $0 \neq Y \subseteq X$, $Gdim(X/Y) < \beta$. Since $Y \neq 0$,

$$K = \{i \in I : Y \cap X_i \neq 0\}$$

is not empty; and

$$X/Y = \cup_{i \in I}(X_i + Y)/Y = \cup_{i \in K}(X_i + Y)/Y$$

with $\{(X_i + Y)/Y : i \in K\}$ a chain of submodules of X/Y. For each $i \in K$, since X_i is β-simple,

$$Gdim((X_i + Y)/Y) = Gdim(X_i/(Y \cap X_i)) \leq \beta - 1.$$

It follows that $Gdim(X/Y) < \beta$. □

6.4.14. LEMMA. Let $N \leq_e M$ such that $Gdim(M/N) = \alpha$ and N is a β-simple module. If $\alpha < \beta$, then M is β-simple.

PROOF. From the hypothesis, it follows that $GdimM = \beta$. It suffices to show that for any $0 \neq X \subseteq M$, $Gdim(M/X) < \beta$. Since $N \leq_e M$, we have $N \cap X \neq 0$; so

$$Gdim((X + N)/X) = Gdim(N/(X \cap N)) < \beta$$

as N is β-simple. Moreover, $Gdim(M/(X+N)) \leq \alpha < \beta$ as $Gdim(M/N) = \alpha$. It follows that $Gdim(M/X) < \beta$. □

6.4.15. LEMMA. [118] Let R be a ring with $\mathcal{T}_r^p(R) = \mathcal{T}_r(R)$.

1. If M is a β-simple module, then $\mu(E(M))$ is β-simple.

2. If M is a semisimple module, then $M = \mu(E(M))$.

PROOF. (1) Since M is a β-simple submodule of $\mu(EM)$, there exists a β-simple submodule X of $\mu(EM)$ maximal with respect to $M \subseteq X$ by (6.4.13). It suffices to show that $X = \mu(EM)$. Suppose $X \neq \mu(EM)$. Then there exists an $N \subseteq \mu(EM)$ such that N/X is α-simple for some ordinal $\alpha > 0$. By (6.4.14), it follows from the maximality of X that $\beta \leq \alpha$. Let $\mathcal{F}$ be the class of all α-simple and all β-simple modules. Then $\sigma[M_{\mathcal{F}}] \in \mathcal{T}_r(R)$, so $N \in \sigma[M_{\mathcal{F}}]$. By (2.2.1), there exist $W \leq K \leq M_{\mathcal{F}}^{(I)}$ for an index set I such that $N \cong K/W \leq_e M_{\mathcal{F}}^{(I)}/W$. Write $M_{\mathcal{F}}^{(I)} = A \oplus B$ where $A = \oplus_{\gamma \in I} A_\gamma$ with each A_γ β-simple, and $B = \oplus_{\gamma \in J} B_\gamma$ with each B_γ α-simple. Thus, $K \leq_e A \oplus B$, and there exists an epimorphism $\theta : K \longrightarrow N$. Let $H = \theta^{-1}(X)$, so $N/X \cong K/H \hookrightarrow (A \oplus B)/H$. Since $X \leq_e N$ (as $M \leq X \leq N \leq EM$), we have $H \leq_e K$ and so $H \leq_e A \oplus B$. Then $H \cap A_\gamma \neq 0$ for each $\gamma \in I$ and $H \cap B_\gamma \neq 0$ for each $\gamma \in J$. Thus,

$$\begin{aligned}
\alpha &= Gdim(N/X) \\
&\leq Gdim((A \oplus B)/H) \\
&\leq Gdim((A \oplus B)/[(\oplus_{\gamma \in I} H \cap A_\gamma) \oplus (\oplus_{\gamma \in J} H \cap B_\gamma)] \\
&= Gdim[(\oplus_{\gamma \in I} A_\gamma/(H \cap A_\gamma)) \oplus (\oplus_{\gamma \in J} B_\gamma/(H \cap B_\gamma))] \\
&< \alpha
\end{aligned}$$

as each A_γ is β-simple and each B_γ is α-simple and $\beta \leq \alpha$. This contradiction shows that $\mu(EM) = X$ is β-simple.

(2) If M is semisimple, then $M \in \mathcal{U}$; so $M \subseteq \mu(EM)$. Suppose $M \neq \mu(EM)$. Then there exists an $N \subseteq \mu(EM)$ such that N/M is α-simple for some ordinal $\alpha > 0$. Since the class of all semisimple modules is a hereditary pretorsion class, it is closed under extensions by hypothesis. Let $\mathcal{F}$ be the class of all α-simple and all 1-simple modules. Then $\sigma[M_{\mathcal{F}}] \in \mathcal{T}_r(R)$, so $N \in \sigma[M_{\mathcal{F}}]$. Next, arguing as in the proof of (1) line by line, one obtains the contradiction $\alpha < \alpha$. □

6.4.16. PROPOSITION. [118] Let R be a ring with $\mathcal{T}_r^p(R) = \mathcal{T}_r(R)$.

1. $\mu(R)$ is a ring direct summand of R.

2. If R is a right V-ring, then every ideal of R is a ring direct summand of R.

PROOF. Let I be an ideal of R. By Zorn's Lemma, there exists an ideal J of R maximal with respect to $I \cap J = 0$. Let $\mathcal{T} = \sigma[X]$ where $X = I \oplus [R/(I \oplus J)]$. Then $\mathcal{T}$ is closed under extensions by hypothesis. So $R/J \in \mathcal{T}$. Thus, there exists a right R-module epimorphism $\theta : K \longrightarrow R/J$ where $K \subseteq A \oplus B$, $A = I^n$, and $B = (R/(I \oplus J))^n$ for some $n > 0$.

Let $L = K \cap B$, and let

$$H/J = \Sigma\{f(L) : f \in \mathrm{Hom}_R(L, R/J)\}.$$

Then H/J is an ideal of R/J and $(H/J)I = (\Sigma f(L))I = \Sigma f(LI) = 0$, as $BI = 0$. But

$$(H/J)I \supseteq [(I \cap H + J)/J](I \cap H) = (I \cap H + J)/J$$

(since $(I \cap H)^2 = I \cap H$ by 6.4.6), so $I \cap H \subseteq J$. Hence $I \cap H = I \cap J = 0$. The maximality of J shows that $H = J$. Thus, $\theta(L) \subseteq H/J = 0$. So θ induces an epimorphism

$$\theta' : K/L \longrightarrow R/J.$$

Let $\pi : A \oplus (B/L) \longrightarrow A$ be the projection. Since $(K/L) \cap (B/L) = 0$ and since $K/L \subseteq A \oplus (B/L)$, the restriction of π to K/L is a monomorphism.

Suppose $I \oplus J \neq R$. We prove that there is a contradiction when R is a right V-ring or $I = \mu(R)$. Choose a maximal right ideal P of R such that $I \oplus J \subseteq P$ and let $\eta : R/J \longrightarrow R/P$ be the natural epimorphism. Then $\eta\theta'\pi^{-1} : \pi(K/L) \longrightarrow R/P$ is an epimorphism. Since $\pi(K/L) \subseteq A$, $\eta\theta'\pi^{-1}$ can be extended to a homomorphism $\phi : A \longrightarrow E(R/P)$.

If $I = \mu(R)$, then $\mu(A) = A$; so $\phi(A) = \phi(\mu(A)) \subseteq \mu(E(R/P)) = R/P$ by (6.4.15)(2), and thus $\phi(A) = R/P$. If R is a right V-ring, then $R/P = E(R/P)$ and so again $\phi(A) = R/P$. Since $I^2 = I$ by (6.4.6), $AI = A$ and so $0 = (R/P)I = \phi(A)I = \phi(AI) = \phi(A) = R/P$, a contradiction. □

6.4.17. COROLLARY. A ring R is a right V-ring with $\mathcal{T}_r^p(R) = \mathcal{T}_r(R)$ iff $R = \Pi_{i=1}^n R_i$, where each R_i is a simple right V-ring with $\mathcal{T}_r^p(R) = \mathcal{T}_r(R)$.

PROOF. "$\Longrightarrow$". By (6.4.11), $R = \Pi_{i=1}^n R_i$ where each R_i is an indecomposable ring and $\mathcal{T}_r^p(R_i) = \mathcal{T}_r(R_i)$. Since R is a right V-ring, each R_i is a right V-ring. Then it follows from (6.4.16) that R_i is a simple ring.

"$\Longleftarrow$". Let $R = \Pi_{i=1}^n R_i$ where each R_i is a simple right V-ring with $\mathcal{T}_r^p(R_i) = \mathcal{T}_r(R_i)$. So it follows from (6.4.10) that $\mathcal{T}_r^p(R) = \mathcal{T}_r(R)$. To see that R is a right V-ring, let N_R be simple and let $N_R \leq_e M_R$. We need to show that $N = M$. Since N_R is simple, there exists some k such that $N = NR_k$ and $NR_i = 0$ for $i \neq k$. It follows from $N_R \leq_e M_R$ that $(NR_j)_R \leq_e (MR_j)_R$ for each j. So $(NR_j)_{R_j} \leq_e (MR_j)_{R_j}$ for each j. Thus, $MR_k = NR_k = N$ and $MR_i = NR_i = 0$ for $i \neq k$. So $M = MR_1 \oplus \cdots \oplus MR_n = N$. □

The implication (3) $\Longrightarrow$ (2) in the next theorem is due to Teply [118].

6.4.18. THEOREM. The following statements are equivalent for a ring R:

1. R is a right QI-ring.
2. R is a right Noetherian ring with $\mathcal{T}_r^p(R) = \mathcal{T}_r(R)$.
3. R has Gabriel dimension and $\mathcal{T}_r^p(R) = \mathcal{T}_r(R)$.

4. $\mathcal{K}_1 \vee \mathcal{K}_2 = E(\mathcal{K}_1, \mathcal{K}_2)$ for all $\mathcal{K}_1, \mathcal{K}_2 \in \mathcal{N}_r^p(R)$.

5. $\mathcal{T}_r^p(R) = \mathcal{T}_r(R)$, and $E(\mathcal{K}_1, \mathcal{K}_2) \in \mathcal{N}_r^p(R)$ for all $\mathcal{K}_1, \mathcal{K}_2 \in \mathcal{N}_r^p(R)$.

6. $\mathcal{T}_r^p(R) = \mathcal{T}_r(R)$ and every singular quasi-injective R-module is injective.

PROOF. (1) $\Longrightarrow$ (2) + (4). Let R be a right QI-ring. By (6.3.9), $\mathcal{N}_r^p(R) = \mathcal{N}_r(R)$ and $\mathcal{T}_r^p(R) \subseteq \mathcal{N}_r(R)$. Thus, (4) follows from (6.1.7), and it also follows from (6.1.7) that every $\mathcal{T} \in \mathcal{T}_r^p(R)$ is closed under extensions. So (2) holds.

(2) $\Longrightarrow$ (3). Right Noetherian rings always have Gabriel dimension.

(3) $\Longrightarrow$ (1). Since R has Gabriel dimension, $\mathcal{U} =$ Mod-R. Thus, for any semisimple module M, $EM \in \mathcal{U}$. So $EM = \mu(EM) = M$ by (6.4.15)(2). So M is injective. A well-known result of Byrd [14] states that if every semisimple right module over a ring is injective, this ring must be right Noetherian. So it follows that R is right Noetherian. Next, we prove that every $\mathcal{T} \in \mathcal{T}_r^p(R)$ is closed under injective hulls. Thus, (1) follows by (6.3.9). Let $M \in \mathcal{T}$. Since R is right Noetherian, $EM = \oplus_\gamma E_\gamma$ where each E_γ is a uniform injective module. So it suffices to show that each $E_\gamma \in \mathcal{T}$. Let $N = M \cap E_\gamma$ and $C = E_\gamma$. Since $N \in \mathcal{U}$, N contains a nonzero β-simple module S for some ordinal β. Since $EN \in \mathcal{U}$, $C = EN = \mu(EN) = \mu(ES)$ is β-simple by (6.4.15)(1). Let τ be the left exact radical corresponding to $\sigma[N]$ (since $\sigma[N] \in \mathcal{T}_r^p(R)$), and let

$$\mathcal{K} = \sigma[\tau(C) \oplus C/\tau(C)].$$

Thus, $\mathcal{K} \in \mathcal{T}_r(R)$ by hypothesis. So $C \in \mathcal{K}$. Then there exists an epimorphism $\theta : K \longrightarrow C$ where $K \subseteq A \oplus B$, $A = \oplus\tau(C)$, and $B = \oplus(C/\tau(C))$. Let $L = K \cap B$. Since C is β-simple and $\tau(C) \leq_e C$, $Gdim(C/\tau(C)) < \beta$ and hence $GdimL < \beta$. Thus, $Gdim\theta(L) < \beta$. It must be that $\theta(L) = 0$ as C is β-simple. Hence θ induces an epimorphism $K/L \longrightarrow C$. But

$$K/L = K/(K \cap B) \cong (K + B)/B \hookrightarrow A$$

and $A \in \sigma[N]$. It follows that $C \in \sigma[N] \subseteq \mathcal{T}$.

(4) $\Longrightarrow$ (1). Let M be a quasi-injective R-module and let $\mathcal{K} = \sigma[M]$. By (4), $\mathcal{K}$ is closed under extensions, so it is a hereditary torsion class. Let τ be the left exact radical corresponding to $\mathcal{K}$. Since M is quasi-injective, $M = E_M(M) = \tau(EM)$. Thus, $\tau(EM/M) = \bar{0}$. Let $\mathcal{L} = d(EM/M)$. Then $EM \in E(\mathcal{K}, \mathcal{L})$. It follows from (4) that $EM \in E(\mathcal{L}, \mathcal{K})$. So there exists $X \leq EM$ such that $X \in \mathcal{L}$ and $EM/X \in \mathcal{K}$. Since $\tau(EM/M) = \bar{0}$, $X \in \mathcal{L}$ implies that $\tau(X) = 0$. But $X \cap M \subseteq \tau(X)$, so it must be that $X = 0$. This shows that $EM \in \mathcal{K} = \sigma[M]$. Since M is quasi-injective, M is EM-injective. This implies that $M = EM$ is injective. So R is right QI.

(4) $\Longrightarrow$ (5) is by (6.4.5).

(1) $\Longrightarrow$ (6). For $\mathcal{T} \in \mathcal{T}_r^p(R)$, $\mathcal{T} \in \mathcal{N}_r(R)$ by (6.3.9). So $\mathcal{T}$ is closed under extensions by (2.3.5). Thus, $\mathcal{T} \in \mathcal{T}_r(R)$.

(5) $\Longrightarrow$ (4). Because of the condition $\mathcal{T}_r^p(R) = \mathcal{T}_r(R)$ and (2.3.5), every pre-natural class is closed under extensions. Thus, for $\mathcal{K}, \mathcal{L} \in \mathcal{N}_r^p(R)$,

$$E(\mathcal{K}, \mathcal{L}) \subseteq E(\mathcal{K} \vee \mathcal{L}, \mathcal{K} \vee \mathcal{L}) = \mathcal{K} \vee \mathcal{L}.$$

But $\mathcal{K} \cup \mathcal{L} \subseteq E(\mathcal{K}, \mathcal{L}) \in \mathcal{N}_r^p(R)$ by (5); so $\mathcal{K} \vee \mathcal{L} \subseteq E(\mathcal{K}, \mathcal{L})$.

(6) $\Longrightarrow$ (1). Note that every quasi-injective module is a direct sum of a nonsingular quasi-injective module and a singular quasi-injective module (since $Z(R_R) = 0$) and hence is injective by (6) and (6.4.7)(2). □

There exist rings R such that $E(-,-)$ is not an operation on $\mathcal{N}_r^p(R)$.

6.4.19. EXAMPLE. For a ring R, let $\mathcal{Z}$ be the class of all singular R-modules and let $\mathcal{F}$ be the class of all nonsingular R-modules. Then $\mathcal{Z}, \mathcal{F}$ are in $\mathcal{N}_r^p(R)$ and $E(\mathcal{Z}, \mathcal{F}) = \{M \in \text{Mod-}R : M/Z(M) \text{ is nonsingular}\}$.

Now let $R = \left\{\begin{pmatrix} a & x \\ 0 & a \end{pmatrix} : a \in \mathbb{Z}, x \in \mathbb{Z}_2\right\}$. Then R is a ring under matrix addition and multiplication. Let

$$I = \left\{\begin{pmatrix} 0 & x \\ 0 & 0 \end{pmatrix} : x \in \mathbb{Z}_2\right\}, \ J = \left\{\begin{pmatrix} a & 0 \\ 0 & a \end{pmatrix} : a \in 2\mathbb{Z}\right\}.$$

Then I and J are ideals and $I \cap J = 0$. Note that a right ideal K of R is essential in R_R iff $I \oplus L \subseteq K$ for some nonzero right ideal L with $L \subseteq J$. It follows that I_R is singular and R/I is nonsingular. So $R \in E(\mathcal{Z}, \mathcal{F})$. Thus, $E(\mathcal{Z}, \mathcal{F})$ is a pre-natural class iff $E(\mathcal{Z}, \mathcal{F})$ is a natural class. But by (3.1.17), the injective hull $E(I)$ is not singular. Thus $I \subseteq Z(E(I)) \neq E(I)$ and so $E(I)/Z(E(I))$ is not nonsingular (is singular indeed). This shows that $E(I)$ is not in $E(\mathcal{Z}, \mathcal{F})$. So $E(\mathcal{Z}, \mathcal{F})$ is not a natural class. Therefore, $E(\mathcal{Z}, \mathcal{F})$ is not a pre-natural class. □

Below is a decomposition theorem of right QI-rings.

6.4.20. THEOREM. A ring R is right QI iff R is a (finite) direct product of simple right QI-rings.

PROOF. Let R be a right QI-ring. Being right Noetherian, R has a decomposition $R = \Pi_{i=1}^n R_i$ where each R_i is an indecomposable ring. Then each R_i is right Noetherian, and $\mathcal{T}_r^p(R_i) = \mathcal{T}_r(R_i)$ by (6.4.10) and (6.4.18). Hence R_i is right QI by (6.4.18). Then it follows from (6.4.16) that R_i is a simple ring.

Conversely, let $R = \Pi_{i=1}^n R_i$ where each R_i is simple right QI-ring. Then R_i is right Noetherian with $\mathcal{T}_r^p(R_i) = \mathcal{T}_r(R_i)$ by (6.4.18). So R is right Noetherian, and it follows from (6.4.10) that $\mathcal{T}_r^p(R) = \mathcal{T}_r(R)$. Then R is right QI by (6.4.18). □

Below is a decomposition theorem of rings R with $\mathcal{T}_r^p(R) = \mathcal{T}_r(R)$.

6.4.21. THEOREM. R is a ring with $\mathcal{T}_r^p(R) = \mathcal{T}_r(R)$ if and only if $R = R_1 \times \cdots \times R_n \times S$ where each R_i is a simple right QI-ring, no nonzero ideal of S has Gabriel dimension, and $\mathcal{T}_r^p(S) = \mathcal{T}_r(S)$.

PROOF. "$\Longleftarrow$". This is by (6.4.10) and (6.4.18).

"$\Longrightarrow$". By (6.4.16), $\mu(R)$ is a ring direct summand of R; the claim then follows from (6.4.10), (6.4.18), and (6.4.20). □

We conclude by giving some examples of rings R with $\mathcal{T}_r^p(R) = \mathcal{T}_r(R)$. In particular, we present rings R with $\mathcal{T}_r^p(R) = \mathcal{T}_r(R)$, but R is neither right V-rings nor right Noetherian rings, as observed by van den Berg [119]. These examples follow from a few results of Golan, A. Viola-Prioli and J. Viola-Prioli, van den Berg and Raftery on chain rings. Recall that a **right chain ring** is a ring all of whose right ideals are linearly ordered by inclusion. We let

$$\eta(I) = \{K_R \subseteq R_R : I \subset K\}$$

for any proper ideal I of R and let

$$\bar{\eta}(I) = \{K_R \subseteq R_R : I \subseteq K\}$$

for any ideal I of R.

6.4.22. PROPOSITION. [121] Let R be a right chain ring. Then $\mathfrak{A} \in$ **fil**-R iff $\mathfrak{A} = \eta(I)$ for some proper ideal I of R or $\mathfrak{A} = \bar{\eta}(I)$ for some ideal I of R.

PROOF. "$\Longleftarrow$". For an ideal I, $\bar{\eta}(I)$ is always a linear topology. For a proper ideal I of R, to see $\eta(I) \in$ **fil**-R, it suffices to show that if $K \in \eta(I)$ and $r \in R$, then $r^{-1}K \in \eta(I)$. This is obviously true if $r \in I$. Hence, assume $r \notin I$ and consider $L = (I + rR) \cap K \in \eta(I)$. Take $s \in L \backslash I$ and write $s = u + rv$ with $u \in I$ and $v \in R$. As $rv \in K$, we have $v \in r^{-1}K$. But, $v \notin I$ since $s \notin I$. So $r^{-1}K \nsubseteq I$. Thus, $r^{-1}K \in \eta(I)$ since R is a right chain ring.

"$\Longrightarrow$". For $\mathfrak{A} \in$ **fil**-R, let $I_0 = \cap\{I : I \in \mathfrak{A}\}$. We first show that I_0 is an ideal of R. Suppose $aI_0 \nsubseteq I$ for $a \in R$ and some right ideal $I \in \mathfrak{A}$. Then $ab \notin I$ for some $b \in I_0$. This implies that $0 \neq (abR + I)/I \cong R/(ab)^{-1}I = R/b^{-1}(a^{-1}I)$ and thus $b^{-1}(a^{-1}I) \neq R$. But, on the other hand, since $a^{-1}I \in \mathfrak{A}$, we have $b \in I_0 \subseteq a^{-1}I$ and hence $b^{-1}(a^{-1}I) = R$, a contradiction. So I_0 is an ideal of R. Thus, if $I_0 \in \mathfrak{A}$, then $\mathfrak{A} = \bar{\eta}(I_0)$. Now assume $I_0 \notin \mathfrak{A}$; we show that $\mathfrak{A} = \eta(I_0)$. To this end, given $K \in \eta(I_0)$ we must have $I_0 \subset K$, so $K \nsubseteq I$ for some $I \in \mathfrak{A}$. Thus, $I \subset K$ since R is a right chain ring. This shows that $K \in \mathfrak{A}$; and it follows that $\mathfrak{A} = \eta(I_0)$. □

The next proposition is well known.

6.4.23. PROPOSITION. Let I be an ideal of R. Then $\bar{\eta}(I)$ is a Gabriel topology iff $I^2 = I$.

PROOF. $\Longleftarrow$. If $I^2 = I$, then $I^2 \in \bar{\eta}(I)$. Let A, B be ideals with $A \in \bar{\eta}(I)$ and $a^{-1}B \in \bar{\eta}(I)$ for all $a \in A$. Then $I \subseteq A$ and $a^{-1}B \supseteq I$ for all $a \in A$. Thus $aI \subseteq B$ for all $a \in A$, that is, $AI \subseteq B$. It follows that $I = I^2 \subseteq AI \subseteq B$, so $B \in \bar{\eta}(I)$.

$\Longrightarrow$. If $I \neq I^2$, then $I^2 \notin \bar{\eta}(I)$, but $I \in \bar{\eta}(I)$. Thus, $a^{-1}I^2 \supseteq I$ for all $a \in I$. So (2.1.8)(4) is not satisfied. □

A proper ideal I of R is said to be **completely prime** if $ab \in I$ always implies $a \in I$ or $b \in I$, or equivalently R/I is a domain. A **reduced** ring is a ring containing no nonzero nilpotent elements.

6.4.24. LEMMA. [63] The following hold for a right chain ring R:

1. If R is reduced, then R is a domain.

2. Every nonzero idempotent proper ideal of R is completely prime.

PROOF. (1) Assume $ab = 0$ where $a, b \in R$. Since R is a right chain ring, either $a = bc$ or $b = ad$. If $a = bc$, then $a^2 = abc = 0$; hence $a = 0$. If $b = ad$, then $(ba)^2 = baba = 0$. Thus, $ba = 0$ and hence $b^2 = bad = 0$; so $b = 0$.

(2) Let $I \neq 0$ be a proper ideal of R with $I^2 = I$. To show I is completely prime, it suffices to show that R/I is reduced by (1). Assume that this is not the case. Then there exists $a \in R$ such that $a^2 \in I$ and $I \subset aR$. We then have $I = I^2 \subseteq (aR)I = aI \subseteq a^2R \subseteq I$. Therefore $I = a^2R$, which in turn implies $I = I^2 = a^2RI = a^2I$. It follows that $a^2 = a^2y$ for a certain element $y \in I$. Since I is proper, $1 - y$ is a unit of R, which forces $I = a^2R = 0$, a contradiction. □

6.4.25. PROPOSITION. [63] Let R be a right chain ring and let I be a completely prime ideal of R. Then $\eta(I)$ is a Gabriel topology.

PROOF. By (6.4.22), $\eta(I)$ is a linear topology. Let U and K be right ideals of R with $K \in \eta(I)$ and $k^{-1}U \in \eta(I)$ for every $k \in K$. We prove that $U \in \eta(I)$. If $U \notin \eta(I)$, then $U \subseteq I$. Take $a \in K \backslash I$. Since $a^{-1}U \in \eta(I)$, we can choose $x \in a^{-1}U \backslash I$. Therefore $ax \in U \subseteq I$. But neither a nor x belongs to I, so I is not completely prime. □

The next theorem follows from (6.4.22), (6.4.23), and (6.4.25).

6.4.26. THEOREM. [119] Let R be a right chain ring all of whose ideals are idempotent. Then $\mathcal{T}_r^p(R) = \mathcal{T}_r(R)$. □

The following theorem of van den Berg and Raftery [120, Theorem 9] shows that there is an abundance of rings R satisfying the hypothesis of (6.4.26). Its proof (omitted here) uses technical methods for constructing noncommutative right chain rings.

6.4.27. THEOREM. [120] The following conditions are equivalent for a chain L:

1. L is an algebraic lattice.

2. There is a right chain domain R such that L is isomorphic to the lattice of proper ideals of R, and all ideals of R are idempotent.

□

Thus, if R is a non-simple ring satisfying the hypothesis of (6.4.26), then $\mathcal{T}_r^p(R) = \mathcal{T}_r(R)$, but R is neither a right V-ring by (6.4.17) nor a right Noetherian ring by (6.4.18) and (6.4.20). We may therefore conclude from (6.4.26) and (6.4.27) that there do exist rings R satisfying $\mathcal{T}_r^p(R) = \mathcal{T}_r(R)$, but R is neither a right V-ring nor a right Noetherian ring.

6.4.28. REFERENCES. Dauns and Zhou [40]; Fenrick [50]; Golan, A. Viola-Prioli and J. Viola-Prioli [63]; Gordon and Robson [69,70]; Handelman [72]; Teply [118]; van den Berg [119]; van den Berg and Raftery [120]; A. Viola-Prioli and J. Viola-Prioli [121,123]; J. Viola-Prioli [126].

6.5 Lattice $\mathcal{N}_r(R)$ and Its Applications

In this section it will be shown that $\mathcal{N}_r(\cdot)$ can be made into a functor. Some consequences of the functoriality of $\mathcal{N}_r(\cdot)$ are explored. In general, natural classes are defined for a given ring R, and depend on R. However, certain very useful classes, the so-called universal natural classes, are defined for every ring, e.g., Goldie torsion, molecular, bottomless, continuous, discrete, types I, II, III. The reason for their universality will be explained by means of the functor $\mathcal{N}_r(\cdot)$.

In this section, abbreviate $\mathcal{N}_r = \mathcal{N}$. The category of all rings $R, S, \dots$ with identity, and identity preserving ring homomorphisms $\phi : R \longrightarrow S$, which are onto, is denoted by **A**. Here the symbol such as **A** for any category refers to the disjoint union of its objects and its morphisms.

Let **B** denote the category whose objects are complete Boolean lattices $L_1, L_2, \dots$ with 0 and 1. And the morphisms $f : L_1 \longrightarrow L_2$ of **B** are zero preserving lattice homomorphisms which are one to one with convex images. Note that $fL_1 = \{y \in L_2 : y \leq f1\}$. Any lattice homomorphism $f : L_1 \longrightarrow L_2$ is said to be **complete** if f preserves arbitrary sups and infs. It follows that the above morphisms $f \in \mathbf{B}$ are complete. For technical reasons, the degenerate Boolean lattice $\{0\}$ with $0 = 1$ must belong to **B**. Above, the map $f : L_1 \longrightarrow L_2$ is a monic ring homomorphism of Boolean rings L_1, L_2 which need not preserve the identity elements.

An arbitrary morphism of **A** will be denoted throughout as $\phi : R \longrightarrow S$ with kernel $\phi^{-1}0 = I \lhd R$, and for simplicity assume that $\phi : R \longrightarrow R/I = S$ is the natural projection. Right singular submodules and right injective hulls with respect to the ring S are denoted by Z^S, Z_2^S, and E^S. (Thus $Z^R = Z$, $Z_2^R = Z_2$, and $E^R = E$). Throughout, let $N = N_S \in$ Mod-S. The induced R-module is denoted by N_ϕ, where $n \cdot r = n(\phi r) = n(r + I)$, $n \in N$. Write $ZN_\phi = Z(N_\phi)$ and $EN_\phi = E(N_\phi)$.

For any subclass $\mathcal{Y}$ of right R-modules as before, $d(\mathcal{Y}) \in \mathcal{N}(R)$ denotes the natural class generated by $\mathcal{Y}$.

6.5.1. DEFINITION. For $\mathcal{K}^S \in \mathcal{N}(S)$, define $\mathcal{K}_\phi^S = \{N_\phi : N \in \mathcal{K}s\}$. The correspondence $N \longrightarrow N_\phi$ induces a covariant functor $\phi^\# :$ Mod-$S \longrightarrow$ Mod-R which on morphisms, is the identity. This functor $\phi^\#$ induces a map

$$\phi^* : \mathcal{N}(S) \longrightarrow \mathcal{N}(R), \ \mathcal{K}^S \longmapsto d(\mathcal{K}_\phi^S).$$

More generally, for any subclass $\mathcal{Y} \subseteq$ Mod-S, $\mathcal{Y}_\phi$ and $\phi^*\mathcal{Y}$ are defined as above.

One of our objectives will be to show that the assignments $S \longrightarrow \mathcal{N}(S)$, $\phi \longrightarrow \mathcal{N}(\phi) = \phi^*$ define a contravariant functor $\mathcal{N} : \mathbf{A} \longrightarrow \mathbf{B}$.

6.5.2. OBSERVATIONS.

1. For any $N \in \text{Mod-}S$, the lattice of S-submodules of N coincides with the lattice of R-submodules of N_ϕ. In particular, essential and complement submodules of the Abelian group $N = N_\phi$ are identical over S and R. The same applies to quotient modules.

2. For $N, N' \in \text{Mod-}S$, $\text{Hom}_S(N, N') = \text{Hom}_R(N_\phi, N'_\phi)$.

3. Since $N \leq_e E^S N$, by (1), also $N_\phi \leq_e (E^S N)_\phi$. Consequently, there exists an embedding $N_\phi \leq_e (E^S N)_\phi \leq_e E(N_\phi)$.

4. There is a natural copy of $E^S N$ in $E(N_\phi)$. As an Abelian group
$$E^S N = (E^S N)_\phi = \{x \in EN_\phi : xI = 0\}.$$
The latter equality follows from the fact that this submodule satisfies Baer's criterion over S, in view of (2).

5. Again by (2), $(E^S N)$ is a quasi-injective R-module. Hence EN_ϕ contains a unique S-injective hull of N as given in (4).

6.5.3. REMARKS. (1) For a pre-natural class $\mathcal{K} \in \mathcal{N}_r^p(S)$, $\mathcal{K}_\phi \in \mathcal{N}_r^p(R) = \mathcal{N}^p(R)$.

(2) The restriction and corestriction $\phi^\# : \mathcal{N}^p(S) \longrightarrow \mathcal{N}^p(R)$, $\phi^\# \mathcal{K} = \mathcal{K}_\phi$, $\mathcal{K} \in \mathcal{N}^p(S)$, is a complete lattice monomorphism with a convex and complete image
$$\phi^\# \mathcal{N}^p(S) = \{\mathcal{K} \in \mathcal{N}^p(R) : \ \forall\, V \in \mathcal{K}, VI = 0\}.$$

(3) Although both $\phi^\# : \mathcal{N}^p(S) \longrightarrow \mathcal{N}^p(R)$ as well as $\phi^* : \mathcal{N}(S) \longrightarrow \mathcal{N}(R)$, are lattice monomorphisms, in general they do not preserve lattice complements, because in general they need not be onto.

PROOF. (1) It suffices to show that for any $N \in \mathcal{K}$, $\text{tr}(\mathcal{K}_\phi, EN_\phi) \in \mathcal{K}_\phi$. Using (6.5.2)(2,4), since $\text{tr}(\mathcal{K}, E^S N) \in \mathcal{K}$, we have
$$\text{tr}(\mathcal{K}_\phi, EN_\phi) = \text{tr}(\mathcal{K}, E^S N)_\phi \in \mathcal{K}_\phi.$$

(2) The function $\phi^\#$ is order preserving and one to one, and finite or infinite suprema are computed by the same formula in either $\mathcal{N}_r^p(S)$ or in $\mathcal{N}_r^p(R)$. Thus (2) holds.

The next straightforward lemma will be used repeatedly later on.

6.5.4. LEMMA. Let $I \lhd R$, $S = R/I$, $\mathcal{K}^S \in \mathcal{N}(S)$, and ϕ, ϕ^* as before. For any $M \in \text{Mod-}R$, define $\text{Ann}_M I = \{m \in M : mI = 0\}$. Then

1. $\phi^* \mathcal{K}^S = \{M \in \text{Mod-}R : \ \exists\, N \in \mathcal{K}^S, N_\phi \leq_e M\}$;

2. $\phi^* \mathcal{K}^S = \{M \in \text{Mod-}R : \ \exists\, N \in \mathcal{K}^S, M \hookrightarrow EN_\phi\}$;

3. $\phi^* \mathcal{K}^S = \{M \in \text{Mod-}R : \text{Ann}_M I \leq_e M, \text{Ann}_M I \in \mathcal{K}^S\}$.

PROOF. Omit (1) and (2). (3) For $M \in \phi^*\mathcal{K}^S$ and $N \in \mathcal{K}^S$ with $M \hookrightarrow EN_\phi$ as in (2), $M \cap N_\phi \leq_e M$, and since $NI = 0$, $M \cap N_\phi \subseteq \text{Ann}_M I$. Thus $\text{Ann}_M I \leq_e M$. Since $M \cap N \in \mathcal{K}^S$, and $(M \cap N)_S \leq_e (\text{Ann}_M I)_S$ by (6.5.2)(1), $\text{Ann}_M I \in \mathcal{K}^S$. □

6.5.5. LEMMA. For $\phi \in \mathbf{A}$, $\phi^* : \mathcal{N}(S) \longrightarrow \mathcal{N}(R)$ as in (6.5.1), and $\mathbf{A}$, $\mathbf{B}$ as at the beginning, the following hold.

1. $\forall\ \mathcal{K}^S, \mathcal{L}^S \in \mathcal{N}(S), \phi^*\mathcal{K}^S \subseteq \phi^*\mathcal{L}^S \iff \mathcal{K}^S \subseteq \mathcal{L}^S$; in particular, ϕ^* is one to one.

2. $\phi^*(\mathcal{N}(S))$ is convex and upper directed in $\mathcal{N}(R)$.

3. For any $\phi \in \mathbf{A}$, $\phi^* \in \mathbf{B}$. In particular, ϕ^* is a complete lattice homomorphism.

PROOF. (1) $\Longleftarrow$. This is trivial. $\Longrightarrow$. Let $\phi^*\mathcal{K}^S \subseteq \phi^*\mathcal{L}^S$. Take any $P \in \mathcal{K}^S$. Then $P_\phi \in \mathcal{K}^S_\phi \subseteq \phi^*\mathcal{K}^S \subseteq \phi^*\mathcal{L}^S$. This means that there exists a $Q \in \mathcal{L}^S$ with $Q_\phi \leq_e P_\phi$ by (6.5.4)(1). Then $Q_S \leq_e P_S$ by (6.5.2)(1), and hence $P \in \mathcal{L}^S$. Thus $\mathcal{K}^S \subseteq \mathcal{L}^S$.

(2) Suppose that $0 \subseteq \mathcal{J} \subseteq \phi^*\mathcal{K}^S$ for some $\mathcal{J} \in \mathcal{N}(R)$ and $\mathcal{K}^S \in \mathcal{N}(S)$. We will show that $\mathcal{J} = \phi^*\mathcal{L}^S$ where $\mathcal{L}^S = \{P \in \mathcal{K}^S : P_\phi \in \mathcal{J}\}$. To show that $\mathcal{L}^S$ is a natural class, it suffices to show that if $P_S \leq_e Q_S$ where $P \in \mathcal{L}^S$ and $Q \in \text{Mod-}S$, then also $Q \in \mathcal{L}^S$. But again, by (6.5.2)(1), $P_\phi \leq_e Q_\phi$. Since $P_\phi \in \mathcal{J}$, also $Q_\phi \in \mathcal{J}$, and hence $Q \in \mathcal{L}^S$. Thus $\mathcal{L}^S \in \mathcal{N}(S)$. Next, it will be proven that $\phi^*\mathcal{L}^S = \mathcal{J}$. By definition of ϕ^*,

$$\begin{aligned}\phi^*\mathcal{L}^S &= \{M \in \text{Mod-}R : \exists\ N \in \mathcal{L}^S, N_\phi \leq_e M\}\\ &= \{M : \exists\ N \in \text{Mod-}S, N_\phi \in \mathcal{J}, N \in \mathcal{K}^S, N_\phi \leq_e M\} \subseteq \mathcal{J},\end{aligned}$$

where the last equality came from replacing "$N \in \mathcal{L}^S$" by the definition of $\mathcal{L}^S$. In order to show that $\mathcal{J} \subseteq \phi^*\mathcal{L}^S$, start with $M \in \mathcal{J}$. Since $M \in \mathcal{J} \subseteq \phi^*\mathcal{K}^S$, there exists $N_\phi \leq_e M$ for some $N \in \mathcal{K}^S$ by (6.5.4)(1). But then $N_\phi \in \mathcal{J}$ since $M \in \mathcal{J}$. By the definition of $\mathcal{L}^S$, $N \in \mathcal{L}^S$. Since $N_\phi \leq_e M$ with $N \in \mathcal{L}^S$, by (6.5.4)(1), $M \in \phi^*\mathcal{L}^S$. Thus $\phi^*\mathcal{L}^S = \mathcal{J}$ and $\phi^*(\mathcal{N}(S))$ is convex in $\mathcal{N}(R)$, and since $\mathcal{N}(S)$ is upper directed, so is $\phi^*(\mathcal{N}(S))$.

(3) By (2), $\phi^*(\mathcal{N}(S))$ is a complete lattice, since $\mathcal{N}(R)$ is complete. The corestriction of ϕ^* to $\mathcal{N}(S) : \mathcal{N}(S) \longrightarrow \phi^*(\mathcal{N}(S))$ is a bijective map of complete lattices which is order preserving with an order preserving inverse, and hence complete. Thus $\phi^* \in \mathbf{B}$. □

6.5.6. DEFINITION. For any ring R and for any class of right R-modules such as $\mathcal{K} \in \mathcal{N}(R) = \mathcal{N}_r(R)$, define

$$\begin{aligned}\mathcal{K}_t &= \{M : M \in \mathcal{K}, Z_2M = M\};\\ \mathcal{K}_f &= \{M : M \in \mathcal{K}, ZM = 0\}.\end{aligned}$$

Note that $\mathcal{K}_t = \{Z_2M : M \in \mathcal{K}\}$, and that $\mathcal{K}_t, \mathcal{K}_f \in \mathcal{N}(R)$. Set

$$\mathcal{N}_t(R) = \{\mathcal{K}_t : \mathcal{K} \in \mathcal{N}(R)\} \text{ and } \mathcal{N}_f(R) = \{\mathcal{K}_f : \mathcal{K} \in \mathcal{N}(R)\}.$$

It is clear that $\mathcal{N}(R) = \mathcal{N}_t(R) \oplus \mathcal{N}_f(R)$ is a lattice direct sum of convex complete sublattices. The following two technical lemmas will be needed later, and for the readers' benefit we provide their proof.

6.5.7. LEMMA. For $\phi : R \longrightarrow S = R/I$ where $I = \phi^{-1}0 \lhd R$, let $N \in$ Mod-S, $N_\phi \in$ Mod-R, and Z^S, Z_2^S as in the beginning. Then the following hold:

(1) $Z^S N \subseteq ZN_\phi$;

(2) $Z_2^S N \subseteq Z_2 N_\phi$.

Now let $I < R$ be a right complement. Then

(3) $Z^S N = ZN_\phi$; hence

(4) $Z_2^S N = Z_2 N_\phi$; and in particular,

(5) $Z^S(R/I) = Z(R/I)$, and $Z_2^S(R/I) = Z_2(R/I)$.

PROOF. (1) Since $N = N_\phi$ as Abelian groups, for $n \in N$, $\text{Ann}_S n = n^\perp/I$, where $n^\perp = \{r \in R : nr = 0\}$. If $n \notin ZN_\phi$, then $n^\perp \oplus B \leq R$ for some $0 \neq B \leq R$. Then $n^\perp/I \oplus [(B+I)/I] \leq R/I$ shows that $n \notin Z^S N$. Hence $Z^S N \subseteq ZN_\phi$.

(2) Since $Z^S N \subseteq N$, $Z^S N$ is an R-module with $(Z^S N)I = 0$. Also $ZN_\phi \subseteq N$, and $(ZN_\phi)I = 0$. Let $\pi : N/Z^S N \longrightarrow N/ZN_\phi$ be the natural quotient map induced by (1). Then π restricts and corestricts to induce the map

$$\pi : Z^S(N/Z^S N) \longrightarrow Z^S(N/ZN_\phi) \subseteq Z(N/ZN_\phi),$$

where the last inclusion follows by applying (1) to the module N/ZN_ϕ. Now (2) follows.

(3) It has to be shown that for any $n \in ZN_\phi$, $\text{Ann}_S n = n^\perp/I$ is an essential right ideal of S. It is here that we need to use the fact that I is a complement right ideal of R to conclude that $n^\perp/I \leq_e R/I$ is essential as a right R-module, and hence by (6.5.2)(1) also essential as a right S-module. Thus (3) holds. Conclusions (4) and (5) now follow from (3). □

6.5.8. LEMMA. For $I \lhd R$ and $S = R/I$, assume in addition that $I < R$ is a complement right ideal. Set $K = I + ZR < R$ and let $\overline{K}$ be the complement closure of K in R. Then

(1) $\overline{K} \lhd R$; and

(2) $Z_2^S(S) = \overline{K}/I$.

PROOF. (1) Note that $\overline{K} = \{r \in R : r^{-1}K \leq_e R\}$. Let $b \in R$, $t \in \overline{K}$, and $r \in t^{-1}K$. Since $K \lhd R$, we have $btr \in K$, $r \in (bt)^{-1}K$, and hence $t^{-1}K \subseteq (bt)^{-1}K \leq_e R$. Thus, $bt \in \overline{K}$.

(2) Let $Z(R/I) = T/I$. Then $T = \{r \in R : r^{-1}I \leq_e R\}$, and hence $K/I \leq Z(R/I) \leq \overline{K}/I$. Since I is a complement right ideal of R, $K/I \leq_e \overline{K}/I$; thus $K/I \leq_e Z(R/I) \leq_e \overline{K}/I$. It follows that $\overline{K}/I \leq_e Z_2(R/I)$. But $\overline{K}$ is a complement right ideal of R, so $\overline{K}/I$ is a complement right ideal of R/I; hence $\overline{K}/I = Z_2(R/I)$. □

6.5.9. For any ring R, any nonsingular cyclic module can be embedded in $E(R/Z_2R)$. For $\mathcal{N}(R) = \mathcal{N}_t(R) \oplus \mathcal{N}_f(R)$, both of the latter are complete Boolean lattices with identities. For $\mathcal{N}_f(R)$, $d(R/Z_2R)$ is the identity element.

6.5.10. COROLLARY. Let $\phi : R \longrightarrow S = R/I$ be a morphism of $\mathbf{A}$ and $K = I + ZR \leq_e \overline{K} \leq R$ as before in the last lemma. Then

$$\phi^*(\mathcal{N}_f(S)) \subseteq \mathcal{N}_f(R),\ \phi^*(d(S/Z_2S)) = d(R/\overline{K}).$$

□

PROOF. Since $Z_2(R/I) = \overline{K}/I$, $Z(R/\overline{K}) = 0$, and hence $R/\overline{K} \in d(R/\overline{K}) \in \mathcal{N}_f(R)$. Since for any $N = N_S$, $Z^S N = 0$ implies by (6.5.7)(3) that $ZN_\phi = 0$, $\phi^*\mathcal{N}_f(S) \subseteq \mathcal{N}_f(R)$. By (6.5.8) and by (6.5.2)(1), $\phi^*(d(S/Z_2S)) = d(R/\overline{K})$. □

6.5.11. DEFINITION. Let $\mathbf{A}^* \subset \mathbf{A}$ have the same objects as $\mathbf{A}$, i.e., all rings $R, S, \ldots$ with identity, but only those surjective identity preserving ring homomorphisms $\phi : R \longrightarrow S \cong R/I$, whose kernels $I = \phi^{-1}0$ are complement right ideals of R.

6.5.12. LEMMA. The morphisms in $\mathbf{A}^*$ are closed under composition, i.e., $\mathbf{A}^*$ is a category.

PROOF. Let $K \lhd R$ and $K \subset L \lhd R$. Note that $L/K < R/K$ is a complement R-submodule if and only if it is a complement right ideal of R/K, since the R- and R/K-submodules of R/K coincide.

It suffices to show that if K is a complement right ideal of R and L/K is a complement R-submodule of R/K, then L is a complement right ideal of R. The proof will hold more generally verbatim the way it is if $K < L < R$ are replaced by any three right R-modules.

Choose $G, H < R$ such that $K \oplus G \leq_e L$ and $L \oplus H \leq_e R$ are essential. It is now that the hypothesis that $K < R$ is a complement is needed to show that $(L/K) \oplus [(H \oplus K)/K] \leq_e R/K$ remains essential. Next, the hypothesis that $L/K < R/K$ is a complement R-submodule is needed to conclude that it is maximal with respect to $(L/K) \cap [(H \oplus K)/K] = 0$. The latter is equivalent to $L \leq R$ being a submodule maximal with respect to $L \cap (H \oplus K) = K$, or since $K \subset L$, by the modular law to $(L \cap H) \oplus K = K$, or $L \cap H = 0$. But the latter implies that $L < R$ is a right complement. □

6.5.13. COROLLARY. For $\mathbf{A}^*$ as above, $\mathbf{A}^*$ is a subcategory of $\mathbf{A}$.

PROOF. Let $K \lhd R$, $L \lhd R$ with $K \subset L$, and let $R \longrightarrow R/K$ and $R/K \longrightarrow (R/K)/(L/K)$ be two morphisms in $\mathbf{A}^*$. Then their composite $R \longrightarrow R/L$ is also in $\mathbf{A}^*$ by the previous lemma. □

Finally all the pieces have been assembled in order to make $\mathcal{N}(\cdot)$ into a functor.

6.5.14. THEOREM. Let $\mathbf{A}^* \subset \mathbf{A}$, $\mathbf{B}$, $\phi : R \longrightarrow S = R/I \in \mathbf{A}$ with kernel $\phi^{-1}0 = I \lhd R$ be as at the beginning, and $\mathcal{N}(S) = \mathcal{N}_t(S) \oplus \mathcal{N}_f(S)$ as in (6.5.6), and $K = I + ZR \leq_e \overline{K} \leq R$ as in (6.5.8). Define $\mathcal{N}(\phi) = \phi^*$ as in (6.5.1). Then the following hold:

(1) $\mathcal{N} : \mathbf{A} \longrightarrow \mathbf{B}$ is a contravariant functor, where $\phi^* : \mathcal{N}(S) \longrightarrow \mathcal{N}(R)$ is a zero preserving monic lattice homomorphism (equivalently ring homomorphism of associated Boolean rings) with a convex (and hence complete) image in $\mathcal{N}(R)$. Consequently, ϕ^* preserves all arbitrary infima and suprema.

(2) $\phi^*\mathcal{N}_t(S) \subseteq \mathcal{N}_t(R)$, and $\mathcal{N}_t(\cdot) \leq \mathcal{N}(\cdot)$ is a subfunctor.

If in (3) and (4) below, in addition $\phi \in \mathbf{A}^*$, then the following hold:

(3) $\phi^*\mathcal{N}_f(S) \subseteq \mathcal{N}_f(R)$, $\mathcal{N}_f(\cdot) \leq \mathcal{N}(\cdot)$ is a subfunctor, and
$$\mathcal{N}(\cdot) = \mathcal{N}_t(\cdot) \oplus \mathcal{N}_f(\cdot)$$
is a direct sum of functors.

(4) $\phi^*\mathcal{N}_f(S) = \{\mathcal{K} \in \mathcal{N}(R) : \mathcal{K} \subseteq d(R/\overline{K})\}$.

6.5.15. NOTATION. As suggested by (5.1.8), let A, B, C, D be functions mapping rings with identity $R, S, \ldots$ to elements of $\mathcal{N}(R), \mathcal{N}(S), \ldots$; $A(R) \in \mathcal{N}(R)$ are the molecular modules, and similarly for bottomless B, continuous (uniform free) C, and discrete D. Let CA be the continuous molecular ones, i.e., $(CA)(S) = C(S) \cap A(S)$. Similarly, I, II, and III are now functions, where $III(R)$ is the natural class of all type III right R-modules as in (5.1.11). Again, the same way for any cardinal $\aleph = 1, \aleph_0, \aleph_1, \ldots$, $\Delta_\aleph(R)$ is the natural class of all right R-modules of local Goldie dimension $\aleph$.

Just what is it about A, B, C, D, I, II, III, $\Delta_\aleph$ and many other such classes that they can be defined for every ring to yield a natural class? The answer is that they are special examples of universal natural classes, which we define and study next. As a corollary we will show that any finite pairwise orthogonal set of universal classes can be extended to a maximal one so $\mathcal{N}$ is a finite direct sum of subfunctors, e.g.,

$$\mathcal{N} = \mathcal{N}_C \oplus \mathcal{N}_D = \mathcal{N}_{CA} \oplus \mathcal{N}_D \oplus \mathcal{N}_B = \mathcal{N}_I \oplus \mathcal{N}_{II} \oplus \mathcal{N}_{III}.$$

This also helps to explain why universal natural classes give direct sum decompositions of modules.

6.5.16. PROPOSITION. Let Δ be a function defined for all rings $S, R, \ldots$ with identity where $\Delta(S) \in \mathcal{N}(S)$, $\Delta(R) \in \mathcal{N}(R), \ldots$, etc. Then (1), (2), and (3) are all equivalent, where each is to hold for all $\phi : R \longrightarrow S = R/I$, with $\phi \in \mathbf{A}$.

(1) $\forall\, N \in \Delta(S)$, $N_\phi \in \Delta(R)$; $\forall 0 \neq P \in c(\Delta(S))$, $P_\phi \notin \Delta(R)$.

(2) $\forall\, N \in$ Mod-S, if $X \oplus Y \leq_e N$ with $X \in \Delta(S)$ and $Y \in c(\Delta(S))$, then $X_\phi \oplus Y_\phi \leq_e N_\phi$ with $X_\phi \in \Delta(R)$ and $Y_\phi \in c(\Delta(R))$.

(3) $\phi^*\Delta(S) \subseteq \Delta(R)$ and $\phi^* c(\Delta(S)) \subseteq c(\Delta(R))$.

PROOF. Note that in (2), $X_\phi \oplus Y_\phi \leq_e N_\phi$ simply because $X \oplus Y \leq_e N$. Thus, immediately (3) $\Longrightarrow$ (2) $\Longrightarrow$ (1). Use of (6.5.4)(1) shows the converse (2) $\Longrightarrow$ (3). (1) $\Longrightarrow$ (3). By hypothesis (1) and again (6.5.4)(1), $\phi^*\Delta(S) \subseteq \Delta(R)$. So if (3) is false, then there is a $V \in c(\Delta(S))$ with $V_\phi \notin c(\Delta(R))$. By definition of $c(\Delta(R))$, there exists a $0 \neq P \leq V_\phi$ with $P \in \Delta(R)$ (and $PI = 0$). But then, since $P \subseteq V \in c(\Delta(S))$, $0 \neq P \in c(\Delta(S))$ with $P_\phi \in \Delta(R)$, contradicting (1). $\square$

6.5.17. DEFINITION. A **universal natural class** is a function Δ mapping rings with identity to natural classes over those rings satisfying the equivalent conditions (1), (2), and (3) of the last proposition. If the last proposition holds only for all ϕ in $\mathbf{A}^*$, then Δ will be called an **$\mathbf{A}^*$-universal natural class**. (Thus an **$\mathbf{A}$-universal natural class** is simply called a universal natural class.)

The key feature in the above concepts is (6.5.16)(2), which says that if you find one essential direct sum $N_1 \oplus N_2 \leq N$ as in (2) for some ring S, then this very same decomposition will also satisfy (2) for all rings R which can be mapped surjectively onto S, i.e., $\phi : R \twoheadrightarrow S \cong R/I$. For an $\mathbf{A}^*$-universal natural class, the above only holds for those rings R, where $I < R$ is a complement right ideal. "Universal natural class" sometimes will be abbreviated to just "universal class."

For universal classes $\Delta_1, \Delta_2, \Delta$, their Boolean combinations are defined componentwise by

$$(\Delta_1 \vee \Delta_2)(R) = \Delta_1(R) \vee \Delta_2(R),$$
$$(\Delta_1 \wedge \Delta_2)(R) = \Delta_1(R) \wedge \Delta_2(R),$$
$$c(\Delta)(R) = c(\Delta(R)),$$

and similarly for more complicated formulas. We will use the next corollary only in the finite case.

6.5.18. COROLLARY. Any (finite or infinite) Boolean combination of universal natural classes is one also.

PROOF. Let Δ, Δ_γ, $\gamma \in \Gamma$, be universal natural classes, where Γ is a set. By (6.5.16)(2) or (3), so is $c(\Delta)$. For any ring R, it follows immediately that $(\bigwedge_{\gamma\in\Gamma}\Delta_\gamma)(R) = \bigwedge_{\gamma\in\Gamma}\Delta_\gamma(R)$ satisfies (6.5.16)(1,2,3). Since $c(\Delta_\gamma)$, $\gamma \in \Gamma$, has been shown to be universal, so is $c(\bigvee_{\gamma\in\Gamma}\Delta_\gamma)(R) = \bigwedge_{\gamma\in\Gamma} c(\Delta_\gamma)(R)$. □

Given any finite set of universal natural classes, upon applying a finite number of the Boolean operations of joins, meets, and complements, this set can be turned into a maximal pairwise disjoint set suitable for direct sum decompositions of modules.

6.5.19. COROLLARY. The functions A, B, C, CA, D, I, II, III, and $\Delta_\aleph$, $\aleph = 1, \aleph_0, \aleph_1, \dots$ are universal natural classes.

PROOF. The functor $\phi^\# : \text{Mod-}S \longrightarrow \text{Mod-}R$, with $\phi : R \longrightarrow S = R/I$ and $\phi^\# N_S = N_\phi$, preserves submodules, essential submodules, direct summands, essential direct sums.

By the above, clearly C, D, and $\Delta_\aleph$ satisfy (6.5.16)(2).

For an S-module N, N_S is atomic if and only if N_ϕ is atomic.

By the above for $N_S \in \text{Mod-}S$, $N \in A(S)$ if and only if $N_\phi \in A(R)$. Next, $c(A) = B$, and let $W_S \in B(S)$. If $W_\phi \notin B(R)$, then it contains an atomic module $0 \neq N_\phi \leq W_\phi$. But then $N_S \leq W_S$ is an atomic submodule, contradicting that $W_S \in B(S)$. Thus A, B, and also CA now have been shown to be universal.

In order to show that (6.5.16)(1) holds for type I, first for any $N_S \in \text{Mod-}S$ the following are equivalent:

(i) $\exists\ EN_\phi = X \oplus Y \oplus C,\ 0 \neq X \cong Y$;

(ii) $\exists\ E^S N = P \oplus Q \oplus D,\ 0 \neq P \cong Q$.

(i) $\Longrightarrow$ (ii). By [66, Cor.2.14], since $(E^S N)_\phi$ is quasi-injective by (6.5.2)(5), (i) induces a decomposition

$$E^S N = (X \cap E^S N) \oplus (Y \cap E^S N) \oplus (C \cap E^S N).$$

Let $\psi : X \longrightarrow Y = \psi X$ be an isomorphism. By (6.5.2)(4),

$$X \cap E^S N = \{x \in X : xI = 0\}$$

and ψ maps the latter set bijectively onto $Y \cap E^S N = \{y \in Y : yI = 0\}$.

(ii) $\Longrightarrow$ (i). Application of "$E(\cdot)$" to (ii) produces (i), i.e.,

$$E(E_S N)_\phi = EN_\phi = EP_\phi \oplus EQ_\phi \oplus ED.$$

For any $N_S \in I(S)$, if $N_\phi \notin I(R)$, then there exist decompositions (i) and (ii), contradicting that $N \in I(S)$. Since

$$\begin{aligned} c(I(S)) &= \{N_S :\ \forall\, 0 \neq L \leq N,\ \ \exists\ E^S L = P \oplus Q \oplus D,\ 0 \neq P \cong Q\} \\ &= \{N_S :\ \forall\, 0 \neq L \leq N,\ \ \exists \text{ decompositions (i) and (ii)} \\ &\qquad \text{for } EL_\phi \text{ and } E^S L\}, \end{aligned}$$

if $0 \neq L \in c(I(S))$, then $L_\phi \notin I(R)$. Hence by (6.5.16)(1), the function I is universal.

For type III, for $N_S \in$ Mod-S, (iii) and (iv) are equivalent.

(iii) $\exists\ EN_\phi = X \oplus Y \oplus C,\ 0 \neq X \cong Y \cong X \oplus Y$;

(iv) $\exists\ E^S N = P \oplus Q \oplus D,\ 0 \neq P \cong Q \cong P \oplus Q$.

(iii) $\Longrightarrow$ (iv). As before, by the quasi-injectivity of $(E^S N)_\phi$,

$$E^S N = (X \cap E^S N) \oplus (Y \cap E^S N) \oplus (C \cap E^S N)$$

with $X \cap E^S N \cong Y \cap E^S N$. Let $f : X \longrightarrow X \oplus Y$ be an isomorphism. Next,

$$\begin{aligned} Z &= \{z = x + y : x \in X, y \in Y, xI = 0\} \\ &= \{z = x + y : x \in X, y \in Y, xI = 0 = yI\} \\ &= (X \cap E^S N) \oplus (Y \cap E^S N). \end{aligned}$$

Then f maps $X \cap E^S N = \{x \in X : xI = 0\}$ bijectively onto Z. Thus

$$X \cap E^S N \cong (X \cap E^S N) \oplus (Y \cap E^S N)$$

holds. That (iv) $\Longrightarrow$ (iii) follows upon applying "$E(\cdot)$" to (iv).

We verify that the function III satisfies (6.5.16)(1). For any S-module W_S, $W \in III(S)$ if and only if for any $0 \neq N \leq W$, N satisfies (iv), and hence N_ϕ satisfies (iii). Thus $N_S \in III(S)$ if and only if $N_\phi \in III(R)$. Suppose that $0 \neq L \in c(III(S))$, but $L_\phi \in III(R)$. Then every $0 \neq N \leq L_\phi$ satisfies (iv) and hence also (iii). But then $0 \neq L_S \in III(S)$, a contradiction. Thus (6.5.16)(1) holds and the function III is universal.

For any ring R, $c(I(R) \vee III(R)) = II(R)$. Hence II is easily seen to be universal by (6.5.19). □

For any ring R, define $\mathfrak{G}(R) = \{M_R : M = Z_2 M\}$. Note that $\mathfrak{G}(R) = \bigvee \mathcal{N}_t(R)$ is the identity element of the Boolean lattice $\mathcal{N}_t(R)$. Then $c(\mathfrak{G})(R) = \bigvee \mathcal{N}_f(R)$ is the identity of $\mathcal{N}_f(R)$.

The next example shows that $\mathfrak{G}$ is not a universal natural class, and that $\mathcal{N}_f : \mathbf{A} \longrightarrow \mathbf{B}$ by $\mathcal{N}_f(\phi) = \phi^*$ for $\phi \in \mathbf{A}$ is not a functor, and $\mathcal{N}_f$ is not a subfunctor of $\mathcal{N}$, although on objects R, $\mathcal{N}_f(R) \subseteq \mathcal{N}(R)$.

6.5.20. EXAMPLE. Let p be a prime and let

$$R = \left\{ \begin{pmatrix} n & x \\ 0 & n \end{pmatrix} : n \in \mathbb{Z}, x \in \mathbb{Z}_{p^\infty} \right\}$$

be the trivial extension of $\mathbb{Z}$ and the $\mathbb{Z}$-module $\mathbb{Z}_{p^\infty}$. Take

$$I = \left\{ \begin{pmatrix} 0 & x \\ 0 & 0 \end{pmatrix} : x \in \mathbb{Z}_{p^\infty} \right\},$$

and $\phi : R \longrightarrow S = R/I \cong \mathbb{Z}$. Set $N_S = S$. Since $I \leq_e ZR \leq_e R$, $Z_2 R = R$. Then the inclusion $Z_2^S N = (0) \subset Z_2 N_\phi = R/I$ in (6.5.7)(1,2) can be proper.

Now $\mathcal{N}_f(S) \neq 0$, while $\mathcal{N}(R) = \mathcal{N}_t(R)$ and $\mathcal{N}_f(R) = 0$, and $\phi^*\mathcal{N}_f(S) \subseteq \mathcal{N}_t(R)$. Thus "$\phi^*\mathcal{N}_f(S) \subseteq \mathcal{N}_f(R)$" is false, and $\mathcal{N}_f$ is not a functor of $\mathbf{A} \longrightarrow \mathbf{B}$.

Here $S \cong \mathbb{Z} \in c(\mathfrak{G})(S)$, and $S_\phi \in \mathfrak{G}(R)$. This violates (6.5.13)(1). So $\mathfrak{G}$ is not a universal natural class.

If we restrict our ring homomorphism to ones with complement kernels, then not only is $\mathcal{N}_f$ a functor, but $\mathfrak{G}$ and $c(\mathfrak{G})(S)$ become $\mathbf{A}^*$-universal. This can be proved by use of (6.5.7)(4) and (6.5.13)(1).

6.5.21. The function $\mathfrak{G}$ which assigns to any ring R the class $\mathfrak{G}(R)$ of Goldie torsion modules and the class $(c\mathfrak{G})(R)$ of nonsingular modules is an $\mathbf{A}^*$-universal natural class.

6.5.22. REFERENCES. Dauns [32,25]; Zhou [136].

6.6 Boolean Ideal Lattice

In the previous section we saw that for any ring R, $\mathcal{N}(R) = \mathcal{N}_t(R) \oplus \mathcal{N}_f(R)$. For any module M, $MZ_2(R) \subseteq Z_2(M)$. Hence $\mathcal{N}_f(R) \cong \mathcal{N}_f(R/Z_2R)$. In this section it will be shown how every ring R contains a Boolean lattice of certain two sided ideals of R isomorphic to $\mathcal{N}_f(R)$. In order to do this, we will have to transfer some information from $E(R/Z_2R)$ to R. For this, the following two lemmas will be needed.

6.6.1. LEMMA. Suppose that $ZM \subseteq K < M$. Then

(1) $\overline{K} = \{m \in M : m^{-1}K \leq_e R\}$ is the unique complement closure of K in M.

(2) $\overline{K}/Z_2M \leq M/Z_2M$ is a complement submodule.

(3) If $Z_2M \subseteq K$ and $K/Z_2M < M/Z_2M$ is a complement submodule, then $K = \overline{K}$.

PROOF. (1) If $K \leq_e C \leq M$ where $C \leq M$ is a complement submodule, then always (even if $ZM \not\subseteq K$) $C \subseteq \overline{K}$. We show $K \leq_e \overline{K}$. For if $m \in \overline{K} \backslash K$, then $m^{\perp} \subset m^{-1}K$ is proper, and hence $0 \neq m(m^{-1}K) \subseteq K$.

(2) For any $N \leq \overline{K}$, $\overline{K}/N \leq M/N$ is a complement submodule, because $\overline{K} \leq M$ is a complement submodule.

(3) Essential extensions modulo complement submodules remain essential. Since by (1), $K \leq_e \overline{K}$, also $K/Z_2M \leq_e \overline{K}/Z_2M$, and by hypothesis (3), $\overline{K} = K$. □

The statement of the next lemma corrects an error in [26, Prop.1.5(2), p.330].

6.6.2. LEMMA. Assume that $Z_2M \subseteq K < M$. Consider the following:

(1) $K < M$ is fully invariant.

(2) $EK < EM$ is fully invariant.

(3) $K/Z_2M < M/Z_2M$ is fully invariant.

(4) $E(K/Z_2M) < E(M/Z_2M)$ is fully invariant.

Then (3) $\Longrightarrow$ (1), and (2) $\Longleftrightarrow$ (4); and if $K < M$ is a complement submodule, then (2) $\Longrightarrow$ (1) and (4) $\Longrightarrow$ (3).

PROOF. (3) $\Longrightarrow$ (1). Let $\phi \in \text{End}_R M$. Since $\phi Z_2M \subseteq Z_2M$, ϕ defines $\widetilde{\phi} \in \text{End}_R(M/Z_2M)$ with $\widetilde{\phi}K/Z_2M \subseteq K/Z_2M$, or $\phi K \subseteq K$.

(4) $\iff$ (2). For any $D \leq M$ with $Z_2M \oplus D \leq_e M$, $Z_2M \oplus (K \cap D) \leq_e K$. Hence $E(K/Z_2M) \cong E(K \cap D) \leq ED \cong E(M/Z_2M)$. Thus (4) holds if and only if $E(K \cap D) \leq ED$ is fully invariant. Since $E(Z_2M) = Z_2(EM)$,

$$EK = Z_2(EM) \oplus E(K \cap D) \leq Z_2(EM) \oplus ED = EM.$$

For any $\psi \in \mathrm{End}_R(EM)$, since $\psi Z_2(EM) \subseteq Z_2(EM)$, (2) holds if and only if

$$\psi E(K \cap D) \subseteq Z_2(EM) \oplus E(K \cap D).$$

Hence (4) $\Longrightarrow$ (2).

Suppose (2) holds. Let $\phi \in \mathrm{End}_R(ED)$. As seen above, to show (4) holds, it suffices to show that $\phi E(K \cap D) \subseteq E(K \cap D)$. Define

$$\psi \in \mathrm{End}_R(EM) = \mathrm{End}_R(Z_2(EM) \oplus ED)$$

by setting $\psi \mid_{Z_2(EM)} = 0$ and $\psi \mid_{ED} = \phi$. Then since

$$EK = Z_2(EM) \oplus E(K \cap D),$$

by hypothesis (2),

$$\phi E(K \cap D) = \psi EK \subseteq Z_2(EM) \oplus E(K \cap D)\ .$$

Since also $\phi E(K \cap D) \subseteq ED$ by definition of ϕ, we have that

$$\phi E(K \cap D) \subseteq ED \cap [Z_2(EM) \oplus E(K \cap D)] = E(K \cap D)$$

by the modular law. Thus (4) holds.

(2) $\Longrightarrow$ (1). For $\phi \in \mathrm{End}_R M$, extend ϕ to $\widehat{\phi} \in \mathrm{End}_R(EM)$. By hypothesis (2), $\widehat{\phi} EK \subseteq EK$. Hence $\phi K \subseteq \widehat{\phi} EK \cap M = EK \cap M = K$. The last equality is because $K \leq M$ is a complement submodule.

(4) $\Longrightarrow$ (3). Apply "(2) $\Longrightarrow$ (1)" to the modules K, M replaced by K/Z_2M, M/Z_2M. □

6.6.3. LEMMA. Let $P < M$ be a complement submodule, and for $P \subseteq I < M$, assume $I/P < M/P$ is a complement. Then $I < M$ is a complement submodule.

PROOF. If $P = I$, we are done. So let $P \oplus B \leq_e I$ for some $0 \neq B < M$. If $I <_e Q \leq M$, then $P \oplus B <_e Q$. Since P is a complement submodule of M, $(P \oplus B)/P <_e Q/P$. Hence $I/P <_e Q/P$ is a contradiction, and $I < M$ must be a complement submodule. □

The next proposition is of independent interest.

6.6.4. PROPOSITION. Let $N < M$ be a complement submodule. Then the following hold:

(1) $EN < EM$ is fully invariant $\Longrightarrow N <_t M$. If $ZM = 0$, then

(2) $EN < EM$ is fully invariant $\iff N <_t M$.

PROOF. (1) If not, let $N < K \leq_t M$ with $K \in d(N)$. In other words, K is a type closure of N in M. Then $N \oplus B \leq_e K$ for some $0 \neq B \leq M$. Since $B \in d(N)$, there exist nonzero submodules V_1 and V_2 such that

$$M \supset N \geq V_1 \cong V_2 \leq B \subseteq M \ .$$

The isomorphism $V_1 \cong V_2$ extends to a map $\psi \in \text{End}_R(EM)$ with $\psi V_1 = V_2$, contradicting the full invariance of $EN < EM$.

(2) This is a consequence of two general facts. For any modules $N < M$ and any endomorphism $\psi : M \longrightarrow M$ whose kernel is a complement, $\psi N \in d(N)$. If $ZM = 0$, then $\psi N \in \Sigma\{V : M \geq V \in d(N)\}$, where the latter is the unique type closure of N in M. Now $N <_t M$ implies that $EN <_t EM$. Simply apply the above two facts to the modules $EN < EM$. □

6.6.5. LEMMA. Assume that $Z_2R \subseteq J < R$ is a right complement and that $EJ < ER$ is fully invariant. Then J/Z_2R is the unique type submodule of R/Z_2R of type $d(J/Z_2R)$.

PROOF. Since $J < R_R$ is a complement submodule, $J/Z_2R < R/Z_2R$ is a complement submodule with $E(J/Z_2R) < E(R/Z_2R)$ fully invariant by (6.6.2), (2) $\Longrightarrow$ (4). But then, by (6.6.4), $J/Z_2R <_t R/Z_2R$ is a type submodule. In nonsingular modules, type submodules of a certain type are unique. □

The previous lemma and next theorem correct an error in [26, 2.5, 2.6, pp.332-333], where the hypothesis that $EJ < ER$ is fully invariant was left out.

6.6.6. THEOREM. For any ring R, let as before $\mathcal{N}_f(R)$ be the set of all natural classes of nonsingular right R-modules. Let $\mathcal{J}(R)$ be the set of all complement right ideals $J \leq R$ such that $Z_2R \subseteq J$ and $EJ < ER$ is fully invariant. (Thus for any $J \in \mathcal{J}(R)$, $J \triangleleft R$ and $J/Z_2R < R/Z_2R$ are both fully invariant.) Then there is a bijection $\mathcal{J}(R) \cong \mathcal{N}_f(R)$, which preserves order in both directions given by

(1) $\mathcal{J}(R) \longrightarrow \mathcal{N}_f(R)$, $J \longmapsto d(J/Z_2R) \in \mathcal{N}_f(R)$;

(2) $\mathcal{N}_f(R) \longrightarrow \mathcal{J}(R)$, $\mathcal{K} \longmapsto J$; where J/Z_2R is the unique type submodule of R/Z_2R of type $\mathcal{K}$.

PROOF. (2) It has to be shown that the map $\mathcal{K} \longmapsto J$ is well defined, onto, and one to one. Since $J/Z_2R \leq_t R/Z_2R$ is a complement, it follows that also $J \leq R$ is a complement by (6.6.3). By (6.6.4), $E(J/Z_2R) \leq E(R/Z_2R)$ is fully invariant. Thus $EJ < ER$ is fully invariant by (6.6.2), (4) $\Longrightarrow$ (2). Hence $J \in \mathcal{J}(R)$. For any $J \in \mathcal{J}(R)$, $d(J/Z_2R) \longmapsto J$ by (6.6.5). Suppose that $\mathcal{K}_1 \neq \mathcal{K}_2 \in \mathcal{N}_f(R)$ are mapped to $J \in \mathcal{J}(R)$, that is, J/Z_2R is the (unique) type submodule of R/Z_2R of both type $\mathcal{K}_1$ and type $\mathcal{K}_2$. Then there exists a cyclic module $0 \neq xR \in \mathcal{K}_1 \cap c(\mathcal{K}_2)$. Thus $xR \cong R/x^\perp$ with

$$Z_2R \subseteq x^\perp < x^\perp \oplus C \leq_e R$$

for some $0 \neq C \leq R$. Then

$$0 \neq xC \cong (C \oplus Z_2R)/Z_2R \subseteq J/Z_2R \in \mathcal{K}_2.$$

Hence $0 \neq xR \in \mathcal{K}_2 \cap c(\mathcal{K}_2)$ is a contradiction.

(1) For any $\mathcal{K} \in \mathcal{N}_f(R)$, define $J \in \mathcal{J}(R)$ by $\mathcal{K} \longmapsto J$ as in (2). By (6.6.5), also $d(J/Z_2R) \longmapsto J$. By what was proved above for part (2), $\mathcal{K} = d(J/Z_2R)$. Thus the map in (1) is the inverse of the function given by (2). Both maps preserve the order which is set inclusion in both $\mathcal{N}_f(R)$ and $\mathcal{J}(R)$. □

6.6.7. COROLLARY. If $ZR = 0$, and ER is regarded as a ring, then $\mathcal{J}(R) \cong \mathcal{J}(ER)$.

PROOF. Since $ZR = 0$, ER is a (von Neumann) regular right self injective ring ([66, Cor.2.31, p.60]). For any $J < R$, $EJ < ER$ is a right ideal of ER.

We will show that $\mathcal{J}(R) \longrightarrow \mathcal{J}(ER)$, $J \longmapsto EJ$ is a bijection whose inverse is $\mathcal{J}(ER) \longrightarrow \mathcal{J}(R)$, $P \longmapsto P \cap R$. By [112, Cor.2.3, p.247], ER is the ring $ER = \mathrm{Hom}_R(ER, ER)$. For $J \in \mathcal{J}(R)$, since $(EJ)_R < (ER)_R$ is fully invariant, $ER \cdot EJ \subseteq EJ \in \mathcal{J}(ER)$.

Conversely, for $P \in \mathcal{J}(ER)$, since $P_R \leq_e (EP)_R$, $P = EP$. But both EP and ER are also injective right ER-modules ([66, Prop.2.9(a), p.45]). The full invariance of $EP < ER$ as right ER-modules implies that P is an ideal in ER. The right R-complement closure of $P \cap R \leq R$ in R is $E(P \cap R) \cap R$. From $EP \cap R = P \cap R \leq R \leq_e ER$, we see that $P \cap R \leq R$ is a right complement ideal. Thus $P \cap R \in \mathcal{J}(R)$, and $\mathcal{J}(R) \cong \mathcal{J}(ER)$.

The next theorem serves as an example of a class of nonsingular rings $R \subset \Lambda = ER$ for which their lattices $\mathcal{J}(R) \cong \mathcal{J}(\Lambda)$ can be found explicitly. Very different kinds of techniques are required for this, which might be applicable to the rings of continuous functions.

6.6.8. THEOREM. If R is a Boolean ring with identity, and Λ its (unique over R) completion, then $\mathcal{N}_f(R) \cong \mathcal{J}(R) \cong \Lambda$. In particular, $\mathcal{N}_f(\Lambda) \cong \mathcal{J}(\Lambda) \cong \Lambda$.

PROOF. Let $R \subseteq \Lambda$ be, as in (1.2.6), the ring of all clopen sets and the field of all regular open sets on X, respectively. View Λ as a lattice of characteristic functions χ_H of regular open sets $H \subseteq X$. Each $J \triangleleft R$ defines the open set

$$S(J) = \bigcup \{B : B \subseteq X, \chi_B \in J\}.$$

Conversely, every open set $\mathcal{O} \subseteq X$ defines an ideal $I(\mathcal{O}) \triangleleft R$ by

$$I(\mathcal{O}) = \{\chi_B : B \subseteq \mathcal{O}, B \text{ is clopen}\}.$$

These are inverse operations $I[S(J)] = J$ and $S[I(\mathcal{O})] = \mathcal{O}$ ([71, Lemma 1, p.82]).

Thus we have a "subset" function S, and an "ideal" function I, which are bijections and mutual inverses:

$$\begin{aligned} S :& \{\text{Ideals of } R\} \longrightarrow \{\text{Open subsets of } X\},\ J \longmapsto S(J), \\ I :& \{\text{Open subsets of } X\} \longrightarrow \{\text{Ideals of } R\},\ \mathcal{O} \longmapsto I(\mathcal{O}). \end{aligned}$$

It is next shown that the complement ideals of R under this bijection correspond precisely to the elements of Λ, i.e., to regular open subsets of X.

For any $J \lhd R$, let H be the interior of the closure of $S(J)$ in X. It can be verified by (1.2.6) that $S(J) \subseteq H$ and that H is regular open, and hence that $\chi_H \in \Lambda$ ([71, pp.12-14]). Regard $\Lambda = \Lambda_R$ as an R-module. Then $J \subseteq \chi_H R$. Suppose that $J < \chi_H R$ is not essential. Then $J \subset J \oplus \chi_G R$ for some $\chi_G \in \Lambda$ and a regular open set $G \subseteq X$ with $\emptyset \neq G \subseteq H$. There exists a clopen subset $A \subseteq X$ with $\emptyset \neq A \subseteq G$.

Consequently, $\chi_A \chi_B = 0$ for any $\chi_B \in J$, that is, for all clopen $B \subseteq S(J)$, $A \cap B = \emptyset$. Thus $A \subseteq \bigcap\{X \backslash B : \chi_B \in J\}$. Upon taking set theoretic complements in the set X we get $S(J) = \bigcup\{B : \chi_B \in J\} \subseteq X \backslash A$. Since A is clopen also, $H \subseteq X \backslash A$. This contradicts that $\emptyset \neq A \subseteq H$. Consequently, $J \leq_e \chi_H R$.

From $\Lambda = \chi_H \Lambda \oplus (1 - \chi_H)\Lambda$ it follows that $\chi_H \Lambda < \Lambda_\Lambda$ and $K \equiv \chi_H R \cap R = \chi_H \Lambda \cap R \leq R$ are complement ideals. Since $J \leq_e K$, K is the complement closure of J. Note that $I(H) = \chi_H R = K$.

In any Boolean ring R, for any $K < R$ and any R-map $f : K \longrightarrow R$, $fK \subseteq K$, i.e., every $K < R$ is fully invariant. By (1.2.6), ER is a Boolean ring if R is, so also $EK < ER$ is automatically fully invariant. Therefore, in view of (6.6.6),

$$\begin{aligned} \mathcal{J}(R) &= \{\chi_H R \cap R : X \supseteq H \text{ is regular open}\}, \text{ and} \\ \mathcal{J}(\Lambda) &= \{\chi_H \Lambda : X \supseteq H \text{ is regular open}\}. \end{aligned}$$

Hence $\mathcal{J}(R) \cong \mathcal{J}(\Lambda) \cong \Lambda$. Thus

$$\mathcal{N}(R) = \{d(\chi_H R \cap R) : X \supseteq H \text{ is regular open}\}$$

and $\mathcal{N}(R) \longrightarrow \Lambda$, $d(\chi_H R \cap R) \longmapsto H$ is a ring and lattice isomorphism. So $\mathcal{N}(R) \cong \Lambda$. □

6.6.9. REFERENCES. Brainerd and Lambek [13]; Dauns [24, 26]; Halmos [71]; Zhou [136].

References

[1] J. Ahsan and E. Enochs, Rings all of whose torsion quasi-injective modules are injective, *Glasgow Math. J.* **25**(1984), 219-227.

[2] J. Ahsan and E. Enochs, Torsionfree injective covers, *Comm. Algebra* **12**(9)(1984), 1139-1146.

[3] J. Ahsan and E. Enochs, Rings admitting torsion injective covers, *Portugal. Math.* **40**(3)(1985), 257-261.

[4] A. Al-Huzali, S.K. Jain and S.R. López-Permouth, Rings whose cyclics have finite Goldie dimension, *J. Algebra* **153**(1992), 37-40.

[5] T. Albu and C. Nástásescu, Relative Finiteness in Module Theory, Marcel Dekker, Inc., 1984.

[6] J.S. Alin and E.P. Armendariz, TTF-classes over perfect rings, *J. Austral. Math. Soc.* **11**(1970), 499-503.

[7] A. Alvarado Garcia, H. Aincón, and J. Rios Montes, On the lattices of natural and conatural classes in R-mod, *Comm. Algebra* **29**(2)(2001), 541-556.

[8] F.W. Anderson and K.R. Fuller, Rings and Categories of Modules, Springer-Verlag, New York-Heidelberg-Berlin, 1974.

[9] A. Beachy and W.D. Blair, Finitely annihilated modules and orders in artinian rings, *Comm. Algebra* **6**(1978), 1-34.

[10] G. Birkhoff, Lattice Theory, *Amer. Math. Soc. Colloq. Pub. XXV*, Providence, RI, 1948.

[11] P.E. Bland, Topics in Torsion Theory, *Mathematical Research* **103** Wiley-VCH Verlag, Berlin, 1998.

[12] A.K. Boyle and K.R. Goodearl, Rings over which certain modules are injective, *Pacific J. Math.* **58**(1975), 43-53.

[13] B. Brainerd and J. Lambek, On the ring of quotients of a Boolean ring, *Canad. Math. Bull.* **2**(1959), 25-29.

[14] K.A. Byrd, Rings whose quasi-injective modules are injective, *Proc. Amer. Math. Soc.* **33**(1972), 235-240.

[15] G. Călugăreanu, Lattice Concepts of Module Theory, *Kluwer Academic Publishers*, Dordrecht, Boston-London, 2000.

[16] V.P. Camillo, Modules whose quotients have finite Goldie dimension, *Pacific J. Math.* **69**(1977), 337-338.

[17] V.P. Camillo and J.M. Zelmanowitz, On the dimension of a sum of modules, *Comm. Algebra* **6**(4)(1978), 345-352.

[18] V.P. Camillo and J.M. Zelmanowitz, Dimension modules, *Pacific J. Math.* **91**(2)(1980), 249-261.

[19] M-C. Chamard, Caractérisations de certaines théories de torsion, *C.R. Acad. Sc. Paris,* Series *A* **271**(1970), 1045-1048.

[20] A.W. Chatters and C.R. Hajarnavis, Rings in which every complement right ideal is a direct summand, *Quart. J. Math. Oxford* **28**(1977), 61-80.

[21] T.J. Cheatham, Finite dimensional torsionfree rings, *Pacific J. Math.* **39**(1)(1971), 113-118.

[22] J. Dauns, Uniform dimensions and subdirect products, *Pacific J. Math.* **126**(1987), 1-19.

[23] J. Dauns, Subdirect products of injectives, *Comm. Algebra* **17**(1989), 179-195.

[24] J. Dauns, Torsion free modules, *Ann. Mat. Pure Appl.* **154**(4)(1989), 49-81.

[25] J. Dauns, Torsion free types, *Fund. Math.* **139**(1991), 99-117.

[26] J. Dauns, Classes of modules, *Forum Math.* **3**(1991), 327-338.

[27] J. Dauns, Direct sums and subdirect products, in: *Methods in Module Theory*, pp. 39-65, Lecture Notes in Pure and Appl. Math., **140**, Marcel Dekker, New York, 1992.

[28] J. Dauns, Modules classifying functors, *Czechoslovak Math. J.* **42**(117)(1992), 741-756.

[29] J. Dauns, Functors and Σ-products, in: *Ring Theory*, pp. 149-171, World Sci. Pub., Singapore, 1993.

[30] J. Dauns, Modules and Rings, Cambridge University Press, Cambridge and New York, 1994.

[31] J. Dauns, Unsaturated classes of modules, *Abelian Groups and Modules* (Colorado Springs, CO, 1995), pp. 211-225, Lecture Notes in Pure and Appl. Math., **182**, Dekker, New York, 1996.

[32] J. Dauns, Module types, *Rocky Mountain J. Math.* **27**(2)(1997), 503-557.

[33] J. Dauns, Intersections of modules, in: *Advances in Ring Theory*, pp. 87-103, Birkhaüser, Boston, 1997.

[34] J. Dauns, Lattices of classes of modules, *Comm. Algebra* **27**(1999), 4363-4387.

[35] J. Dauns, Goldie dimensions of quotient modules, *J. Australian Math. Soc.* **71**(2001), 11-19.

[36] J. Dauns, Generalized pure submodules, *J. Algebra* **242**(2001), 1-19.

[37] J. Dauns, Natural classes and torsion theories, *J. Algebra Appl.* **2**(1)(2003), 85-99.

[38] J. Dauns and L. Fuchs, Infinite Goldie dimensions, *J. Algebra* **115**(2)(1988), 297-302.

[39] J. Dauns and Y. Zhou, Some non-classical finiteness conditions of modules, in: *Algebra and Its Applications*, pp. 133-159, AMS Contemporary Mathematics **259**(2000).

[40] J. Dauns and Y. Zhou, Sublattices of the lattice of pre-natural classes, *J. Algebra* **231**(2000), 138-162.

[41] J. Dauns and Y. Zhou, On simple, primitive and prime rings relative to a torsion theory, *Bull. Austral. Math. Soc.* **62**(2000), 297-301.

[42] J. Dauns and Y. Zhou, Type dimension of modules and chain conditions, *Houston J. Math.* **29**(1)(2003), 15-23.

[43] J. Dauns and Y. Zhou, Type submodules and direct sum decompositions of modules, *Rocky Mountain J. Math.* **35**(1)(2005), 83-104.

[44] V. Dlab, A characterization of perfect rings, *Pacific J. Math.* **33**(1970), 79-88.

[45] V. Dlab, On a class of perfect rings, *Canad. J. Math.* **22**(1970), 822-826.

[46] N.V. Dung, D.V. Huynh, P.F. Smith, and R. Wisbauer, Extending Modules, Longman Scientific and Technical, 1994.

[47] E.E. Enochs, Injective and flat covers, envelopes and resolvents, *Israel J. Math.* **39**(3)(1981), 189-209.

[48] C. Faith, Algebra: Rings, Modules and Categories I, Springer-Verlag, New York-Heidelberg-Berlin, 1973.

[49] C. Faith and E.A. Walker, Direct-sum representations of injective modules, *J. Algebra* **5**(1967), 203-221.

[50] M. Fenrick, Conditions under which all preradical classes are perfect hereditary torsion classes, *Comm. Algebra* **2**(4)(1974), 365-376.

[51] J.W. Fisher, On the nilpotency of nil subrings, *Canad. J. Math.* **22**(1970), 1211-1216.

[52] J.W. Fisher, Nil subrings of endomorphism rings of modules, *Proc. Amer. Math. Soc.* **34**(1972), 75-78.

[53] K.R. Fuller, On direct representations of quasi-injectives and quasi-projectives, *Arch. Math.* **20**(1969), 495-502.

[54] P. Gabriel, Des catégories abélennes, *Bull. Soc. Math. France* **90**(1962), 323-448.

[55] A. García, H.Rincón, and R.J. Ríos, On the lattices of natural and conatural classes in R-mod, *Comm. Algebra* **24**(2001), 541-556.

[56] J.L. García Hernández and J.L. Gómez Pardo, V-rings relative to Gabriel topologies, *Comm. Algebra* **13**(1985), 58-83.

[57] B.J. Gardner, Rings whose modules form few torsion classes, *Bull. Austral. Math. Soc.* **4**(1971), 355-359.

[58] J. Golan, Localization of Noncommutative Rings, Marcel Dekker, New York, 1975.

[59] J. Golan, Torsion Theories, *Pitman Monographs and Surveys in Pure and Applied Mathematics*, **29**. Longman Scientific & Technical, Harlow; John Wiley & Sons, Inc., New York, 1986.

[60] J.S. Golan, Linear Topologies on a Ring: an Overview, Longman Scientific & Technical, New York, 1987.

[61] J.S. Golan, Embedding the frame of torsion theories in a larger context - some constructions, in : *Rings, Modules and Radicals*, pp. 61-71, Longman Scientific & Technical, 1989.

[62] J.S. Golan and S.R. Lopéz-Permouth, QI-filters and tight modules, *Comm. Algebra* **19**(8)(1991), 2217-2229.

[63] J.S. Golan, A. Viola-Prioli and J. Viola-Prioli, Ducompact filters and prime kernel functors, *Comm. Algebra* **22**(12)(1994), 4637-4651.

[64] J.L. Gómez Pardo, The rational Loewy series and nilpotent ideals of endomorphism rings, *Israel J. Math.* **60**(1987), 315-332.

[65] K.R. Goodearl, Singular torsion and the splitting properties, *Memoirs Amer. Math. Soc.* **124**(1972).

[66] K.R. Goodearl, Ring Theory: Nonsingular Rings and Modules, Marcel Dekker Inc., 1976.

[67] K.R. Goodearl, Direct sum properties of quasi-injective modules, *Bull. Amer. Math. Soc.* **82**(1976), 108-110.

[68] K.R. Goodearl and A.K. Boyle, Dimension theory for nonsingular injective modules, *Memoirs Amer. Math. Soc.* **177**(1976).

[69] R. Gordon and J.C. Robson, Krull Dimension, *Mem. Amer. Math. Soc.* **133** (1973).

[70] R. Gordon and J.C. Robson, The Gabriel dimension of a module, *J. Algebra* **29**(1974), 459-473.

[71] P. Halmos, Lectures on Boolean Algebras, Math. Studies No. 1, D. Van Nostrand Co., Princeton, NJ, 1963.

[72] D. Handelman, Strongly semiprime rings, *Pacific J. Math.* **60**(1)(1975), 115-122.

[73] J.J. Hutchinson, Quotient full linear rings, *Proc. Amer. Math. Soc.* **28**(1971), 375-378.

[74] S.K. Jain and S.R. López-Permouth, Rings whose cyclics are essentially embeddable in projective modules, *J. Algebra* **128**(1990), 257-269.

[75] A.V. Jategaonker, Localization in Noetherian rings, Cambridge University Press, 1986.

[76] R.E. Johnson, Quotient rings of rings with zero singular ideal, *Pacific J. Math.* **11**(1960), 710-717.

[77] M.A. Kamal and B.J. Müller, Extending modules over commutative domains, *Osaka J. Math.* **25**(1988), 531-538.

[78] I. Kaplansky, Rings of Operators, W.A. Benjamin, Inc., New York-Amsterdam, 1968.

[79] H. Katayama, On the lattice of left linear topologies on a ring, *J. Fac. Liberal Arts Yamaguchi Univ. Natur. Sci.* **29**(1995), 1-4 (1996).

[80] K. Koh, Quasisimple modules and other topics in ring theory, in: *Lectures in Rings and Modules*, pp. 322-433, LNM 246, Springer, New York, 1972.

[81] E.R. Kolchin, Galois theory of differential fields, *Amer. J. Math.* **75**(1953), 753-824.

[82] K. Kunen, Set Theory, an Introduction to Independence Proofs, Studies in Logic and the Foundations of Mathematics, **102**, North-Holland Publishing Co., Amsterdam-New York, 1980.

[83] A. Levy, Basic Set Theory, Perspectives in Mathematical Logic, Springer, Berlin-Heidelberg-New York, 1979.

[84] E. Matlis, Cotorsion Modules, *Mem. Amer. Math. Soc.* **49**(1964).

[85] E. Matlis, Decomposable modules, *Trans. Amer. Math. Soc.* **125** (1966), 147-179.

[86] R.W. Miller and M.L. Teply, The descending chain condition relative to a torsion theory, *Pacific J. Math.* **83**(1979), 207-219.

[87] S.H. Mohamed and B.J. Müller, Continuous and Discrete Modules, *London Math. Soc. Lectures Notes* **147**, Cambridge Univ. Press, 1990.

[88] J.R. Montes and G.T. Sanchez, A general theory of types for nonsingular injective modules, *Comm. Algebra* **20**(8)(1992), 2337-2364.

[89] B.J. Müller and S.T. Rizvi, On the decomposition of continuous modules, *Canad. Math. Bull.* **25**(1982), 296-301.

[90] B.J. Müller and S.T. Rizvi, On injective and quasi-continuous modules, *J. Pure Appl. Algebra* **28**(1983), 197-210.

[91] W.K. Nicholson and B. Sarath, Rings with a largest linear topology, *Comm. Algebra* **13**(1985), 769-780.

[92] D. Northcott, Injective envelopes and inverse polynomials, *J. London Math. Soc.* **8**(1974), 290-296.

[93] J.D. O'Neill, An unusual ring, *J. London Math. Soc.* **44**(2)(1991), 95-101.

[94] B.L. Osofsky, Rings all of whose finitely generated modules are injective, *Pacific J. Math.* **14**(1964), 645-650.

[95] B.L. Osofsky, Nonsingular cyclic modules, *Proc. Amer. Math. Soc.* **19**(1968), 1383-1384.

[96] B.L. Osofsky and P.F. Smith, Cyclic modules whose quotients have all complements submodules direct summands, *J. Algebra* **139**(1991), 342-354.

[97] S.S. Page and M.F. Yousif, Relative injectivity and chain conditions, *Comm. Algebra* **17**(4)(1989), 899-924.

[98] S. Page and Y. Zhou, When direct sums of singular injectives are injective, *Ring Theory* (Granville, OH, 1992), pp. 276-286, World Sci. Publishing, River Edge, NJ, 1993.

[99] S. Page and Y. Zhou, On direct sums of injective modules and chain conditions, *Canad. J. Math.* **46**(1994), 634-647.

[100] S. Page and Y. Zhou, Direct sums of quasi-injective modules, injective covers and natural classes, *Comm. Algebra* **22**(1994), 2911-2933.

[101] M.M. Parmenter and Y. Zhou, Relative injectivity of modules and excellent extensions, *Quaestiones Math.* **22**(1999), 101-107.

[102] R.S. Pierce, A note on complete Boolean algebras, *Proc. Amer. Math. Soc.* **9**(1958), 892-896.

[103] E. Popescu, A characterization of Π-reducible rings, *Rev. Roum. Math. Pures et Appl.* **22**(1977), 537-540.

[104] F. Raggi, J.R. Montes and R. Wisbauer, The lattice structure of hereditary pretorsion classes, *Comm. Algebra* **29**(2001), 131-140.

[105] M. Saleh, On q.f.d. modules and q.f.d. rings, *Houston J. Math.* **30**(3)(2004), 629-636.

[106] F. Sandomierski, Semisimple maximal quotient rings, *Trans. Amer. Math. Soc.* **128**(1967), 112-120.

[107] R.C. Shock, Polynomial rings over finite dimensional rings, *Pacific J. Math.* **42**(1972), 251-258.

[108] R.C. Shock, The ring of endomorphisms of a finite dimensional module, *Israel J. Math.* **11**(1972), 309-314.

[109] P.F. Smith, Modules for which every submodule has a unique closure, *Ring Theory* (Granville, OH, 1992), pp. 302-313, World Sci. Publishing, River Edge, NJ, 1993.

[110] P.F. Smith and A. Tercan, Generalizations of CS-modules, *Comm. Algebra* **21**(1993), 1809-1847.

[111] W. Stephenson, Modules whose lattice of submodules is distributive, *Proc. London Math. Soc.* **28**(1974), 291-310.

[112] B. Stenström, Rings of Quotients, Springer-Verlag, New York, 1975.

[113] H, Storrer, On Goldman's primary decomposition, in: *Lectures in Rings and Modules*, pp. 617-661, LNM 246, Springer, New York, 1972.

[114] M.L. Teply, Torsionfree injective modules, *Pacific J. Math.* **28**(2)(1969), 441-453.

[115] M.L. Teply, Some aspects of Goldie's torsion theory, *Pacific J. Math.* **29**(2)(1969), 447-459.

[116] M.L. Teply, Homological dimension and splitting torsion theories, *Pacific J. Math.* **34**(1970), 193-205.

[117] M.L. Teply, Torsionfree covers II, *Israel J. Math.* **23**(2)(1976), 132-136.

[118] M.L. Teply, On the idempotence and stability of kernel functors, *Glasgow Math. J.* **37**(1995), 37-43.

[119] J.E. van den Berg, When every torsion preradical is a torsion radical, *Comm. Algebra* **27**(11)(1999), 5527-5547.

[120] J.E. van den Berg and J.G. Raftery, Every algebraic chain is the congruence lattice of a ring, *J. Algebra* **162**(1)(1993), 95-106.

[121] A. Viola-Prioli and J. Viola-Prioli, Rings whose kernel functors are linearly ordered, *Pacific J. Math.* **132**(1988), 21-34.

[122] A. Viola-Prioli and J. Viola-Prioli, Asymmetry in the lattice of kernel functors, *Glasgow Math. J.* **33**(1)(1991), 95-97.

[123] A. Viola-Prioli and J. Viola-Prioli, Rings arising from conditions on preradicals, in: *Ring Theory* (Granville, OH, 1992), pp. 343-349, World Sci. Publishing, River Edge, NJ, 1993.

[124] A. Viola-Prioli, J. Viola-Prioli and R. Wisbauer, Module categories with linearly ordered closed subcategories, *Comm. Algebra* **22**(1994), 3613-3627.

[125] A. Viola-Prioli, J. Viola-Prioli and R. Wisbauer, A description of closed subcategories, *Comm. Algebra* **23**(1995), 4173-4188.

[126] J. Viola-Prioli, When is every kernel functor idempotent?, *Canad. J. Math.* **27**(1975), 545-554.

[127] C.L. Walker and E.A. Walker, Quotient categories and rings of quotient, *Rocky Mountain J. Math.* **2**(1972), 513-555.

[128] R.W. Wilkerson, Finite dimensional group rings, *Proc. Amer. Math. Soc.* **41**(1973), 10-16.

[129] R. Wisbauer, Localization of modules and the central closure of rings, *Comm. Algebra* **14**(1981), 1455-1493.

[130] R. Wisbauer, Generalized co-semisimple modules, *Comm. Algebra* **18**(1990), 4235-4253.

[131] R. Wisbauer, Foundations of Module and Ring Theory, *Gordon and Breach Science Publishers*, 1991.

[132] R. Wisbauer, On module classes closed under extensions, in: *Rings and Radicals*, pp. 73-97, Pitman Res. Notes Math. Ser., **346**, Longman, Harlow, 1996.

[133] Q. Wu and J.S. Golan, On the endomorphism ring of a module with relative chain conditions, *Comm. Algebra* **18**(1990), 2595-2609.

[134] J. Xu, Flat Covers of Modules, *Lecture Notes in Mathematics* **1634**, Springer, 1996.

[135] Y. Zhou, Direct sums of M-injective modules and module classes, *Comm. Algebra* **23**(1995), 927-940.

[136] Y. Zhou, The lattice of natural classes of modules, *Comm. Algebra* **24**(5)(1996), 1637-1648.

[137] Y. Zhou, Notes on nilpotency of nil subrings of endomorphism rings of modules, *Quaestiones Math.* **19**(1996), 1-5.

[138] Y. Zhou, Relative chain conditions and module classes, *Comm. Algebra* **25**(2)(1997), 543-557.

[139] Y. Zhou, Weak injectivity and module classes, *Comm. Algebra* **25**(8)(1997), 2395-2407.

[140] Y. Zhou, Nonsingular rings with finite type dimension, in: *Advances in Ring Theory*, pp. 323-333, Birkhäuser, Boston, 1997.

[141] Y. Zhou, Decomposing modules into direct sums of submodules with types, *J. Pure Appl. Algebra* **138**(1999), 83-97.

[142] Y. Zhou, The lattice of pre-natural classes of modules, *J. Pure Appl. Algebra* **140**(2)(1999), 191-207.

Index

ACC, ascending chain condition, 34

class of modules
  $c(\mathcal{F})$, 19
  $d(\mathcal{F})$, 19
  $c(P)$, 19
  $d(P)$, 19
  $\mathfrak{F}$, 9
  $\mathfrak{G}$, 9
  $\mathfrak{G}_M$, 16
  hereditary pretorsion, 8
  hereditary torsion, 8
  hereditary torsion free, 8
  $i_M(\mathcal{F})$, 65
  Mod-R, 2
  M-natural, 24
  natural, 20
  $\mathcal{P}(\infty)$, 111
  pre-natural, 29
  pretorsion, 8
  pretorsion free, 8
  saturated, 20
  $\sigma[M]$, 12
  $\mathcal{S}(R)$, 43
  stable, 8
  torsion, 8
  torsion free, 8
  universal natural, 196
cogenerator of $\mathcal{K}$, 51
cofinal ordinal indexed set, 26
complement closure, 2
completely prime ideal, 187
completion, 10
condition $(*)$, 26
convex subgroup, 96
convex sublattice, 160
convex subsemigroup, 96
cover, 65

DCC, descending chain condition, 50
dense right ideal, 3, 11
Dedekind domain, 127
dimension
  Gabriel dimension, $G.\dim M$, 181
  local Goldie dimension, $Gd(M)$, 113
  type, $t.\dim M$, 75
  uniform, $u.\dim M$, 75
direct sum decomposition
  complement direct summands, 37
  complement maximal direct summands, 37
  complement type summands, 91
direct limit, 66
directed partially ordered set, 66
direct system of modules, 66

$E(\cdot)$, 1
$E_M(\cdot)$, 13
$\aleph$-product, 114
$\text{Ext}_R(\cdot,\cdot)$, 7
extension operator $E(\cdot,\cdot)$, 177

field of sets, 6
filter, 10
free algebra, 114, 115
$\mathcal{F}$-cover, 65
$\mathcal{F}$-precover, 65

$G_{\mathfrak{A}}$, 30
Gabriel topology, 10

$H_{\mathcal{K}}(M)$, 21
$H_{\mathcal{K}}(R)$, 21
hereditary filter, 10

injective hull, 1

inverse equivalence, 104

$\mathcal{K}$-composition series, 54

lattice
- algebraic, 154, 173
- atom in, 163
- Boolean, 5
- coatom in, 163
- compact element in, 154
- complemented, 5
- complete, 5
- distributive, 173
- $\mathcal{F}_r(R)$, 151
- **fil**-R, 152
- M-ptors, 153
- modular, 5
- $\mathcal{N}(R)$, $\mathcal{N}_r(R)$, 20
- $\mathcal{N}_l(R)$, 20
- $\mathcal{N}^p(R)$, $\mathcal{N}_r^p(R)$, 29
- $\mathcal{N}_l^p(R)$, 29
- $\mathcal{N}(R, \mathcal{K})$, 157
- $\mathcal{N}(R, M)$, 24
- $\mathcal{T}_r(R)$, 151
- $\mathcal{T}_r^p(R)$, 151
- $\mathcal{T}_r^p(\mathcal{K}, R)$, 153
- **Tor**-R, 152
- uniquely complemented, 165
- upper continuous, 156

linear topology, 10
$L(R)$, 21

maximal set of pairwise orthogonal types, 109
$M_{\mathcal{F}}$, 12
module
- (A_2), 145
- Abelian, 111
- α-simple, 181
- Artinian, 61
- atomic, 21
- bottomless, 110
- (C_1), 39
- (C_{11}), 118
- (C_2), 39
- (C_3), 39
- c-generated, 40
- c-limited, 46
- compressible, 110
- condition (A_2), 145
- continuous, 110
- countably decomposable, 117
- CS, 107
- dimension, 137
- directly finite, 111
- discrete, 110
- extending, 39
- $\mathcal{F}$-cover, 65
- $\mathcal{F}$-precover, 65
- finitely decomposable, 117
- fully decomposable, 117, 129
- fully invariant, 119
- Goldie torsion, 2, 10
- idempotent square, 111
- idempotent square free, 111
- $\mathcal{K}$-cocritical, 25
- $\mathcal{K}$-injective, 13
- large, 163
- local (direct) summand of, 39
- local type (direct) summand of, 89
- locally idempotent square free, 111
- locally Noetherian, 48
- maximal direct summand of, 37
- M-cogenerated, 2
- M-generated, 2
- M-injective, 2
- molecular, 110
- monoform, 17
- M-singular, 15
- n-decomposable, 117
- Noetherian, 48, 60
- non M-singular, 15
- nonsingular, 2
- orthogonal, 72
- parallel, 72
- purely infinite, 111
- quasi-continuous, 39
- quasi-injective, 2

quasi-injective hull, 17
ρ-torsion, 8
ρ-torsion free, 8
Σ-quasi-injective, 40
singular, 2
square, 111
square free, 111
square full, 111
strongly uniform, 17
(T_3), 118
τ-closed, 60
τ-cocritical, 60
t-ACC, 88
t-DCC, 101
t-indecomposable, 89
TS, 94
type I, 111
type II, 111
type III, 111
uniform, 1
unique complement closure (UC), 133
unique type closure (UTC), 132
weakly injective, 87
weakly R-injective, 87
Morita equivalence, 104
Morita invariant property, 104

natural set of right ideals, 21

orthogonal projection argument, 90
orthogonality, 72

parallelity, 72
partial endomorphism, 133
partial homomorphism, 133
$\mathcal{P}(L(R))$, 20
prefilter, 10
precover, 65
pre-natural set of right ideals, 30
preradical, 8
idempotent, 8
left exact, 8
radical, 8
Projection argument, 20

rational extension, 3
ring
Artinian, 171
Boolean, 5
chain, 187
distributive, 173
formal triangular matrix, 82
indecomposable, 89
Laurent polynomial, 84
Noetherian, 46
maximal right quotient, 4
of right quotients, 3, 86
opposite, R^{op}, 83
perfect, 167, 169, 175
polynomial, 84
QFD, 87
QI, 186
reduced, 187
(Von Neumann) regular, 99
right full linear, 99
right self-injective, 99
semiartinian, 166, 167, 169
semisimpe (Artinian), 171
V- , 181
progenerator, 104

semigroup, 96
subfunctor, 8
submodule
closed, 2
complement, 2
essential, 1
$\mathcal{K}$-critical, 25
fully invariant, 119
maximal type, 99
rational, 17
rationally closed, 62
ρ-closed, 8
ρ-dense, 8
singular, 2
torsion, 9
type, 72
subsemigroup, 97
superspectivity, 109, 118
τ-composition series, 60

torsion theory, 9
  cogenerated by $\mathcal{K}$, 9
  generated by $\mathcal{K}$, 9
  Goldie, 10
  hereditary, 9
  Lambek, 11
T-nilpotency, 62
torsion free set of right ideals, 22
totally ordered Abelian group, 96
trace, $\operatorname{tr}(M, N)$, $\operatorname{tr}(\mathcal{K}, N)$, 13
trivial extension, 23
type dimension, 75
$\mathcal{T}_{\mathfrak{A}}$, 10
type, 20
type complement, 73
type closure, 73

valuation, 95
valuation domain, 95
value group, 96